Oracle
Database In-Memory
架构与实践

萧宇 编著

清华大学出版社
北京

内 容 简 介

Database In-Memory 是 Oracle 内存计算技术家族中的核心成员，也是 Oracle 数据库企业版最重要的数据库选件之一。Database In-Memory 通过独有的双格式数据库架构及一系列内存计算软硬件技术的结合，可以使传统的运营系统即刻开启 HTAP（混合事务/分析处理）能力，或者为传统的数据仓库系统提供实时分析的能力。本书全面深入介绍 Database In-Memory 架构和技术，涵盖了从 12.1.0.2 版本 Database In-Memory 诞生到 2021 年 8 月最新发布的 21c 版本之间所有重要的 Database In-Memory 特性。

本书面向对 Oracle 数据库和内存计算技术感兴趣的读者，适合的对象包括企业架构师、数据库管理员、数据分析师和应用开发人员，不仅是读者全面学习 Database In-Memory 内存计算技术的入门指南，同时也是深入了解 Oracle Database In-Memory 的极有价值的参考书籍。

本书封面贴有清华大学出版社防伪标签，无标签者不得销售。
版权所有，侵权必究。举报：010-62782989，beiqinquan@tup.tsinghua.edu.cn。

图书在版编目（CIP）数据

Oracle Database In-Memory 架构与实践 / 萧宇编著. —北京：清华大学出版社，2022.4
ISBN 978-7-302-60088-6

Ⅰ. ①O… Ⅱ. ①萧… Ⅲ. ①关系数据库系统 Ⅳ. ①TP311.132.3

中国版本图书馆 CIP 数据核字（2022）第 018750 号

责任编辑：王　芳
封面设计：刘　键
责任校对：徐俊伟
责任印制：丛怀宇

出版发行：清华大学出版社
　　　　网　　址：http://www.tup.com.cn, http://www.wqbook.com
　　　　地　　址：北京清华大学学研大厦 A 座　　邮　编：100084
　　　　社 总 机：010-83470000　　　　　　　　邮　购：010-83470235
　　　　投稿与读者服务：010-62776969, c-service@tup.tsinghua.edu.cn
　　　　质量反馈：010-62772015, zhiliang@tup.tsinghua.edu.cn
印 刷 者：北京富博印刷有限公司
装 订 者：北京市密云县京文制本装订厂
经　　销：全国新华书店
开　　本：185mm×260mm　　印　张：23　　字　数：547 千字
版　　次：2022 年 4 月第 1 版　　印　次：2022 年 4 月第 1 次印刷
印　　数：1～2000
定　　价：99.00 元

产品编号：092292-01

序言

虎年春节来临之际,欣闻萧宇关于内存数据库的新书要出版了,他和我打趣说,你要不要写个序?我自是答应了。

内存数据库并不是一个新的概念,早在 20 世纪八九十年代,就已经有学者提出了将所需数据存放在内存中提高处理速度的概念原型。随着近年来硬件技术革新性的快速发展以及终端实时应用对超低响应时间超大吞吐量的需求,内存数据库从无到有,从概念模型到广泛而成熟的产品应用,仅仅经过了不到二十年的时间。各大数据库厂商和开源社区纷纷推出了自己的内存数据库产品,来适应新的市场变化。其中比较典型的是针对事务型的内存加速数据库和分析型应用的列存储内存数据库。近年来,列存储内存数据库吸引了更多的关注,其研究的重点也在于软硬件相结合的设计和开发技术。

天下武功,唯快不破,对于内存数据库而言更是如此,高性能是它最早的设计初衷,也是内存数据库的代名词。内存数据库并非简单的将数据加载在内存中,为了同时进行事务处理和分析处理,需要巧妙地设计数据存储格式以及更新算法;为了在不影响性能的前提下尽量节约空间,需要使用针对不同数据的不同应用特性的压缩算法;为了有效利用硬件的新性能,查询处理引擎需要做针对性的重构和修改;同时还要应对纷纷涌现出的新硬件技术,例如,协同处理器(co-processor)、专用集成电路(ASIC)、图形处理单元(GPU)、现场可编程门阵列(FPGA)、持久内存、远程直接内存访问(RDMA)等。可以说,内存数据库脱胎于传统磁盘数据库,却有自己独特的技术挑战和优势。

除此之外,做研究的学者们可能会轻视产品在实际应用中的融合性和易用性。我们常常笑说快乐总是短暂的,做一个产品的原型开发是最开心的时候,可这部分大概只占了整个产品研发不到 10%的时间,产品化的过程却占据了工程最繁复、最头疼的 90%。对于数据库用户而言,他们想要的是一个不需要任何人为干预自动选择最优技术策略的产品,自治数据库应运而生。简而言之,自治数据库依旧是关系型数据库,但它能够根据用户的应用场景自动定义数据不同的存储层级,自动选择适用于内存处理的数据,自动选择最优的压缩算法,自动进行后台的数据更新等。这些小而美的特点才是让内存数据库更加通用更加流行的法宝。

本书从内存技术的起源聊起,详细地讨论了 Oracle 内存数据库产品。其中,包括对高层次的结构体系的讨论,以及对所有特性的非常细节的介绍,甚至手把手地介绍使用窍门,对 Oracle

内存数据库的使用者来说，这是一本不可多得的好工具书，对所有对 Oracle 内存数据库感兴趣的学者和用户而言，也是最为详尽的中文书，值得一读。

龚玮薇

Oracle 内存技术研发总监

前言

2015年，由于项目的原因，我逐渐开始关注内存计算技术。2017年，我出版了第一本关于内存计算的书：《TimesTen 内存数据库架构与实践》。实际上，Oracle 内存计算技术家族还有另一个重要的成员：Database In-Memory。2018 年底，在与某物流客户的联合测试中，我第一次亲身体验了 Database In-Memory 的易用性，用户启用 Database In-Memory 实现了查询提速 108 倍，而应用和查询未做任何修改。2020 年初，Oracle 中国公司推出了 Oracle 19c 公益课堂，通过网络为公众介绍 Oracle 产品和技术。在第一批拟定的课程主题中，我最终选择了 Database In-Memory。经过一个多月的精心准备，阅读了大量产品手册、白皮书、博客文章、网络视频与论文，也做了大量的实验。公开课吸引了超过 500 人参加，课后我在甲骨文云计算公众号上发表了文章"加速度：走进 Oracle Database In-Memory"，作为课程的文字版，同时也作为这一个月来学习的总结。2020 年底，我开始有了出书的想法，希望可以完整和系统地介绍 Database In-Memory。目前市面上仅有的一本介绍 Database In-Memory 的书是基于 Oracle 12cR2 版本的，由 Oracle Press 于 2017 年出版。Database In-Memory 在 12cR2 版本之后又发布了很多新特性，所以我觉得一些理论可以更深入展开，也可以增加动手操作的内容，以加深对此项技术的理解。相信我学习 Database In-Memory 的这些心得体会、经验与教训是值得与大家分享的，我想做的正如我非常喜欢的散文家王鼎钧所说："我一边赤脚行走，一边把什么地方有荆棘、什么地方有甘泉写下来，放在路旁让后面走过来的人拾去看看"。简而言之，我为什么写这本书？一是觉得 Database In-Memory 值得写，二是我有信心能把它写好。

本书适用于不同类型的读者。如果你是数据库管理员，可以通过此书了解内存计算的基本概念、Database In-Memory 的体系架构和内部运行机制、监控与调优。如果你是架构师，可以了解如何利用新型内存计算技术简化整体架构设计，将 Database In-Memory 应用于企业级混合负载业务系统或数据仓库系统，助力构建实时企业。如果你是应用开发人员，可以学习如何充分利用数据库自身的能力，简化应用开发，聚焦于真正创造业务价值的核心问题。

希望本书能帮助读者编写出更高效的应用，设计出更敏捷的架构。阅读本书的唯一先决条件是对关系型数据库有基础了解。如果读者具备 Oracle 数据库管理、关系型数据库应用开发或系统架构设计能力中的一种，或对其他类型的内存数据库有所了解，对于理解本书内容也将大有裨益。

第 1 章首先介绍了内存计算技术的发展历史以及推动其发展的软硬件技术和新型企业应用需求，对内存计算技术的概念及分类进行了阐述。最后对当前主流的内存数据管理产品做了简要介绍,包括商业和开源领域的内存数据库和内存数据网格产品。

第 2 章介绍如何搭建 Database In-Memory 实验环境,包括安装虚拟化引擎 VirtualBox、虚拟机环境管理工具 Vagrant、下载随书示例、安装示例表和生成并导入测试数据,后续章节中的示例和概念讲解均基于此实验环境展开。接下来介绍了便于开发和测试的管理工具 SQL Developer 及命令行编辑工具 rlwrap。最后介绍了 Database In-Memory 相关的学习资源,包括文档、博客、视频和动手实验资源等。

Database In-Memory 是 Oracle 数据库企业版的选件之一。如果说 Database In-Memory 是一片树叶,Oracle 数据库则是一棵参天大树。在探索树叶的脉络之前,对这棵大树有一全局的概念是非常必要的。因此第 3 章首先对 Oracle 数据库相关的重要基本概念做了简要介绍,包括数据库版本和版本号、选件和管理包体系以及升级和更新的概念。重点介绍了 Database In-Memory 中最核心的概念,包括行列双格式存储、数据库内存管理、Database In-Memory 架构和内存压缩单元架构。

第 4 章介绍了两个重要的 Database In-Memory 操作:配置与发布。配置部分包括如何为数据库启用和禁用 Database In-Memory,如何开启和关闭数据库对象的 INMEMORY 属性,如何指定 INMEMORY 对象的压缩级别和发布优先级。发布方面则涉及发布的基本概念和工作原理、不同的发布方式和发布对象与过程的监控等。

第 5 章全面介绍了与 Database In-Memory 相关的管理工具。首先介绍了一些传统的图形化和命令行管理工具,如 SQL Developer、Oracle Enterprise Manager、SQL Plus 和 SQLcl。然后介绍了如何生成优化器统计信息和 SQL 执行统计信息,以及生成、显示和解读执行计划,这些概念对于理解 Database In-Memory 的执行都是非常重要的。Oracle 数据库建议器部分介绍了 In-Memory 建议器和压缩建议器,并解答了两个常见问题,即哪些数据库对象适合使用 Database In-Memory,以及如何估算不同压缩级别对于内存的消耗。本章最后介绍了用于控制 Database In-Memory 行为的初始化参数和优化器提示,以及监控 Database In-Memory 状态的系统视图。

第 6 章介绍了 Database In-Memory 最基础同时也是最核心的性能优化技术,包括内存列式存储格式、压缩、存储索引和 SIMD 向量处理。然后介绍了操作下推的概念及 Database In-Memory 所支持的各类下推操作。最后介绍了优化器在改进分析查询上的两类增强:In-Memory 联结和 In-Memory 聚合。本章介绍的技术、概念和算法是 Database In-Memory 提供极速分析性能的关键,也充分体现了 Oracle 在数据库软件领域的长期和深厚积累。

第 7 章介绍了 Database In-Memory 高级性能优化功能,包括通过预计算实现性能提升的 In-Memory 表达式,通过全局字典避免解压缩和哈希计算的联结组,通过与 SIMD 指令紧密配合实现的 In-Memory 深度向量化,针对 JSON、NUMBER、全文本和地理空间多种数据类型的优化,以及结合多线程和并行处理等技术实现的 In-Memory 扫描优化。

高可用性是关键业务系统最重要的非功能性需求之一。第 8 章首先介绍了 Oracle 最高可用性架构,然后详细介绍了 Database In-Memory 如何与 RAC 集群环境结合,包括数据的分布、复制和数据库服务等重要概念。接下来介绍了 Database In-Memory 与 ADG 灾备组件紧密集成的多种场景,以及如何利用数据库服务实现 Database In-Memory 与灾备角色的绑定。最后介绍了 FastStart,这是可以在数据库启动时加速内存列式数据发布的功能。

第 9 章介绍了 Database In-Memory 与数据可管理性相关的一些功能。首先介绍了自动数据优化，可以将内存列式存储作为新的数据层级，根据制定的策略自动将对象发布到或移出内存列式存储。然后介绍了 Automatic In-Memory，可以在内存不足时自动移除"冷"对象，以保证"热"对象的发布；在最新的 21c 版本中，即使未设置 INMEMORY 属性的数据库对象也可以通过其进行管理。

　　大数据有数据量大、数据类型多、产生速度快 3 个主要特征。第 10 章主要关注 Database In-Memory 与大数据的"量"和"速"的结合。本章首先介绍了 Database In-Memory 如何与多种形式的外部表协同工作，然后介绍了内存优化行存储。最后介绍了 Exadata 对 In-Memory 列格式的支持，此功能将 Database In-Memory 的好处延伸到 Exadata 存储层的大容量闪存，同时可以保证良好的性能。

　　本书的一大特色是理论与动手实践相结合，在计算机旁阅读此书是最佳的方式，这样可以对书中的概念进行验证和探索。全书附带大量脚本和代码，以加深读者对产品特性和基本概念的理解，所有脚本和代码均按章节组织，可以扫描文末二维码自助下载。

　　衷心感谢我的朋友和同事们在出书过程中给予的帮助与指导，感谢家人对我写作的支持。感谢清华大学出版社的王芳编辑，尽管未曾谋面，但这已经是我们配合的第 3 本书。感谢 Oracle 公司的同事 Joyce Li、叶大海、段辉、朱俊峰、孙骏，Database In-Memory 产品经理 Andy Rivenes，Database In-Memory 高级开发经理 Gong Weiwei，以及 SAP 中国公司的 Jennifer Jia 和 Long Sheng，感谢你们通过 Oracle 内部论坛、电话、邮件及现场对我提出问题的耐心指导和答疑。感谢海信集团 IT 与数据管理部总经理许敏、副总经理袁俊卿、基础架构及云服务部部长曲栋，网鼎明天科技有限公司的李兆，在 Oracle 第一届 Database In-Memory 高峰会议上无私分享了海信在 SAP 核心系统使用 Database In-Memory 的经验和教训，这些来自用户视角的不同观点也给予了我新的启发，并融入本书的内容当中。最后，我想对我的父母说，这本书是献给你们的，感谢你们从小培养了我阅读和思考的习惯，给予我克服困难的勇气和动力，谢谢你们！

萧　宇
2021 年 9 月

代码脚本下载

目录

第 1 章 内存计算概述 ··· 1
  1.1 内存计算的兴起 ··· 1
    1.1.1 硬件的发展 ··· 2
    1.1.2 软件的发展 ··· 3
    1.1.3 企业应用的需求 ··· 4
  1.2 内存计算技术分类 ·· 5
    1.2.1 内存数据库 ··· 5
    1.2.2 内存数据网格 ·· 6
  1.3 典型内存数据管理产品 ·· 7
    1.3.1 Oracle TimesTen ··· 7
    1.3.2 Oracle Database In-Memory ·· 9
    1.3.3 MySQL HeatWave ·· 11
    1.3.4 Oracle Coherence ··· 13
    1.3.5 SAP HANA ·· 15
    1.3.6 Redis ·· 17

第 2 章 搭建 Database In-Memory 实验环境 ·· 20
  2.1 虚拟化引擎 Oracle VM VirtualBox ··· 21
  2.2 版本控制系统 Git ··· 23
  2.3 虚拟机环境管理工具 Vagrant ·· 23
  2.4 安装示例表与数据 ·· 26
    2.4.1 随书示例 ··· 26
    2.4.2 Star Schema Benchmark 示例 ··· 27
    2.4.3 Oracle 数据库标准示例 ·· 30
    2.4.4 TPC-H 基准示例 ·· 30
  2.5 命令行编辑工具 rlwrap ··· 31
  2.6 数据库开发与管理工具 SQL Developer ······································· 32
    2.6.1 Windows 下的 SQL Developer 安装 ································· 32

	2.6.2	Linux下的SQL Developer安装 ·········· 33
	2.6.3	使用SQL Developer连接数据库 ·········· 33
2.7	Database In-Memory 学习资源 ·········· 35	
	2.7.1	文档资源 ·········· 35
	2.7.2	Database In-Memory官方博客 ·········· 36
	2.7.3	动手实验资源 ·········· 36
	2.7.4	My Oracle Support ·········· 37
	2.7.5	视频学习资源 ·········· 39

第3章 Database In-Memory 基本概念与架构 ·········· 40

3.1	Oracle 数据库基本概念 ·········· 40	
	3.1.1	Oracle数据库版本简介 ·········· 40
	3.1.2	Oracle数据库企业版选件 ·········· 42
	3.1.3	数据库管理包 ·········· 43
	3.1.4	数据库版本号 ·········· 44
	3.1.5	数据库升级与更新 ·········· 44
3.2	Database In-Memory 体系架构 ·········· 48	
	3.2.1	双格式存储 ·········· 48
	3.2.2	Oracle数据库内存结构与管理 ·········· 49
	3.2.3	Database In-Memory架构 ·········· 51
	3.2.4	内存压缩单元架构 ·········· 52
3.3	Exadata 作为最佳 Database In-Memory 平台 ·········· 53	
3.4	何时使用 Database In-Memory ·········· 55	
	3.4.1	Database In-Memory适用场景 ·········· 55
	3.4.2	分析型查询与应用的考虑 ·········· 56

第4章 Database In-Memory 配置与发布 ·········· 58

4.1	配置与发布基本流程 ·········· 58	
4.2	启用 Database In-Memory ·········· 58	
	4.2.1	In-Memory Area初始设置 ·········· 58
	4.2.2	多租户下的In-Memory Area设置 ·········· 60
	4.2.3	调整In-Memory Area大小 ·········· 61
	4.2.4	Base Level特性 ·········· 62
	4.2.5	常见错误与处理 ·········· 63
4.3	指定 Database In-Memory 对象 ·········· 65	
	4.3.1	哪些对象可以发布到内存列式存储 ·········· 65
	4.3.2	开启和关闭INMEMORY属性 ·········· 65
4.4	发布 INMEMORY 对象 ·········· 67	
	4.4.1	确认对象已发布 ·········· 67
	4.4.2	发布优先级与自动发布 ·········· 69

  4.4.3 通过查询手工发布 ·· 71
  4.4.4 通过PL/SQL子程序手工发布 ·· 71
  4.4.5 通过初始化参数控制发布 ·· 76
4.5 重新发布 ·· 76
4.6 发布进程与发布速度 ·· 79
4.7 指定内存压缩级别 ·· 82
4.8 内存列式存储与行式存储映射 ·· 84
4.9 移除数据库对象 ·· 91
4.10 禁用 Database In-Memory ·· 93

第 5 章 Database In-Memory 管理工具 ·· 96

5.1 SQL Developer ·· 96
5.2 Oracle Enterprise Manager ·· 97
  5.2.1 In-Memory Central ·· 98
  5.2.2 SQL性能分析器 ·· 101
5.3 命令行管理工具 ·· 103
  5.3.1 SQL Plus ·· 103
  5.3.2 SQLcl ·· 107
  5.3.3 Data Pump ·· 112
5.4 统计信息与执行计划 ·· 113
  5.4.1 优化器统计信息 ·· 113
  5.4.2 SQL执行统计信息 ·· 115
  5.4.3 解读执行计划 ·· 117
  5.4.4 生成和显示执行计划 ·· 118
5.5 Oracle 数据库建议器 ·· 121
  5.5.1 In-Memory建议器 ·· 121
  5.5.2 压缩建议器 ·· 126
5.6 实时 SQL 监控 ·· 128
5.7 Database In-Memory 初始化参数 ·· 132
5.8 Database In-Memory 视图 ·· 134
  5.8.1 数据字典视图 ·· 134
  5.8.2 动态性能视图 ·· 135
  5.8.3 In-Memory视图使用示例 ·· 137
5.9 优化器提示 ·· 140

第 6 章 Database In-Memory 基础性能优化 ·· 142

6.1 列格式组织 ·· 142
6.2 内存存储索引 ·· 145
6.3 SIMD 向量处理 ·· 148
6.4 数据压缩 ·· 152

		6.4.1 行级压缩	152
		6.4.2 混合列压缩	157
		6.4.3 内存列压缩	160
	6.5	操作下推	168
		6.5.1 过滤谓词下推	169
		6.5.2 聚合下推	170
		6.5.3 下推与内存存储索引	173
	6.6	In-Memory 联结	173
		6.6.1 联结方法	173
		6.6.2 In-Memory联结与布隆过滤器	175
		6.6.3 部分表发布时的In-Memory联结	183
	6.7	In-Memory 聚合	184
		6.7.1 In-Memory聚合基本概念	184
		6.7.2 In-Memory聚合性能比较	186
	6.8	索引优化	191

第 7 章 Database In-Memory 高级性能优化 194

7.1	In-Memory 表达式	194
	7.1.1 In-Memory虚拟列	194
	7.1.2 In-Memory Expression	198
7.2	In-Memory 联结优化	203
	7.2.1 联结组（Join Group）	203
	7.2.2 In-Memory深度向量化	208
7.3	In-Memory 数据类型优化	210
	7.3.1 In-Memory JSON列	210
	7.3.2 In-Memory优化运算	216
	7.3.3 In-Memory全文本列	218
	7.3.4 In-Memory Spatial支持	222
7.4	In-Memory 扫描优化	226
	7.4.1 In-Memory动态扫描	226
	7.4.2 In-Memory混合扫描	229

第 8 章 Database In-Memory 与高可用性 233

8.1	Oracle 最高可用性架构	233
8.2	Database In-Memory 与 RAC	235
	8.2.1 利用OCI搭建RAC实验环境	235
	8.2.2 利用Vagrant搭建RAC实验环境	237
	8.2.3 In-Memory数据分布	243
	8.2.4 In-Memory复制	250
	8.2.5 In-Memory与并行执行	252

8.2.6　In-Memory与实例子集发布 ……………………………… 257
8.2.7　实例失效时的In-Memory重新发布 …………………………… 266
8.3　Database In-Memory与ADG …………………………………… 270
8.3.1　利用OCI搭建ADG实验环境 …………………………………… 270
8.3.2　利用Vagrant搭建ADG实验环境 ……………………………… 272
8.3.3　ADG基本概念与Database In-Memory参数 ………………… 275
8.3.4　ADG常用管理和监控命令 ……………………………………… 276
8.3.5　主备数据库发布相同的对象 …………………………………… 279
8.3.6　仅在备数据库发布对象 ………………………………………… 280
8.3.7　主备数据库发布不同的对象 …………………………………… 282
8.3.8　Database In-Memory与ADG主备切换 ……………………… 289
8.4　In-Memory FastStart …………………………………………… 293

第9章　Database In-Memory与可管理性 …………………………… 296
9.1　In-Memory自动数据优化 ………………………………………… 296
9.1.1　自动数据优化基本概念 ………………………………………… 296
9.1.2　In-Memory自动数据优化 ……………………………………… 300
9.2　Automatic In-Memory …………………………………………… 307
9.2.1　自动In-Memory管理 ……………………………………………… 308
9.2.2　自治In-Memory管理 ……………………………………………… 311
9.3　Database In-Memory与分区 ……………………………………… 314
9.3.1　分区发布 …………………………………………………………… 315
9.3.2　分区裁剪 …………………………………………………………… 316
9.3.3　智能分区联结 …………………………………………………… 317
9.3.4　分区交换 ………………………………………………………… 319

第10章　Database In-Memory与大数据 …………………………… 322
10.1　Database In-Memory与外部表 ………………………………… 322
10.1.1　外部表基本概念 ………………………………………………… 322
10.1.2　In-Memory普通外部表 ……………………………………… 324
10.1.3　In-Memory分区外部表 ……………………………………… 333
10.1.4　In-Memory混合分区表 ……………………………………… 336
10.2　内存优化行存储 ………………………………………………… 338
10.2.1　行存储快速查询 ………………………………………………… 338
10.2.2　行存储快速摄入 ………………………………………………… 342
10.3　Exadata In-Memory列格式支持 ………………………………… 345
10.3.1　In-Memory列式缓存基本操作 ………………………………… 346
10.3.2　RAC环境下的In-Memory列式缓存 ………………………… 349
10.3.3　In-Memory列式缓存性能比较 ………………………………… 351

第1章

内存计算概述

1.1 内存计算的兴起

早在 2000 年,一位年轻人在 Intel 开发者论坛上代表他所在的企业发言:"我想把整个互联网放在内存,放在随机存取内存中。"[①] 他的发言引起了一阵笑声,但很快这个想法就被人淡忘了。这位青年人名叫 Larry Page,当时他的公司并不知名,而后来则逐渐被人们熟知,这就是大名鼎鼎的 Google。2006 年,曾获图灵奖的微软科学家 Jim Gray 提出了一个新的观点:"磁带已死,磁盘即磁带,闪存即磁盘,内存为王"。如今,越来越多的互联网企业将数据放置在内存中,很多企业也将内存计算广泛应用于数据缓存、流式数据处理、大数据分析、物联网等领域以提升用户体验,加速产品价值实现。2012 年,内存计算首次出现在 Gartner 的 10 大战略技术趋势中。2013 年,Gartner 将内存计算列为影响信息架构的十大技术趋势之一。2011 到 2014 年,内存数据库管理系统和内存分析连续四年出现在 Gartner 新兴技术成熟度曲线中。隶属于 IDG 的 ComputerWorld 也在 2014 年将内存分析评为大数据分析八大趋势之一。毋庸置疑,内存计算如今已成为主流技术趋势并广泛应用于信息技术各领域中。

那么什么是内存计算?按照 Forrester 的定义,内存计算是利用内存来加速数据访问和应用的性能,并降低应用开发复杂度的技术。Gartner 将内存计算定义为一种革命性的技术,使得企业应用可以在海量数据集上运行高级查询或复杂的事务处理,不仅在速度上快至少一个数量级,并且具有更好的可扩展性,同时仍保持数据的可用性、一致性和完整性。Gartner 公司也强调,这里的内存指的是计算机的主存,即掉电后不再保留数据的内存,而并非闪存之类的非易失存储。GridGain 网站将内存计算定义为一种中间层软件,将数据分布于一组集群服务器的内存中并实现并行处理。GridGain 的定义虽然不尽全面,但也突出了内存计算的一个重要特点,即作为一个新的层次,通过部署在传统的中间层和数据库层之间,来提升整个架构的可扩展性、可

[①] http://www.intel.com/pressroom/archive/speeches/ag021500.htm.

用性和性能。

内存计算并不是一项新的和未经验证的技术。早在 20 世纪 90 年代，IBM 公司的 S/360 大型机就已使用了内存计算技术，目前已经有超过 50 家软件厂商发布了内存计算相关产品。内存计算不仅应用于分析领域，也广泛应用在软件即服务、社交网络、金融、电信、在线游戏等领域。内存计算技术并非只有大型企业才消费得起，许多内存计算软件最大的客户即为中小企业。内存计算生逢其时，但还有很长的路要走，内存计算的兴起将对用户体验、应用设计原则、产品架构和企业战略产生巨大的影响。

1.1.1 硬件的发展

摩尔定律是信息技术领域最知名的分析预测之一，指用一美元所能买到的计算机性能，每隔 18 个月便会增加一倍，这一定律揭示了信息技术进步的速度。内存计算时代的到来也与处理器、内存、磁盘等硬件技术的发展息息相关。

CPU 被称为计算机的大脑，第一个单芯片的微处理器 Intel 4004 由 Intel 公司于 1971 年发布。1978 年，Intel 发布了第一个 16 位的处理器 8086，可寻址 1MB 内存，从此引入了 x86 架构。1985 年，Intel 发布了 32 位处理器 80386，可寻址 4GB 内存。2004 年，Intel 发布了第一个 64 位多核处理器 Xeon。在整个 CPU 的发展进程中，不仅 CPU 时钟频率越来越高，而且芯片密度越来越大，出现了多 CPU 和多核架构。AMD 于 2020 年发布的锐龙 Threadripper 3990X CPU 拥有 64 核。Intel 于 2021 年发布的至强 8358 处理器已经具备 32 核。随着 CPU 及核数的不断增长，CPU 与内存之间的通信出现了瓶颈，于是出现了新的 NUMA 架构以改进传统对称多处理（SMP）架构的扩展性，非一致内存访问（NUMA）架构将 CPU 分为多个模块，每个模块都具备本地和远端内存，这也是其称为非一致的原因。NUMA 架构一方面提升了系统的扩展性，可以组合出更多 CPU、更强处理能力的系统，如 SGI UV、Oracle SPARC T5-8、HP Superdome 及 16 路及以上的 x86 服务器；另一方面，NUMA 架构不仅是硬件架构上的改变，同时也影响应用和数据的设计，特别是针对并行计算应用，数据和处理应尽量遵循本地化的原则，毕竟访问远端内存开销很大。分析型内存计算可以很好地利用 NUMA 架构，一方面分析型数据通常以列式存储，方便进行垂直分区后并行处理；另一方面，分析型数据数值重复度和稀疏度较高，通过与数据压缩技术结合，可进一步缓解内存访问速度和 CPU 处理速度之间的差异。

在数据库发展的过程中，数据库设计者和使用者一直在尽量避免磁盘 I/O。从硬件的角度看，数据处理分为三个部分，即处理器进行计算、存储存放数据以及两者之间的数据传输系统，而这三者中磁盘始终是瓶颈。尽管磁盘的密度越来越高，容量越来越大，但磁盘的 I/O 处理能力却提升得不多。例如一般磁盘的吞吐量在 200MB/s，而内存的数据吞吐量可达到 100GB/s。基于闪存的 SSD 磁盘出现后，读取性能有了较大的提升，但由于其工作原理以及擦写寿命，并不适合写密集型工作负载。机械磁盘的读写延迟一般为 10ms，SSD 的读和写延迟分别为 25μs 和 300μs，而 DRAM 的读写延迟为 60ns，比机械磁盘快 5000 倍，即使是与 SSD 相比，读写速度也要快 400 倍。最初，由于价格高昂和容量有限，内存仅用于缓存部分数据，随着内存容量的不断增加和价格的大幅下降,将所有数据置于内存成为可能,这时,磁盘不再作为数据的主存储,而是作为数据备份和持久化的设备。根据 Gartner 的报告,内存价格每年下降幅度为 32%。2006 年,1GB 内存的价格近 100 美元,2016 年只需要不到 4 美元。内存容量方面,微软 Windows Server 2022 在 x86 服务器上最多可支持 48TB 内存,Red Hat 企业版 Linux 8 在 x86 服务器上可支持 24TB 内存。

1.1.2 软件的发展

在硬件发展的同时,一些软件技术的普及也促进了内存计算技术的发展,其中主要的技术为列式存储或行列双格式存储、内存压缩、分区等技术。

传统关系型数据库主要面向在线事务处理应用,这些应用需要在频繁的小规模更新时仍可以保证很好的性能,因此行式存储成为主流的记录组织格式。而对于在线分析处理和数据挖掘类查询密集型应用,访问模式完全不同,这类应用通常只访问记录中少量属性,通过对大量数据进行计算后得到汇总结果集。如果使用行式存储组织数据,即使只访问少数几列,也需要读取记录中的所有数据,大量的 I/O 只能产生少量有效数据。因此对于在线分析处理类应用,产生了两种应对的方法,即预计算和垂直分区方法。预计算方法将预先计算的结果存放于立方体或物化视图中,这种方法可以对汇总的结果进行切片、下钻和上卷等操作。但这种方法的缺点在于只能加速模式已知的查询,而对灵活的即席查询并不适用,另外,还需要考虑事实表和预计算结果集之间的同步问题。而垂直分区则是将数据按列组织,由于需要查询的属性值集中存放在连续的数据块中,因此对于合计、分组、排序、映射等操作,通过较少的访问即可获取更多的有效数据,这极大提高了数据分析类应用的效率。列式存储提高了 I/O 效率,通过将列式存储完全加载到内存则进一步避免了磁盘 I/O 访问,并且可以充分发挥多处理器多核的并行处理优势,对于即席查询也可以获得很好的性能。内存中列式存储还可以结合单指令多数据(SIMD)CPU 指令进行高效扫描,利用布隆过滤器(Bloom filter)进行高效表联结操作。

最初的内存列式数据库仅适用于数据分析类应用,一些新型的内存数据库通过同时支持事务类应用增加了数据库的普适性,如 Oracle Database In-Memory 和 SAP HANA,这些数据库通过同时支持行列双格式实现对不同特征数据操作的统一支持,同时通过后台任务保持两种格式间数据的一致性。

与列式存储经常结合的一个软件特性是压缩,由于列属性值集中存放,并且由于分析的列取值通常重复度较高,或者大量的取值为空,因此压缩的效率非常高。压缩不仅可以节省昂贵的内存空间,同时也可以提升性能。内存中压缩减少了 CPU 和内存间数据的移动,并且由于在等量空间中存储或缓存了更多的数据,因此可以提高数据处理的吞吐量和缓存命中率。内存数据库最常使用的压缩算法为字典压缩,这是一种轻量级的无损压缩。字典压缩可以将源数据转换为等宽的值,这非常适合于结合 SIMD 实现批量和并行处理,同时也支持无须解压直接在压缩格式数据上进行操作。也可以选择叠加压缩算法或选择压缩比更高的算法,如 RLE 游程编码或 ZIP 算法,但需要在空间节省和处理开销间进行平衡。

1.1.3 企业应用的需求

除了软件和硬件的不断进步，另一个促使内存计算发展的重要因素是企业应用的需求。在云计算、大数据、社交化和移动化趋势的推动下，企业应用需要处理的应用模式越来越复杂，数据量越来越大，对数据处理实时性的要求也越来越高。换句话说，越来越多的传统企业正在向实时企业转变。

实时企业应具有数据驱动、敏捷和高效三种特质，也就是从数据中获得实时洞察的能力，快速调整以适应内外部变化的能力，以及持续流程改进以最大化收益的能力。所有这些能力都需要依赖企业最重要的资产：数据。企业需要在更短的时间内处理更多的数据，更快地对外部市场环境和客户需求做出反应，以提升业务运营效率和竞争力。内存计算技术非常适合于构建实时企业这一目标。

云计算的本质是共享的基础设施、共享的服务，云计算使得传统的孤岛式系统建设转向基于共享资源池的建设模式。在集中整合的环境下，应用需要处理更多的数据和并发访问，因此底层架构需要具备更强的性能和可扩展性。可扩展性不仅意味着当工作负载变化时可以提供足够的存储、网络和计算能力，同时还必须保证相应的服务水准，包括性能和高可用性。大数据是另一个主要驱动力。大数据的三个特征：数据量大、产生速度快和类型多样决定了新型的大数据应用不仅需要处理传统的企业内部数据，还需要应对来自移动端、社交媒体、传感器等多种数据源的海量数据，同时必须快速处理这些数据，以支持实时预警监控、实时辅助决策等需求。这无疑对整个系统架构的性能和可扩展性提出了更高的要求。

无论是公有云还是私有云，大多数应用的运行都依赖于关系型数据库。而数据库恰恰是应用架构层级中最难扩展的部分。传统的数据库扩展方法包括集群、复制和分片。数据库集群技术的典型代表是 Oracle RAC，通过为一个数据库配置多个实例实现可扩展性，每一个实例都可以读写数据。集群技术由于需要共享存储，并且写数据时需要加锁来同步内存缓存中的数据，因此集群的扩展性是有限的，并且在配置和管理上更加复杂。复制技术通过提供一个或多个数据副本来实现业务分离和读写分离，通常副本是只读的，可用于开发测试、数据备份、分析报表等只读应用。复制的问题在于主从数据库之间可能存在数据差异，并非完全一致。另外由于从数据库通常为只读，因此对于需要可写的应用并不能扩展。另一个常用的扩展方法是分片，分片即将一个大数据库拆分为多个小数据库，拆分可以依据业务规则，如按区域、按时间范围，或根据技术规则，如哈希值等。相对于集群和复制，分片具有最佳的可扩展性，但管理维护也最为复杂。分片要求应用能感知数据位置，这通常需要在代码中嵌入路由逻辑或添加新的路由应用来重定向访问路径。当系统扩展时，应用需要能适应分片拓扑的变化，并需要考虑数据的重新分布和迁移。对于一些主数据和参照数据，需要复制到每一个分片中，出于性能考虑，通常采用异步复制，这可能会出现数据一致性的问题。跨多库的聚合操作通常执行效率不高，实现更为困难。相当于单一数据库，分片系统的高可用保护更为复杂，需要为每一个分片实现单独保护。

通过内存计算技术可以提高整个数据库架构的可扩展性，目前可以采用两种方式。第一种是使用独立内存数据库直接替代传统关系型数据库，这类新型数据库性能上优于传统磁盘数据库，功能上与传统关系型数据库类似，支持 SQL 处理，支持事务和分析混合负载，支持集群与复制等功能。但总地来说，成熟的产品还较少。另一种方式是在应用层和数据库层间添加新的内存层，应用层仍保持无状态，而将状态存放于内存层。新的内存层通常为分布式架构，具有良好的可扩展性。内存层自动加载数据库中的数据并将更新自动持久化到后端数据库。对于应用，内存层可保证性能和数据库层的最终持久。利用内存的性能，可提升应用的响应时间和减轻后端数据库的压力。

除了可扩展性的提升，内存计算还极大简化了整个应用软件的架构。传统架构中事务处理和分析处理是分离的，不仅需要复杂的数据抽取、转换和加载，并且需要通过构建立方体或物化视图来加速分析。这种分离的架构不仅带来了数据冗余，并且可能导致数据不一致，维护和更改也比较困难。内存计算通过将所有数据置于内存中，可统一满足不同特征的工作负载，去除了架构中冗余的层次，保证了性能和数据一致性，相应的应用逻辑和硬件架构更加简单，运维更方便，整体拥有成本更加节省，IT 可以更快速地满足不断变化的业务需求，促进业务敏捷和业务创新。

1.2 内存计算技术分类

通过内存计算技术可以支撑内存应用平台，包括内存分析平台、复杂事件处理和内存应用服务器。Gartner 将内存计算技术分为两类，即内存数据管理和高性能消息基础架构。内存数据管理是内存计算技术的主流，又细分为内存数据库管理系统和内存数据网格两类。内存数据库和内存数据网格这两个概念经常容易混淆，虽然两者都是基于内存，并且随着技术的发展也相互借鉴了对方的一些特性，但两者仍然有着明显的区别。简单来说，内存数据库是基于磁盘的关系型数据库的延伸，可用来替代传统的数据库；而内存数据网格技术上更近似于基于内存的 NoSQL 或 NewSQL 数据库，因此也具有更好的可扩展性。两者在应用架构中的位置也不一样，内存数据库位于数据层，而内存数据网格位于中间层，通常作为数据库或其他应用的前置来提高访问速度和减轻后端的压力。

1.2.1 内存数据库

内存数据库的前身是基于磁盘的关系型数据库，随着技术的发展以及内存价格的不断下降，逐渐出现了内存为主存的数据库系统。这里的内存指的是 DRAM，而不是固态硬盘或闪存。通常来说，内存数据库的数据全部驻留在内存中，因此数据访问的速度远高于基于磁盘的系统，并且数据库引擎的实现更简洁，执行指令的时间更快。后来又逐渐出现了内存与磁盘混合的数据库，充分结合了内存的高性能和磁盘的高容量低成本优势。

内存数据库的另一个主要特征是支持数据库的关系型特性，包括事务的原子性、一致性、隔离性和持久化、SQL 数据访问、JDBC 和 ODBC 访问接口、存储过程和函数等高级特性。因此，内存数据库通常可以替换传统的基于磁盘的关系型数据库，并且应用只需要做很少的改动就可以与新的内存数据库兼容。

传统的磁盘数据库通常是纵向扩展的系统，在横向扩展上不具备优势，这主要是由于跨节点的 SQL 联结操作会带来很大的性能开销。利用内存的性能优势，新型的分布式内存数据库可以在一定程度上提高数据库的可扩展性，同时保证数据的一致性。内存数据库另一种横向扩展的方法是通过数据分片技术，这种方式通常需要应用定制的数据路由以及基于业务特征实现数据划分。

典型的内存数据库产品包括 Oracle TimesTen、Redis、SAP HANA 和 Oracle Database In-Memory。

1.2.2 内存数据网格

内存数据网格是另一种主流的内存数据处理技术，与内存数据库一样，数据也是加载到内存中处理，与内存数据库的 SMP 架构不同，内存数据网格通常采用 MPP 架构，SMP 和 MPP 架构如图 1-1 所示。同时内存数据网格的目标是实现更高的可用性和可扩展性，因此非常适合于大数据量高性能并行处理。

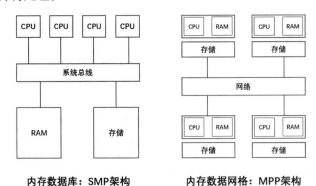

图 1-1　SMP 和 MPP 架构

内存数据网格通常采用分布式对象存储方式实现，应用以键值对的方式存储和访问对象。传统的数据库通常只能使用字符串作为键或值，而内存数据网格的键值对取值可以使用任意的对象，这带来了更大的灵活性，同时由于无须额外的数据组织和分解，应用在业务逻辑的处理上更加简捷。可以直接操作对象也是内存数据网格和内存数据库的主要区别之一。与内存数据库可以独立运行不同，内存数据网格的后端通常需要传统磁盘数据库的支撑，以实现对象关系映射和数据的持久化存储。

内存数据网格的另一个特点是极佳的横向可扩展性，这得益于其底层的分布式架构。通过添加数据节点，内存数据网格可以实现近线性的处理性能和容量扩展，数据通常只在数据节点本地处理，节点之间几乎没有通信，只有在数据复制或增加节点引发数据重新分布时，节点间才有大量的数据交互。

内存数据网格通过锁机制可以实现事务处理和对数据的并发访问，尽管一些数据网格产品也实现了类 SQL 和简单的 SQL 处理，但总地来说，对于 SQL 的支持通常不如传统数据库完备和强大。正因为如此，从应用开发的角度看，如果是从传统数据库移植到内存数据库，应用的改动比较少，而迁移到内存数据网格时，通常需要更多的工作量。

典型的内存数据网格产品包括 Oracle Coherence、Pivotal GemFire、Apache Geode、Hazelcast、GridGain 和 Apache Ignite。

1.3 典型内存数据管理产品

以下介绍的典型内存数据管理产品，每一种都各有特点，例如 TimesTen 的优势在于 SQL、PL/SQL 的支持以及与 Oracle 的缓存集成；Oracle Database In-Memory 的优势在于海量数据的统计分析及应用透明性；Coherence 的优势在于良好的扩展性，并行处理能力和与应用逻辑的贴合；SAP HANA 不仅是一个内存数据库，同时也是一个应用平台，可以很好地与 SAP ERP、CRM、BW 等应用结合；Redis 具有明显的互联网基因，也具备丰富的数据结构，简单易用，一些企业级特性如安全等在不断完善。总之，选择哪一种产品与实际的业务需求、产品的成本与特性，以及开发者的技能和产品熟悉程度等因素紧密相关，还需要具体情况具体分析。

1.3.1 Oracle TimesTen

TimesTen 是一种关系型内存数据库产品，也是第一个商用的关系型内存数据库。TimesTen 最初来自惠普实验室的内存驻留数据库项目 Smallbase。1995 年，TimesTen 被嵌入到惠普的电信产品 OpenCall 中，这是其第一次正式的商业应用。1996 年，TimesTen 从惠普分拆并在加利福尼亚州成立独立公司。1998 年，TimesTen 发布第一个商用的关系型内存数据库，产品在电信和金融行业得到广泛应用。2005 年 6 月，TimesTen 被 Oracle 正式收购，成为其旗下实时数据管理产品家族中的一员。

图 1-2 为 TimesTen 数据库的发展简史，从中可以看出，在被 Oracle 收购之后，TimesTen 与 Oracle 数据库的集成更加紧密，在应用开发、产品管理、数据库功能方面都吸收和借鉴了 Oracle 数据库的很多特性，这大大降低了 TimesTen 数据库的学习、管理和应用的成本，例如 TimesTen 支持 ANSI SQL 和 PL/SQL，支持 Oracle Clusterware，支持 JDBC、ODBC、OCI、Pro*C 开发接口，兼容 Oracle 数据类型，支持 SQL Developer 开发以及 Enterprise Manager 管理等。

图 1-2 TimesTen 数据库的发展简史

TimesTen 支持基于 SQL 的数据访问以及事务的原子性、一致性、隔离性和持久化。除了标准的客户端服务器驱动，TimesTen 还支持性能优化的直连驱动访问 TimesTen 数据库。使用直连驱动时，TimesTen 数据库可以直接关联到应用程序的进程内存空间，应用可以直接访问内存数据，避免了进程间通信，从而极大提升了应用访问的性能。TimesTen 通过检查点文件和日志机制保证事务完整性和数据持久化。当事务提交时，事务日志可以按同步或异步的方式写入日志文件，同时后台进程定期将内存数据库完整写入检查点文件，检查点和日志文件结合可以保证数据库的可恢复性。

如图 1-3 所示，典型的 TimesTen 应用模式可分为 3 类，即传统模式、应用层缓存模式和网格模式。传统模式下的 TimesTen 为保证事务一致性的关系型内存数据库，同时可通过单向或双向数据库复制提供高可用性。应用层缓存模式下的 TimesTen 为 Oracle 数据库企业版的选件，可以用来缓存 Oracle 数据库的数据子集，以提高应用层的响应时间。缓存的数据可以是只读的，如频繁访问的热点数据；也可以是可修改的，如物联网应用中的高速数据摄入。同一份数据可以缓存到多个 TimesTen 数据库用于读取，也可以从多个 TimesTen 数据库采集数据然后最终汇入后端 Oracle 数据库。应用程序使用标准 SQL 读取和更新缓存，并自动执行缓存和 Oracle 数据库之间的数据同步。

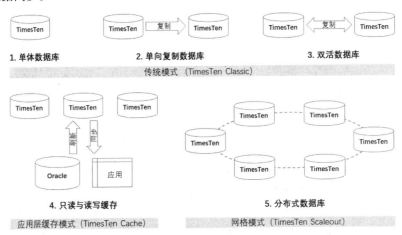

图 1-3 典型的 TimesTen 应用模式

网格模式下的 TimesTen 即 18.1 版本推出的 TimesTen Scaleout，是一个由运行 TimesTen Scaleout 实例的互连主机组成的网格，这些实例共同提供内存中数据的快速访问、容错和高可用性。一个网格中可以创建多个数据库，这些数据库相互独立。数据库中的数据可以创建 1~2 个副本，从而在实例失效时仍可以保证数据可用性。TimesTen Scaleout 的最大好处在于，内存容量和处理能力可随实例的增加而横向扩展，从而可以满足更大规模内存计算应用的需求。

除以上集中典型的应用模式外，利用 TimesTen 的内存计算速度和 SQL 处理能力，还可以构建提供实时响应的数据集成中心（Data Integration Hub）。如利用 Oracle GoldenGate 等实时数据集成工具将 Oracle 或其他数据库中的数据实时复制到 TimesTen，利用 API 将应用数据写入 TimesTen，然后基于这些内存中的数据对外提供服务。这不仅可以提供实时的数据响应以提升用户体验，同时可以减轻后端数据库或应用的压力，让后端释放出更多的能力来处理其他任务。

1.3.2 Oracle Database In-Memory

2014 年 6 月 10 日，Oracle CEO 拉里·埃里森在总部宣布了一项突破性的创新技术：Oracle Database In-Memory。和 Multitenant 一样，Database In-Memory 是 12c 版本数据库中一个全新选件，是继 Exadata 之后最重要的数据库创新之一。在第三方研究机构 Forrester 于 2017 年第一季度发布的报告中[①]，Database In-Memory 和 Oracle TimesTen 一起被归类为内存数据库的范畴，并在市场和产品战略方面均处于首位。

在架构层面，Database In-Memory 可以在数据仓库，或在混合负载的源 OLTP 系统上直接启用。在源端直接启用 Database In-Memory，可避免额外再搭建一套 ODS 系统，减少或免除 ETL 任务。同时由于 Database In-Memory 无须建立专门的分析型索引，还会带来简化管理和交易性能提升等好处。临时性的数据使用并不能从 Database In-Memory 中受益，如 ETL 中的暂存区域。但由于可以直接从内存中读取数据以及避免了更新分析型索引的开销，数据抽取和加载的速度仍可以得到提升。Oracle 提供 In-Memory 建议器，可以通过分析现有数据库的工作负载，判断其是否可以从 Database In-Memory 中受益，哪些表适合于发布到内存列式存储中，同时获取其大致的压缩比和性能提升倍数。

传统的数据库以行的方式存放数据，例如 Oracle 数据库的磁盘存储和内存缓存。行式存储比较适合在线事务处理类应用，因为应用可以快速访问一行记录中的多列。而数据分析类应用则不同，它往往只需要访问部分列中的大量数据，因此相应产生了使用列式存储的数据库，如 KDB、Sybase IQ、Vertica 等。列式数据库非常适合统计分析和数据挖掘类应用，但是当数据发生修改时会引发很大的开销。

Database In-Memory 采用了独特的行列双格式存储，结合了两种格式的优点。应用访问数据库时，查询优化器会根据工作负载特征自动选择行式或列式存储，所有这一切对于应用是透明的，应用代码无须改变。事务日志处理、数据持久化仍由传统的行式存储处理，列式存储则可以专注于加速内存中数据的扫描和过滤。通过更新日志、后台刷新和实时合并机制，从列式存储中数据查询的数据可以始终保持最新，并提供事务一致性。

不同于 TimesTen 和 SAP HANA 需要将所有数据加载到内存中，Database In-Memory 可以选择加载部分表、表的部分分区或表的部分列，因此提供更大的灵活性。可以通过优先级调整表加载的顺序，如指定在数据库启动时加载，或是在实际访问时即时加载。Database In-Memory 支持多种压缩算法，如字典压缩、RLE 压缩和 ZIP 压缩。由于列中的取值重复度相对较高，并且可能存在空值，因此通常可以达到 2～20 倍的压缩比，实际对总内存的增长通常只在 20%左右。如果需要准确的内存需求估算，Database In-Memory 提供压缩建议器，可以对不同压缩级下的压缩比进行评估。

---

① http://www.oracle.com/us/corporate/analystreports/forrester-imdb-wave-2017-3616348.pdf

Database In-Memory 的另一个优势在于与 Oracle 数据库企业版功能和选件的集成，这意味着其可以通过 RAC 实现高可用和横向扩展，结合 ADG 实现灾备和生产端负载转移，以及通过分区实现数据生命周期管理和分区优化的联结和并行处理。

内存计算技术绝不仅仅是将数据放入内存中，从图 1-4 所示的 Database In-Memory 的整个发展历程中就可以看出这一点。在 12.1.0.2 版本的首次发布中，纯内存列式格式、In-Memory 压缩、存储索引和 SIMD 向量处理构成了 Database In-Memory 最核心和最基本的功能，同时也是即使将所有数据置入行式的内存缓存中，内存列式存储也会更快的根本原因。在这 4 项基本功能中，前 3 项均为软件实现，SIMD 向量处理是软件和硬件指令集的结合。In-Memory 联结和 In-Memory 聚合均为优化器的改进，基于布隆过滤器和 Key Vector 数据结构。

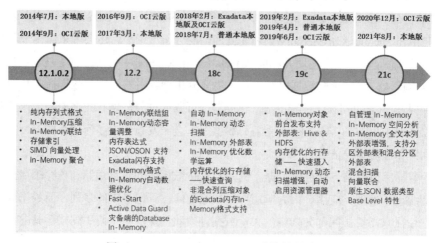

图 1-4　Database In-Memory 功能特性演进

2017 年 3 月 1 日，Oracle 数据库 12.2.0.1 版本正式发布。在性能方面，Database In-Memory 推出了 In-Memory Expression、In-Memory JSON 和 In-Memory 联结组三项新功能。In-Memory Expression 可以将用户定义的表达式虚拟列，或由优化器自动捕获到的频繁执行的表达式物化到内存（IMEU）中，通过避免重复计算提高查询效率。In-Memory JSON 使用 Oracle 优化的二进制 OSON 格式存放 JSON 数据，从而可以与 SIMD 向量计算结合，实现性能提升。In-Memory 联结组为两个表的联结列共享了相同的字典，从而避免了解压带来的额外开销。在容量方面，Database In-Memory 支持 Active Data Guard 和 Exadata，使内存计算延伸到灾备端和超大容量的 Exadata Flash Cache。Exadata 本身支持混合列压缩，Database In-Memory 使 Exadata 支持新的纯列式格式，超大的表可以将不同分区分布在内存和闪存中，但同一个段对象不能在两者间分布。在可用性方面，FastStart 可将内存中的数据定期刷新到磁盘中，这样就省去了发布数据时进行列式转换和压缩的开销，加快了发布速度。在易管理性方面，Database In-Memory 支持与自动数据优化（ADO）策略结合，实现对内存对象的生命周期管理。ADO 根据热图信息，自动将对象加载到内存或从内存中清除，或者调高对象的压缩级以节省空间。

2018 年 2 月 16 日，Oracle 正式发布了数据库 18c 版本，18c 是 Oracle 数据库首个按年度发布的版本。Database In-Memory 在此版本中的增强主要体现在性能、大数据和自动化管理三方面。分析型任务会大量使用数学计算，Oracle 的 NUMBER 类型是用软件实现的，而且宽度不一。In-Memory Optimized Arithmetic 使用优化的紧凑等宽格式存储 NUMBER 类型数据，使得数学计算可以利用 SIMD 指令加速。In-Memory 动态扫描可以利用空闲的 CPU 资源，自动启动轻量级的线程进行并行扫描以提升性能，并且不会对其他任务造成影响。In-Memory 动态扫描与 Oracle 已有的并行执行机制并不冲突，两者是互补关系。Oracle 18c 是 Database In-Memory 支持外部表的首个版本，这使得用户的数据领域延伸到 Oracle 数据库之外。通过将外部表发布到内存，可以避免对外部存储的访问，提升分析性能，同时可以充分利用到 Oracle 数据库的安全性和丰富的 SQL 能力。在 12.2 版本自动数据优化支持的基础上，Automatic In-Memory 将内存自动化管理又向前推进了一步。在内存紧张导致对象无法完整发布时，Automatic In-Memory 会将最不活跃的对象从内存中驱逐，为需要发布的对象留出空间，所有这些都是自动完成的。

2019 年 2 月 13 日，数据库 19c 版本率先在 Exadata 上得到支持。在此版本中，Database In-Memory 对之前部分特性做了增强。例如 In-Memory 动态扫描无须手动启用资源计划即可自动运行，可通过 POPULATE_WAIT PL/SQL 函数支持前台发布，这样应用可以等到对象完全发布后再启动，带给用户良好的使用体验。在大数据方面，新增对 Hive 和 HDFS 两类外部数据的支持，并支持对外部表的并行查询。

2021 年 1 月 13 日，Oracle 在其公有云上率先发布了 21c 版本。在易管理性方面，自治 In-Memory 在自动 In-Memory 基础上更进一步，可实现数据库对象的完全自动化管理。在大数据方面，Database In-Memory 增加了分区外部表和混合分区表的支持，以及全文本列、原生 JSON 列和空间数据的支持。性能方面新增了对向量化联结与混合扫描的支持。最后，Base Level 特性允许用户无须许可即可使用最多 16GB 内存列式存储，此特性可以让更多用户体验 Database In-Memory 的功能和好处。

1.3.3　MySQL HeatWave

MySQL 是最流行的开源数据库，在 2021 年 8 月的 DB-Engines 大排名中仅次于商用数据库 Oracle，位居第二。MySQL 是面向 OLTP 优化的数据库，支持不同种类的存储引擎。在 MySQL HeatWave 发布之前，其中和内存相关的存储引擎包括 MEMORY 和 NDB Cluster。MEMORY 存储引擎可以将整个表的数据置入内存中，提供快速低延时的访问。但由于其不提供持久化，数据在硬件异常或数据库重启时会丢失，因此只适合于存放以读取为主、临时访问的非关键数据，如缓存或会话管理数据。和 MEMORY 一样，NDB Cluster 也是内存数据库引擎，通过定期磁盘检查点支持数据的持久化，提供更好的事务支持和可扩展性。

如果要对 MySQL 中的数据进行分析，通常需要定义复杂的业务逻辑将 MySQL 中的数据抽取到专门的分析型数据仓库。其中涉及数据传输的安全保障问题、数据实时性和增量抽取的问题，以及维护多个系统的管理开销和成本。2020 年 4 月，Oracle 在公有云上发布了 MySQL 数据库服务，紧接着在年底又推出了配套的 MySQL 分析型引擎，2021 年 1 月正式更名为 MySQL HeatWave。

MySQL HeatWave 是基于内存的分析型引擎，由一个 MySQL 数据库服务和多个分析节点组成。HeatWave 安装于 MySQL 数据库服务器节点，负责集群管理，加载数据到 HeatWave 节点的内存中，查询调度和查询执行。HeatWave 支持所有的 MySQL 语法，因此现有使用标准 ODBC/JDBC 连接器的应用和分析工具均可不加改变地使用。HeatWave 使得 MySQL 可以同时支持交易型和分析型负载，MySQL 自动根据工作负载的特征将查询转发到 OLTP 引擎或分析型引擎。

需要分析型处理的数据以内存列格式存放于 HeatWave 节点的内存中，节点的数量取决于需要分析的数据量和内存压缩比。HeatWave 分析集群最少为两个节点，最多可扩展到 64 个节点，节点类型支持虚拟机和物理机。单个虚拟机节点配置为 16 CPU 核和 512GB 内存，可以存放约 400GB 数据。10TB 以上数据建议采用物理机节点，单个物理机节点配置为 128CPU 核和 2TB 内存。

分析型数据在整个 HeatWave 节点中进行分区，在分区内部可以结合多核和 SIMD 向量处理，整体架构提供良好的可扩展性。HeatWave 集群中的每个节点，节点中每一个 CPU 核都支持对分区的数据进行并行处理，包括扫描、过滤、联结、聚合等操作。HeatWave 集群可在线添加和减少节点，但由于数据是在所有节点中进行分布的，因此集群伸缩后需要重新加载数据以重新分布。

利用 HeatWave 实现 MySQL 分析型负载处理的典型流程如下。

（1）确定需要加载到 HeatWave 中的表。分析型查询涉及的表只有加载到 HeatWave 中后，相应的处理才能卸载到 HeatWave 中进行处理。

（2）指定需要加载到 HeatWave 中的列。可以排除不支持的数据类型以及查询不涉及的列，如注释列。

（3）选择列的压缩类型。对于字符串列支持变长压缩（VARLEN）和字典压缩（DICTIONARY）。变长压缩支持联结操作、字符串函数和 LIKE 谓词。如果不需要支持这些操作并且列中的唯一值较少，可以选择字典压缩。

（4）定义数据分布键。默认情况下，表中的数据按主键水平分布到所有 HeatWave 节点。如果其不能达到性能预期，也可以自定义数据分布键，如常用查询中的联结键和分组键。

（5）指定分析型引擎。MySQL 数据库服务使用 InnoDB 作为主引擎，对于需要加载到 HeatWave 中的表，还需指定分析型引擎，即 RAPID 引擎。

（6）加载数据到 HeatWave 集群。加载完成后，在 MySQL 数据库服务中的数据更改自动复制到 HeatWave。复制每 200ms 或当变化数据积累超过 64MB 时进行一次。

（7）当需要分析的数据变化时，按需增减 HeatWave 节点的数量。扩展或减少节点的操作是在线进行的，完成后需重新加载需要分析的数据。

目前的 MySQL HeatWave 数据库系统只能在 Oracle 公有云（OCI）上提供，并且其中的 MySQL 数据库只能是独立数据库。MySQL 数据库服务在 2020 年 12 月支持入站复制，可以将应用场景扩展到本地或集群 MySQL 数据库，如图 1-5 所示。本地数据库可以是独立 MySQL 数据库，或 MySQL NDB Cluster。源端的集群数据库也可以是建立在 Oracle 公有云上的 MySQL HA 数据库系统。

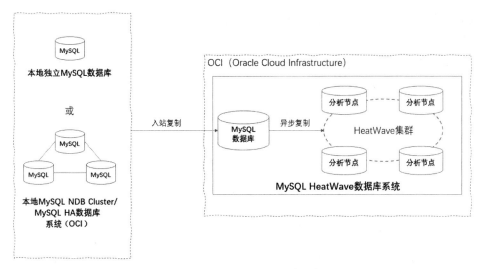

图 1-5　MySQL HeatWave 应用场景

MySQL HeatWave 是一项比较新的服务，功能也在不断完善中。2021 年 8 月 10 日，MySQL HeatWave 同时发布了两项新功能：横向扩展数据管理和 MySQL Autopilot。

横向扩展数据管理为 HeatWave 新增了对象存储持久化层，以列格式保持同步到 HeatWave 中的数据。这既可以保证一致和更快速的恢复速度，也可以在 HeatWave 集群中节点失效时利用其进行恢复。

MySQL Autopilot 利用机器学习实现了配置、数据加载、查询执行和故障处理等多项重要任务的自动化。MySQL Autopilot 包括以下 4 类自动化功能。

（1）系统自动设置。通过对需要加载到 HeatWave 中的表进行采样估算 HeatWave 节点的数量，避免了人工估算的不准确和反复调整。

（2）数据加载。可以提供 3 种功能，①自动并行加载：自动生成脚本并使用最佳并行度来加载数据到 HeatWave；②自动数据放置：推荐分区键以实现最佳数据分布和查询性能；③自动编码：为列推荐最适合的编码方式，以保证查询性能和节省存储空间。

（3）查询执行。可以提供 4 种功能，①自动查询计划改进：利用历史查询的统计信息优化未来查询的执行计划；②自动查询时间估计：在查询运行之前预测查询执行时间；③自动更改传播：自动确定数据变化传播到横向扩展数据管理层的最佳时间；④自动调度：通过队列中各查询预估运行时长自动调度，实现总体等待时间最短。

（4）错误处理。当 HeatWave 集群中的节点失效时，可以进行自动错误恢复，配置新节点并重新加载数据。

1.3.4　Oracle Coherence

和 TimesTen、DBIM 一样，Coherence 也是 Oracle 的内存数据处理产品，不过前者属于内存数据库，而 Coherence 属于内存数据网格。Coherence 最初属于 Tangosol 公司，于 2007 年 3 月被 Oracle 收购。在第三方研究机构 Forrester 于 2015 年第三季度发布的报告[①]中，Coherence 位于所有内存网格产品的第一位。

Coherence 是 Oracle 融合中间件的组成部分，是一个基于 Java 的，高度可伸缩和容错的分布式缓存引擎，提供数据缓存、数据复制、实时事件处理和分布式计算服务，从而确保可伸缩性、可用性、可靠性和性能。和内存数据库一样，Coherence 将频繁访问的数据缓存到内存中，不仅提高了数据处理的速度，同时极大降低了后端数据源的压力。Coherence 内存网格架构如图 1-6 所示。

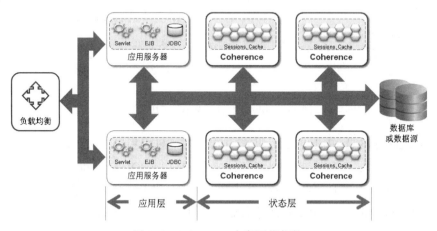

图 1-6 Coherence 内存网格架构

和关系型数据库不同，Coherence 中存放的是无模式数据，这点非常类似于 NoSQL 数据库。数据以键值对的形式分布式存放于 Coherence 节点中，所有节点形成一个大的集群，并通过节点间复制实现缓存的高可用性。Coherence 作为内存数据网格，比内存数据库具有更高的可扩展性和并行处理能力，通过添加节点可以实现内存容量和处理能力的线性增长，当添加或删除 Coherence 节点时，数据会自动在所有节点间重新分布，以实现负载均衡。

Coherence 分布式架构的另一个好处是可利用数据网格的并行处理功能，使应用程序可在整个数据网格中并行查询和分析数据。Coherence 对数据搜索、聚合和排序提供现成的支持，包括对自定义分析函数的支持。

Coherence 支持丰富的缓存拓扑架构，包括全复制缓存、分区缓存、混合缓存等。全复制缓存的特点是，每一个节点上的数据都被复制到内存网格中余下的所有节点，因此网格中的每一个节点都拥有完整的数据集，不过网格的数据容量也受限于网格中最小节点的内存。全复制缓存具有最佳的读取性能，但由于数据需要复制到所有节点，因此会带来一定的开销。分区缓存将数据分布于网格中的所有节点，每一个节点只有整个数据集合的部分数据，为保证高可用性，这些数据被复制到其他网格中的节点。分区缓存的读性能低于全复制缓存，如果数据在本节点不存在，只需要额外的一步即可命中数据，写性能需要引发额外的开销，具体开销取决于后端

---

① https://www.forrester.com/report/The+Forrester+Wave+InMemory+Data+Grids+Q3+2015/-/E-RES120420

数据副本的数量。分区缓存的数据容量为所有节点容量之和，因此可以处理较大的数据集。混合缓存是多层次缓存，近端采用本地容量有限的缓存保证访问性能，后端使用分区缓存保持数据容量的线性扩展。

Coherence 后端通常需要数据库作为支撑，以实现数据的持久化。Coherence 中的对象可实时或异步持久化到数据库，数据库中的数据默认以异步方式刷新到 Coherence 中的对象，通过 Oracle GoldenGate HotCache，数据库中的变化可以实时同步到 Coherence 中的对象，从而保证数据的时效性。

Coherence 通过锁机制或特定的 API 来支持内存中的事务，保证数据一致性和完整性，直到这些事务被永久保存到外部数据源。

Coherence 支持内存中流式数据处理，通过预先定义规则，结合内存的高效性，Coherence 可以在海量的数据中快速过滤和发现关联事件，从而实现实时决策和事件驱动的处理。

Coherence 可以支持在 Java、C/C++ 和 .NET 上开发的应用程序，以及其他语言支持的 RESTful API 接口。此外，Coherence 支持在以不同语言编写的应用程序之间共享数据。

作为 Oracle 产品家族中的一员，Coherence 与 Oracle 的其他许多产品进行了紧密集成，包括 WebLogic Server、ATG、PeopleSoft，以及通过 Oracle Enterprise Manager 进行统一管理。

1.3.5　SAP HANA

SAP HANA 是一个基于列式的内存数据库和应用平台，将事务处理、分析处理以及业务逻辑处理功能组合至内存中，突破了传统交易型数据库架构中，开发支持实时分析应用的限制。高性能分析装置（High Performance ANalytic Appliance，HANA）基于 C++语言编写，其中一些业务和预测分析功能使用了 SAP 内部的 L 语言开发，L 语言也是 C++语言的变种。HANA 采用软硬件一体的形式，硬件由 IBM、Dell、华为等认证的合作伙伴提供。

HANA 于 2009 年开始开发，于 2011 年正式发布。HANA 由 SAP 完全重新构建，虽然诞生的时间并不长，但产品的构想则早已有之，并且 HANA 通过并购以及与研究机构的合作实现了快速发展。最初的产品中已经融入了隶属于波茨坦大学的 Hasso Plattner 学院的研究成果、TREX 搜索引擎、2005 年收购的关系型行式内存数据库 P*TIME，以及与 MySQL AB 合作的关系型数据库 MaxDB 等技术。Hasso Plattner 是 SAP 的创始人，HANA 最初的命名实际上表示 Hasso's New Architecture，也说明了 HANA 是基于多种长期积累技术基础上的新构想。此外，通过对 Business Objects 和 Sybase 的收购，HANA 也融合了这些产品的知识产权，如 Sybase IQ 列式数据库、Business Objects 数据联邦等。之后，产品经过了不断的发展和演进。2015 年，SAP 发布了集成了 HANA 的新一代 ERP 产品 S/4HANA，2016 年，SAP 发布 HANA 2.0，以及集成了 HANA 的新一代数据仓库 BW/4HANA。2020 年，SAP 发布 HANA Cloud。

SAP HANA 集成了一个完整的数据库管理系统，使用标准的 SQL 接口，支持事务的原子性、一致性、隔离性和持久化。SAP HANA 支持大多数入门级 SQL92 语句，使用 Open SQL 的 SAP 应用程序可以不经改变直接在 SAP HANA 平台上运行。SQL 是 SAP HANA 的标准接口，并以

SQL 扩展的方式实现了自由搜索功能，此外通过使用多维表达式（MDX）直接支持了商务智能客户端的使用。和其他数据库系统相比，HANA 的一个显著优势在于与 SAP 应用的紧密集成，包括 SAP 商务套件中的 ERP、CRM、PLM、SCM 以及 SAP BW 数据仓库平台。

SAP HANA 架构如图 1-7 所示，核心服务包括 Index 服务器、预处理服务器和名称服务器。Index 服务器包含实际的数据和数据处理引擎；预处理服务器用于分析文本数据，并提取文本搜索功能所需要的信息；名称服务器存放整个 HANA 系统的拓扑信息，如租户数据库的位置。

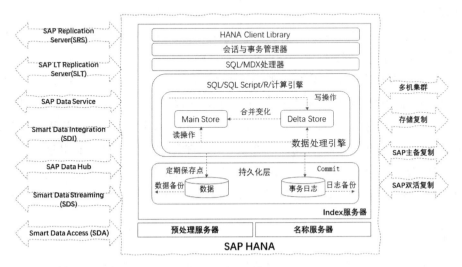

图 1-7　SAP HANA 架构

和 Oracle Database In-Memory 一样，HANA 综合了行列两种格式，并且都是通过联合这两种格式实现了多版本并发控制（MVCC）。不同的是，由于 Oracle 在行格式上已有长期积累，因此以行格式为主，以列格式为辅，变化量记录在列格式一端。而 HANA 以列格式为主，变化记录在行格式中。这是由于其最初的设计是从列格式起步，用于加速分析系统。在图 1-7 所示的数据处理引擎部分，数据以列压缩格式存放于 Main Store 中，写操作不会直接修改 Main Store 中的数据，而是以追加方式插入到行格式的 Delta Store 中，后台在适当时候会将 Delta Store 中的变化合并到 Main Store 中，形成最新版本的数据。Oracle 的 MVCC 实现需要操作两次，包括修改实际数据，并将修改前的数据写入回滚段中。HANA 的实现采用 Insert Only 方式，只需一次操作，而且 Delta Store 中的数据是未压缩的，比较适合高速的插入、删除和更新，但同时需要占用更多的内存空间。HANA 的持久化层采用了定期保存点（Savepoint）和事务日志机制。在事务提交时记录日志，然后后台定期地将内存中的数据以快照的方式存放于 Savepoint 文件中。Savepoint 类似于 TimesTen 的检查点文件，通过 Savepoint 结合日志文件，即可保证数据库的数据一致性和正常恢复。

HANA 是无共享架构，每个节点都有自己的数据和日志，支持内存的纵向和横向扩展。横向扩展时需要考虑数据和应用的分布，其中包括：数据类型，例如主数据、交易数据、参照数据、数据的参照一致性等；数据依赖性、数据访问路径；工作负载，例如读写比例、交易还是查询、更新频率等。数据分布的手段相应可以将相互依赖的表放在一起，将相互关联的表的分区放在同一节点，对于主数据、参照数据，可以复制到每一个节点，以实现数据属地化访问，避免节点间通信。

HANA 的高可用性通过集群和复制实现。支持的集群文件系统包括 IBM 的 GPFS、EMC 的 MPFS、Oracle 的 OCFS2、XFS 和 NFS。通过集群文件系统，备用节点可以访问其他所有活动节点上的底层存储，因此故障时可以按一对多的方式接管失效节点，继续运行业务。HANA 的灾备通过复制实现，可以基于存储的底层数据复制、基于集群文件系统的复制，或基于传输日志的 SAP 系统复制。复制传输可以基于同步或异步方式，备端可处于备用状态，或设为只读以分担主端的工作负载。

数据集成方面，HANA 可以与 SAP ERP、SAP BW 或非 SAP 的第三方数据源进行集成。SLT（SAP Landscape Transformation replication server）和 SRS（SAP Replication Server）均能提供实时数据集成，前者基于触发器，后者基于数据库日志。SAP Data Service 和 SDI（Smart Data Integration）是基于 ETL 的数据复制工具，主要用于准实时的批量数据集成。SAP Data Hub 并不拥有自己的存储，其主要用于多数据源集中管理、编排调度与治理。SDS（Smart Data Streaming）非常适合于处理实时产生的事件数据，并迅速分析结果并给出响应，如社交媒体、物联网传感器和智能设备的数据。无须将数据复制到 HANA，SDA（Smart Data Access）允许访问远程数据，就好像数据存储在本地一样。这使得应用程序可以实时访问和集成来自多个系统的数据。

1.3.6 Redis

Redis 是 NoSQL 类型的内存数据库，最初是由意大利的 Salvatore Sanfilippo 在 2006 年用 C 语言编写，之后得到了 VMware 和 Pivotal 的资助。Redis 是 REmote DIctionary Server 的缩写，正如其名字所示，Redis 的主要用途是内存中的快速信息检索。Redis 是最流行的键值数据库之一，在 Stack Overflow 网站的调查中，Redis 在 2017 到 2020 年连续 4 年位居开发者最喜爱的数据库榜首。典型的 Redis 应用场景包括高性能数据缓存、游戏排行榜、消息发布和订阅、流式数据处理、实时数据分析、物联网高速数据摄入等。Redis 提供免费的开源版本，也提供功能增强的商用版 Redis Enterprise。Redis 支持在本地部署，也可以在 Azure、Google 和 AWS 上以云服务形式提供。

和前面介绍的商用数据库不同，除了内存中运行的高性能外，Redis 还支持丰富的数据类型和算法，如 Strings、Hashes、Lists、Sets、Sorted Sets、Bitmaps 和 HyperLogLogs。这是 Redis 非常重要的特性，只有理解和选择适合的数据类型，才能设计出更好的应用。例如 Strings 类型除了可以表示整型、浮点型、字符串和位图外，还可以存放 XML、JSON、HTML 等文本。Lists 可用来表示队列、堆栈和集合。HyperLogLogs 作为一种常用的算法，可以估算出集合中唯一元素的数量，计算速度快，消耗内存少。这些算法和数据结构可以与互联网应用紧密结合，因此 Redis 也被称为数据结构服务器。在灵活的数据类型基础上，Redis 也着力于将自身打造为多模数据库，以减少跨模式数据开发、集成和运维的成本。Redis 通过动态添加模块的方式来支持这些新的模式，包括 RediSearch 对全文搜索的支持、RedisJSON 存储和查询 JSON 文档、RedisGraph

对 Graph 数据结构的支持、RedisTimeSeries 对时序数据的支持，以及 RedisAI 对人工智能的支持。

Redis 是单线程的，这点非常重要，在任何时候 Redis 只会执行一个命令，因此 Redis 在执行事务的中途是不可能同时执行其他事务的，这也保证了 Redis 事务的原子性和隔离性，不会出现并发访问冲突和数据一致性问题。

作为开源软件，Redis 支持的开发语言比商业软件丰富，除了常用的 C、C++、C#、Java，还支持 Node.js、Ruby、Perl、R、Perl 等近 50 种开发语言。从 Redis 2.6 版本开始，在服务器端也支持执行 Lua 脚本。

内存是临时的易失性存储，为防止数据丢失，Redis 提供了两种持久化的方法：RDB（Redis DataBase）和 AOF（Append-Only File），两种方法可独立或结合使用。RDB 类似于 SAP HANA 的 Savepoint 或 TimesTen 的检查点文件，定期生成 Redis 实例的时间点快照，单个 RDB 文件即可用来恢复数据库。AOF 采用类似于日志的机制，所有的操作都以追加的方式记录在日志文件中，在 Redis 重新加载时，通过重演这些日志即可实现数据库恢复。

RDB 非常适合于备份，例如每小时、每天、每月一次的备份，然后可以按时间点恢复，RDB 也支持压缩，可以方便地传输到远端或公有云上实现灾备。在数据量较大时，RDB 的恢复速度比 AOF 更快。由于快照比较消耗资源，因此生成快照的频率不能过高，这也使得 RDB 不太适合于 RPO（恢复点目标）较小的情形。AOF 采用追加的方式记录日志，记日志的频度可以低至秒级，因此效率非常高，对系统影响小，并大大降低了数据丢失的可能性。AOF 以原始格式记录了所有的操作，更易于理解和解析，不过对于同样的操作集，AOF 文件通常比 RDB 要大。为保证 Redis 的最高可用性和最少数据丢失，最佳建议是结合使用 RDB 和 AOF 两种方式，以同时保证粗粒度和细粒度的数据恢复。

复制被广泛用于提高读扩展性，这样所有的副本可以处理读，而主数据库只需处理写。复制还可以通过数据冗余实现数据保护。如图 1-8 所示，Redis 支持主从复制和双活复制。主从复制支持一个主数据库复制到多个从数据库，也支持从数据库的级联复制。Redis 双活复制使用 CRDT（无冲突复制数据类型）技术实现了地理分布的双活应用，使得两个站点都可以实现本地读写，并保持数据的最终一致性。

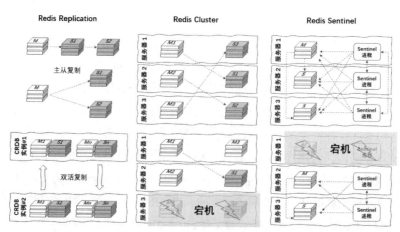

图 1-8　Redis 的复制与扩展

Redis 最初被设计运行在单机上，但单机的内存和能力有限，可能无法容纳所有数据和应对更高的并发访问，此时可以通过分片来实现数据容量和处理能力的扩展。最初的分片方案是通过客户端或第三方的分片中间件来实现的，例如 Twitter 在 2012 年发布的工具 twemproxy，也称为 nutcracker，可以实现哈希等分片算法，并可以在增加节点时自动实现数据重新分片。2015 年，Redis 在 3.0 版本推出了服务器端的分片方案：Redis Cluster。Redis Cluster 可以将数据自动分布到多个节点，同时通过为主节点配置 1 到多个从节点，可以提供一定程度的可用性保护。如图 1-8 中间部分，当服务器 3 故障时，服务器 1 上的从数据库 S3 升级为主数据库 M3，保证了整个集群的数据完整和业务连续。Redis Cluster 内部提供主从数据复制保护，但本质上还是一个用于线性扩展的数据分片方案。

Redis 官方的高可用方案为 Redis Sentinel，可实现 Redis 节点的故障自动切换。Redis Sentinel 是一个分布式系统，目标是提供可靠的主从自动切换，在主节点失效时，可以自动将从节点升级为主节点，在 Sentinel 出现之前，这些都需要手工处理。但 Sentinel 不能分布数据，主节点必须具有所有数据，而其他都是副本。Sentinel 是一个独立的进程，在指定的端口监听。Sentinel 进程间相互保持通信，同时 Sentinel 可以监控所有的 Redis 实例。Sentinal 相当于连接代理，监控后端 Redis 实例的健康状态，因此客户端需要首先连接到 Sentinel 进程，然后根据配置转发到相应 Redis 实例。

第 2 章

搭建 Database In-Memory 实验环境

在实际环境中操作练习是非常有效的学习方式，可以获得直观的体验并加深对概念的理解。本章将介绍如何搭建一个 Database In-Memory 实验环境，后续章节中绝大部分演示都将基于此环境，读者也可以在此环境中执行操作以对照演示结果。

本书所使用的演示环境包括 3 类，即单实例数据库环境、集群数据库（RAC）环境和主备数据库复制（ADG）环境。单实例数据库环境需要主机至少有 8GB 内存，日常使用的笔记本电脑或台式机通常可以满足要求。其余两种环境需要至少 32GB 内存，可以考虑在虚拟化服务器或公有云中搭建。本章仅介绍单实例环境的搭建，RAC 和 ADG 环境的搭建请参见第 8 章。

图 2-1 为实验环境所包含的软硬件组件。通过这些软件可以快速创建标准化的实验环境，并可实现自动化的实验环境创建和管理。

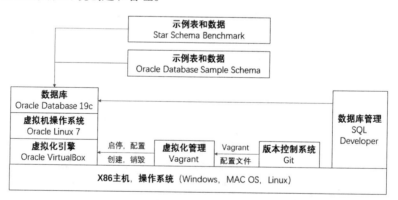

图 2-1 Database In-Memory 实验环境软硬件组件

以下为作者所搭建实际实验环境中包含的软硬件配置及版本，供大家参考。

（1）Dell Latitude 7410、Windows 10、16G 内存、英特尔 i7-10610U 4 核处理器。

（2）Oracle VM VirtualBox 6.1。

（3）Git 2.30.0。

（4）Vagrant 2.2.14。

(5) SQL Developer 20.2.0.175。

(6) Oracle Linux 7 及 Oracle 数据库 19c（由 Vagrant 自动创建）。

基于以上配置，接下来将介绍 Database In-Memory 实验环境的详细搭建过程，这里假设主机硬件环境已就绪，主机操作系统已安装完毕。

2.1 虚拟化引擎 Oracle VM VirtualBox

VirtualBox 是一款非常流行的开源、免费的 x86 虚拟化引擎。虚拟化引擎分为裸金属和寄居式两类，VirtualBox 属于后者，即其需要安装在宿主操作系统之上，而非直接安装于物理主机。通常我们称安装 VirtualBox 的主机为宿主机，而通过 VirtualBox 创建的虚拟机为客户机。由于 VirtualBox 属于 Oracle 公司，因此很多 Oracle 产品的演示虚拟机均基于 VirtualBox 构建，例如 Oracle 数据库和 Oracle Linux。

VirtualBox 的安装包括两部分，即平台特定的基础包和适合所有平台的扩展包。基础包包含了虚拟化引擎及管理虚拟机、计算、存储和网络资源的图形界面。如图 2-2 所示，在 VirtualBox 的下载页面[①]可以看到这两部分软件的下载地址。

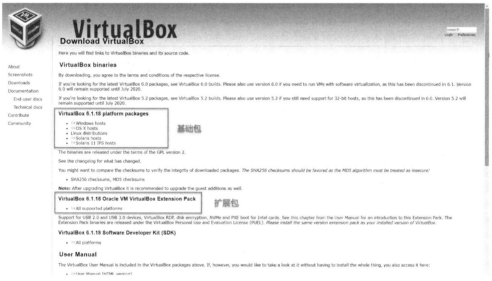

图 2-2　Oracle VM VIrtualBox 下载页面

首先下载与实际环境匹配的基础包并安装，本例为 Windows hosts。VirtualBox 6.1.16 的安装包大小约为 105MB，安装后占用空间约 217MB。虚拟机占用空间通常为几 GB 到十几 GB，为防止硬盘空间满，可以将默认虚拟机目录设定为大容量外接磁盘或其他合适位置，如图 2-3 所示。在 VirtualBox 主界面，选择 File>Preferences... 可进入此设置界面。

---

① https://www.virtualbox.org/wiki/Downloads

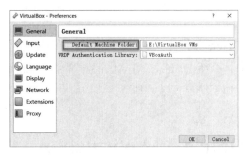

图 2-3　修改 VirtualBox 虚拟机默认目录

基础包安装完毕，接下来单击 All supported platforms 下载并安装扩展包（Extension Pack）。扩展包的版本必须与基础平台包一致，后续如果基础包升级，扩展包也需要做相应的更新。扩展包提供了以下增强支持：

（1）USB 设备。

（2）VirtualBox 远程桌面协议。

（3）Intel PXE 启动。

（4）使用宿主机摄像头。

（5）使用宿主机 PCI 设备。

（6）透明磁盘数据加密。

通过以下命令可确认扩展包安装成功：

```
$ VBoxManage list extpacks
Extension Packs: 1
Pack no. 0:   Oracle VM VirtualBox Extension Pack
Version:      6.1.16
Revision:     140961
Edition:
Description:  Oracle Cloud Infrastructure integration, USB 2.0 and USB 3.0 Host
Controller, Host Webcam, VirtualBox RDP, PXE ROM, Disk Encryption, NVMe.
VRDE Module:  VBoxVRDP
Usable:       true
Why unusable:
```

除基础包和扩展包外，VirtualBox 还包括客户机附加包，即名为 VBoxGuestAdditions.iso 的光盘镜像，需要在虚拟机内部安装。客户机附加包使用户对虚拟机的操作更友好更方便，其包含的主要功能如下所述。

（1）鼠标光标集成。

（2）与宿主机间的共享目录。

（3）更好的图形显示支持。

（4）时间同步。

（5）与宿主机共享的剪贴板。

（6）自动登录。

建议先不安装客户机附加包，因为后续会通过虚拟机环境管理工具 Vagrant 自动安装。

2.2 版本控制系统 Git

Git 是最为流行的版本控制系统之一，是开源和免费的软件。由于其在开发者群体中的高度认可和普遍采用，很多企业都将其代码项目存放在 Git 对应的平台 GitHub 上。GitHub 于 2018 年被微软以 75 亿美元收购。

访问 Git 官方网址[①]，如图 2-4 所示，单击 64-bit Git for Windows Setup 下载安装文件。安装文件大小约为 47MB，安装完成后占用空间约 259MB。以下命令可显示 Git 的安装版本：

```
$ git --version
git version 2.30.0.windows.2
```

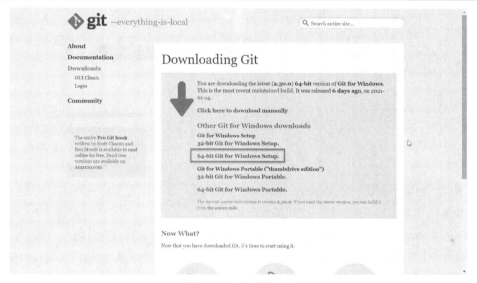

图 2-4　Git 下载页面

2.3 虚拟机环境管理工具 Vagrant

Vagrant 是用于构建和管理虚拟机环境的工具。和 Ansible、Terraform、Chef 一样，Vagrant 属于基础设施即代码（IaC）类工具。IaC 指通过机器可识别的配置文件供应和管理基础设施的流程。由于目标基础设施可以被代码化或以文本表示，因此可以实现版本管理和供应过程的自动化。自动化进而带来了简单易用、可重复部署等好处，使得开发和运维之间协同更顺畅，合作更紧密。

Vagrant 提供了自动化和可移植的虚拟机管理。虚拟机的创建、额外软件的安装、配置更改、启停和销毁等操作均可以通过 vagrant 命令完成。Vagrant 必须与虚拟化引擎等基础设施供应软件相结合，如 VirtualBox、Hyper-V 和 Docker。

---

① https://git-scm.com/download

登录 Vagrant 官网①，下载 Windows 版软件并安装，下载页面如图 2-5 所示。安装文件约 232MB，安装后空间占用约 1GB。Oracle 在 Github 上提供了官方的 Vagrant 项目②，可以很方便地创建 Oracle 软件环境，如 Oracle 数据库、Oracle Linux、Oracle RAC 和 Oracle ADG。执行以下命令，下载 Oracle Vagrant 项目，然后进入目录 OracleDatabase/19.3.0。

```
git clone https://github.com/oracle/vagrant-projects.git
cd vagrant-projects/OracleDatabase/19.3.0
```

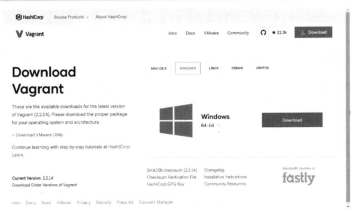

图 2-5　Vagrant 下载页面

此目录下包含了创建一套 Oracle 19.3 单实例数据库所需的配置文件和脚本。在 Vagrant 正式供应数据库环境之前，还需要完成两部分准备工作。

（1）下载 Oracle 数据库相应版本安装介质。Oracle 提供的 Vagrant 项目均以 Oracle Linux 作为底层操作系统，因此需要预先从官网③下载 Linux 版数据库安装介质 LINUX.X64_193000_db_home.zip（约 3GB），然后存放到对应的 Vagrant 项目目录，本例为 OracleDatabase/19.3.0。

（2）修改实验主机配置文件。此项目虚拟机内存默认设置（VM_MEMORY）只有 2300MB，可以分配给数据库的内存（totalMemory），默认设置为 1521MB。为了后续演示需要，需要将这两个参数分别调整到 4096MB 和 3072MB。其中虚拟机内存设置位于文件 Vagrantfile 中，数据库内存设置位于 ora-response 目录下的文件 dbca.rsp.tmpl 中。

```
# OracleDatabase/19.3.0/Vagrantfile 文件中的修改
VM_MEMORY = default_i('VM_MEMORY', 4096)
# OracleDatabase/19.3.0/ora-response/dbca.rsp.tmpl 文件中的修改
totalMemory=3072
```

实际上，众多配置都可以在这两个文件中进行修改和设置，例如数据库字符集、数据库监听端口、SYS 用户口令等。然后执行以下代码开始供应虚拟机：

```
vagrant up
```

---

① https://www.vagrantup.com/downloads

② https://github.com/oracle/vagrant-projects.git

③ https://www.oracle.com/database/technologies/oracle19c-linux-downloads.html

第一次执行 vagrant up 命令的时间会比较长,因为需要更新操作系统和安装数据库软件。取决于网速和主机配置,整个过程需要约 30 分钟甚至更长时间。虚拟机就绪后,可通过 vagrant ssh 命令登录数据库主机。如果习惯用 PuTTY[①],可以安装插件并以 putty 登录。

```
vagrant plugin install vagrant-multi-putty
vagrant putty
```

通过 Vagrant 建立的虚拟机,默认登录的操作系统用户为 vagrant,具有 sudo 权限。虚拟机与宿主机之间可通过 /vagrant 目录共享文件:

```
[vagrant@oracle-19c-vagrant ~]$ id -un
Vagrant

[vagrant@oracle-19c-vagrant ~]$ sudo -l
...
User vagrant may run the following commands on oracle-19c-vagrant:
    (ALL) ALL
(ALL) NOPASSWD: ALL

[vagrant@oracle-19c-vagrant ~]$ df /vagrant
Filesystem       1K-blocks      Used Available Use% Mounted on
vagrant          242218284 148278156  93940128  62% /vagrant
```

然后可以查看主机配置和数据库的版本等信息:

```
[vagrant@oracle-19c-vagrant ~]$ cat /etc/oracle-release
Oracle Linux Server release 7.9

[oracle@oracle-19c-vagrant ~]$ free -m
              total        used        free      shared  buff/cache   available
Mem:           3646         502         807        1146        2336        1770
Swap:          4095           7        4088

[vagrant@oracle-19c-vagrant ~]$ sudo su - oracle

[oracle@oracle-19c-vagrant ~]$ sqlplus / as sysdba
SQL> select banner_full from v$version;

BANNER_FULL
--------------------------------------------------------------------------------
Oracle Database 19c Enterprise Edition Release 19.0.0.0.0 - Production
Version 19.3.0.0.0
SQL> show pdbs

    CON_ID CON_NAME                       OPEN MODE  RESTRICTED
---------- ------------------------------ ---------- ----------
         2 PDB$SEED                       READ ONLY  NO
         3 ORCLPDB1                       READ WRITE NO
```

在首次运行 vagrant 命令时,其运行日志的最后部分会显示随机生成的 SYS、SYSTEM 和

---

① https://www.putty.org

PDBADMIN 用户的口令。如果当时没有记录或忘记了口令，在主机内部也提供了口令重置脚本：

```
cd ~oracle
./setPassword.sh "Welcome1"
```

以下为常用的 vagrant 管理命令：

```
$ vagrant status              # 查看虚拟机状态
Current machine states:
oracle-19c-vagrant            running (virtualbox)

$ vagrant halt                # 停止虚拟机
==> oracle-19c-vagrant: Attempting graceful shutdown of VM…

$ vagrant up                  # 如虚拟机存在则启动虚拟机，否则创建虚拟机
Bringing machine 'oracle-19c-vagrant' up with 'virtualbox' provider…

$ vagrant destroy             # 销毁虚拟机
    oracle-19c-vagrant: Are you sure you want to destroy the 'oracle-19c-vagrant' VM?
[y/N] N
```

2.4　安装示例表与数据

目前为止，已经拥有了一台安装了 Oracle 数据库的虚拟机，接下来将在数据库中安装示例表和导入测试数据。

以下将介绍 Star Schema Benchmark 示例和 Oracle 数据库标准示例的安装。对于本书中大部分的性能测试和功能演示，将使用 Star Schema Benchmark 示例来完成；Oracle 数据库标准示例将用于其他扩展功能的演示。

2.4.1　随书示例

所有随书示例和代码可从 GitHub 下载，在虚拟机中运行以下命令：

```
sudo -s
cd /etc/yum.repos.d
wget http://yum.oracle.com/public-yum-ol7.repo
yum install git -y
su - oracle
git clone https://github.com/XiaoYu-HN/dbimbook
```

进入 dbimbook 目录，其中包括每个章节的示例脚本：

```
$ cd dbimbook
$ ls -l
total 52
drwxr-xr-x. 3 oracle oinstall 4096 Jul 26 17:30 chap02
drwxr-xr-x. 2 oracle oinstall 4096 Jul 26 17:30 chap04
drwxr-xr-x. 3 oracle oinstall 4096 Jul 26 17:30 chap05
drwxr-xr-x. 3 oracle oinstall 4096 Jul 26 17:30 chap06
drwxr-xr-x. 2 oracle oinstall 4096 Jul 26 17:30 chap07
```

```
drwxr-xr-x. 2 oracle oinstall   25 Jul 26 17:30 chap08
drwxr-xr-x. 2 oracle oinstall   56 Jul 26 22:25 chap09
drwxr-xr-x. 2 oracle oinstall 4096 Jul 26 17:41 chap10
-rw-r--r--. 1 oracle oinstall  702 Jul 26 17:30 imstats.sql
-rw-r--r--. 1 oracle oinstall   10 Jul 26 17:30 README.md
-rw-r--r--. 1 oracle oinstall  247 Jul 26 17:30 sample_profile
-rw-r--r--. 1 oracle oinstall   79 Jul 26 17:30 showplan.sql
-rw-r--r--. 1 oracle oinstall   42 Jul 26 17:30 SSB_TABLES.list
-rw-r--r--. 1 oracle oinstall   30 Jul 26 17:30 userlogin.sql
-rw-r--r--. 1 oracle oinstall   78 Jul 26 17:30 xplan.sql
```

2.4.2 Star Schema Benchmark 示例

这里将使用标准的 SSB（Star Schema Benchmark）[①]用于后续测试。SSB 也称为星形模型基准，是由麻省大学波士顿分校开发，专门用于评估数据库系统中星形分析型查询的性能。星形模型通常由一个事实表和多个维度表组成，事实表中包含业务相关可衡量的定量数据，维度表则包含与事实数据相关的描述性属性。多个维度表围绕事实表呈星形放射状，星形模型因此而得名。SSB 模型包含 1 个事实表和 4 个维度表，其说明如表 2-1 所示。

表 2-1 SSB 模型中的事实表与维度表

表 名	类 型	说 明
LINEORDER	事实表	订单表。每一订单由多个订单项组成，每一订单项包括价格、数量、折扣、运输方式及关联的客户、供应商、订单日期和部件号
CUSTOMER	维度表	客户维度。包括客户姓名、城市、国家和地区等
SUPPLIER	维度表	供应商维度。包括供应商名称、地址、城市、国家和地区等
PART	维度表	部件维度。包括部件名称、制造商、尺寸、颜色和所属分类等
DATE_DIM	维度表	时间维度。每一日期一行，包括此日期所属的年、季度、月和星期等

运行以下脚本创建 SSB 用户及表结构，指定新建 SSB 用户的口令：

```
# su - oracle
$ cd ~/dbimbook/chap02
$ ./create_ssb_schema.sh
Please input password for user SSB you will create:<新建 SSB 用户口令>
```

将上面设置的 SSB 用户口令更新到配置文件，后续脚本会使用此用户和口令：

```
$ vi ~/dbimbook/userlogin.sql
connect ssb/<新建 SSB 用户口令>@orclpdb1
```

接下来生成 SSB 示例数据：

```
cd ~
git clone https://github.com/electrum/ssb-dbgen.git
cd ssb-dbgen
sed -i 's/^MACHINE.*=.*/MACHINE=LINUX/' makefile          # 将平台改为 Linux
```

[①] https://www.cs.umb.edu/~xuedchen/research/publications/StarSchemaB.PDF

```
make                                                          # 生成可执行程序 dbgen

# 利用 dbgen 生成示例数据，-T 指定表，-s 指定数据放大系数
# 以下依次生成 CUSTOMER 表、PART 表、SUPPLIER 表、DATE_DIM 表和 LINEORDER 表
./dbgen -s 8 -T c
./dbgen -s 24 -T p
./dbgen -s 8 -T s
./dbgen -s 1 -T d
./dbgen -s 2 -T l
```

建议使用容器数据库架构，并建立可插拔数据库来运行示例。随书示例默认使用可插拔数据库 orclpdb1 运行，如果 orclpdb1 不存在，可使用类似于以下的语句来创建：

```
CREATE PLUGGABLE DATABASE orclpdb1 ADMIN USER pdbadmin IDENTIFIED BY welcome1
DEFAULT TABLESPACE users DATAFILE 'user01.dbf' SIZE 100M AUTOEXTEND ON;
```

确保网络服务名 orclpdb1 可以解析，以下为解析失败的示例：

```
$ tnsping orclpdb1

TNS Ping Utility for Linux: Version 21.0.0.0.0 - Production on 21-JUN-2021 10:17:22

Copyright (c) 1997, 2020, Oracle.  All rights reserved.

Used parameter files:
/u01/app/oracle/homes/OraDB21000_home1/network/admin/sqlnet.ora

TNS-03505: Failed to resolve name
```

如果解析失败，进入输出信息中显示参数文件所在目录，添加网络服务名条目即可：

```
# 查询主机名
$ hostname
dbim

# 查询服务名
$ lsnrctl status|grep orclpdb1
Service "orclpdb1.sub12092311540.training.oraclevcn.com" has 1 instance(s).

$ cd $ORACLE_HOME/network/admin

# 根据主机名和服务名建立网络服务名条目
$ cat <<-EOF >> tnsnames.ora
ORCLPDB1 =
  (DESCRIPTION =
    (ADDRESS = (PROTOCOL = TCP)(HOST = dbim)(PORT = 1521))
    (CONNECT_DATA =
      (SERVER = DEDICATED)
      (SERVICE_NAME = orclpdb1.sub12092311540.training.oraclevcn.com)
    )
  )
EOF

# 确保网络服务名可以连通
```

```
$ tnsping orclpdb1
...

Used TNSNAMES adapter to resolve the alias
Attempting to contact (DESCRIPTION = (ADDRESS = (PROTOCOL = TCP)(HOST = dbim)(PORT =
1521)) (CONNECT_DATA = (SERVER = DEDICATED) (SERVICE_NAME = orclpdb1.sub12092311540.
training.oraclevcn.com)))
OK (0 msec)
```

最后将示例数据导入数据库：

```
export SSBPWD='<SSB 用户口令>'
export CTLDIR=~/dbimbook/chap02/SSB/controlfiles
cd ~/ssb-dbgen

sqlldr ssb/$SSBPWD@orclpdb1 control=$CTLDIR/supplier.ctl
sqlldr ssb/$SSBPWD@orclpdb1 control=$CTLDIR/part.ctl
sqlldr ssb/$SSBPWD@orclpdb1 control=$CTLDIR/customer.ctl
sqlldr ssb/$SSBPWD@orclpdb1 control=$CTLDIR/date.ctl
sqlldr ssb/$SSBPWD@orclpdb1 control=$CTLDIR/lineorder.ctl
```

确认数据已经导入：

```
$ cd ~/dbimbook/chap02
$ ./query_ssb_schema.sh
TABLE_NAME                COUNT
------------------------  ------------
CUSTOMER                   240,000
DATE_DIM                     2,556
LINEORDER               11,997,996
PART                     1,000,000
SUPPLIER                    16,000
```

SSB 示例中共有 4 类共 13 个测试 SQL 语句。从第 1 类到第 4 类，每一类 SQL 涉及的维度表逐渐增多。这些 SQL 语句存放在随书示例代码中，后续的演示中会使用代码来引用这些 SQL，例如，1_1 表示 SSB 示例 SQL 文件 1_1.sql。另外，在 SQL 中通过注释添加了 SSB_SAMPLE_SQL 关键字，以便 Oracle 在监控时能通过关键字捕捉到这些 SQL。

```
$ cd ~/dbimbook/chap02/SSB/sql
$ ls
1_1.sql  1_3.sql  2_2.sql  3_1.sql  3_3.sql  4_1.sql  4_3.sql
1_2.sql  2_1.sql  2_3.sql  3_2.sql  3_4.sql  4_2.sql

$ cat 1_1.sql
SELECT /* 1.1 SSB_SAMPLE_SQL */
    SUM(lo_extendedprice * lo_discount) AS revenue
FROM
    lineorder,
    date_dim
WHERE
        lo_orderdate = d_datekey
    AND d_year = 1993
    AND lo_discount BETWEEN 1 AND 3
```

```
       AND lo_quantity < 25;
```

2.4.3 Oracle 数据库标准示例

Oracle 数据库标准示例[1]由官方提供，包括示例数据库对象和示例数据，广泛用于 Oracle 官方文档，用来演示和说明 SQL 语言概念和数据库特性。Oracle 数据库标准示例目前共包括 7 个用户，分别是 HR（人力资源）、OE（订单条目）、PM（产品媒体）、IX（信息交换）、SH（销售历史）、BI（商务智能）和 CO（客户订单），其中，CO 出现较晚，概念与 OE 类似，包含更多的新特性如 JSON 支持。Oracle 数据库标准示例数据量比 SSB 示例少，但涵盖的对象类型、关联关系更丰富和复杂，因此非常适合于功能说明和演示。

假设数据库已就绪，以下为 Oracle 数据库标准示例的安装过程，详情可参考官方文档[2]。HR 到 BI 的用户数据可以一次性安装完成，CO 用户和数据需要最后单独安装。

```
cd ~
git clone https://github.com/oracle/db-sample-schemas.git
cd db-sample-schemas
perl -p -i.bak -e 's#__SUB__CWD__#'$(pwd)'#g' *.sql */*.sql */*.dat

export SYSPW=<数据库 SYSTEM 用户口令>      # 例如 export SYSPW='ABcd123_#'
export DB=<需要安装示例的目标数据库>         # 例如 export DB=ORCLPDB1
export USERPW=<新建 Schema 用户口令>       # 假设用户 HR、OE、PM、IX、SH、BI 和 CO 口令一样
# 例如 export USERPW='Welcome1'

# 安装从 HR 到 BI 共 6 个用户的示例
sqlplus system/$SYSPW@$DB @mksample $SYSPW $SYSPW $USERPW $USERPW $USERPW $USERPW
$USERPW $USERPW users temp $ORACLE_HOME/demo/schema/log/ $DB

# 单独安装 CO 用户示例
cd customer_orders
sqlplus sys/$SYSPW@$DB as sysdba @co_main $USERPW $DB users temp

sqlplus hr/$USERPW@$DB                    # 用 HR 用户登录，确认安装成功
```

2.4.4 TPC-H 基准示例

TPC-H 是决策支持基准，由面向业务的即席查询和示例数据组成。TPC-H 被广泛应用于数据仓库、数据集市等决策支持系统的性能测试。前面介绍的 Star Schema Benchmark 即基于 TPC-H，可以认为是 TPC-H 的简版。下面将简单介绍 TPC-H 示例的安装过程，尽管本书并没有使用这些示例，但读者可以基于这些示例进行性能比较和对照学习。

---

[1] https://docs.oracle.com/en/database/oracle/oracle-database/21/comsc/index.html

[2] https://github.com/oracle/db-sample-schemas

第一步需要下载 TPC-H 工具包[①]，此软件需要注册下载。最新版本为 3.0.0，大小约为 27MB。下载完成后解压软件，并进入 dbgen 目录：

```
$ unzip /vagrant/tpc-h-tool.zip
$ cd TPC-H_Tools_v3.0.0/dbgen/
```

生成并修改 makefile，编译生成可执行的示例数据生成文件：

```
$ cp makefile.suite makefile

# 按平台和数据库类型修改 makefile 中的环境变量
$ vi makefile
...
CC      = gcc
DATABASE= ORACLE
MACHINE = LINUX
WORKLOAD = TPCH
...

$ make

$ ls -l dbgen
-rwxr-xr-x. 1 oracle oinstall 106520 Aug  7 18:35 dbgen
```

使用 dbgen 生成 8 张表的示例表数据，指定放大系数为 1，其总数据量约为 1GB。

```
$ ./dbgen -s 1

$ du -sh *tbl
24M     customer.tbl
725M    lineitem.tbl
4.0K    nation.tbl
164M    orders.tbl
114M    partsupp.tbl
24M     part.tbl
4.0K    region.tbl
1.4M    supplier.tbl
```

示例数据文件的格式为以竖线分割的文本文件，在 Oracle 数据库中建立表结构后，可以使用后续介绍的 SQL Developer、SQLcl 或外部表方式将数据导入。

2.5 命令行编辑工具 rlwrap

Oracle 数据库命令行实用程序 SQL*Plus 是最常用的 Database In-Memory 配置管理工具，但其本身不支持命令历史和命令行在线编辑功能，使用起来很不方便。通过与 rlwrap（readline wrapper）结合，SQL*Plus 即可补充这些能力，使用效率大为提高。以下为 rlwrap 的安装过程：

```
sudo -s
```

[①] http://tpc.org/tpc_documents_current_versions/current_specifications5.asp

```
yum install readline-devel readline -y
yum install autoconf automake -y
git clone https://github.com/hanslub42/rlwrap.git
cd rlwrap
autoreconf --install
./configure; make install
```

最后，为方便快捷调用，建议为 sqlplus 等常用命令设置别名，然后添加到用户配置文件中。以下为设置别名和添加配置示例，注意需要相应修改 ssb 和 sys 用户的口令。

```
$ cd ~/dbimbook
$ cat sample_profile
export SSBPWD=Welcome1
export SYSPWD=Welcome1
alias sqlplus='rlwrap sqlplus'
alias ssb="sqlplus ssb/$SSBPWD@orclpdb1"
alias sys="sqlplus sys/$SYSPWD@orclpdb1 as sysdba"
alias sqlssb="sql ssb/$SSBPWD@orclpdb1"
alias sqlsys="sql sys/$SYSPWD@orclpdb1 as sysdba"

$ cat sample_profile >> ~/.bash_profile
$ source .bash_profile
```

使用 sqlplus 登录数据库，即可通过上下键浏览命令历史，或使用丰富的 Emacs 快捷键实现命令查看和行内编辑。例如 Ctrl+A 移动到行首，Ctrl+E 移动到行尾。详细的 Emacs 快捷键组合可查看 GNU Emacs 参考卡片[①]。

2.6 数据库开发与管理工具 SQL Developer

SQL Developer 是由 Oracle 提供的免费的图形化数据库开发与管理工具，可简化数据库开发和提高数据库管理效率。在 SQL Developer 中可以浏览数据库对象，运行 SQL 语句和脚本，编辑和调试 PL/SQL，导入和导出数据，创建和查看报告。除 Oracle 数据库外，SQL Developer 还提供了与 Oracle Data Miner、Oracle OLAP、Oracle TimesTen 内存数据库、SQL Developer Data Modeler 和其他第三方数据库的集成。

2.6.1 Windows 下的 SQL Developer 安装

如图 2-6 所示，从 Oracle 官网下载[②]安装介质，平台选择 Windows 64-bit with JDK 8 included。此安装包中已嵌入了 JDK 软件，因此无须额外的设置，只需直接解压即可完成安装。可执行程序为解压后最上一级目录中的 sqldeveloper.exe。SQL Developer Windows 安装包约 422MB，安装后约 655MB。

[①] https://www.gnu.org/software/emacs/refcards/pdf/refcard.pdf

[②] https://www.oracle.com/tools/downloads/sqldev-downloads.html

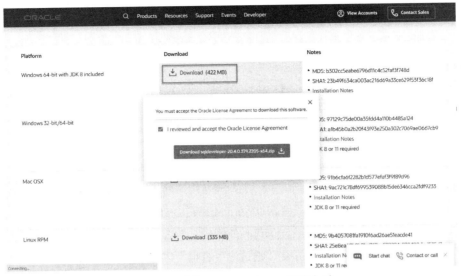

图 2-6　下载 SQL Developer 安装介质

2.6.2　Linux 下的 SQL Developer 安装

在 Linux 下不提供 SQL Developer 和 JDK 合一的安装包，因此这两部分需要单独安装。Linux 下 SQL Developer 的安装步骤如下：

（1）确保 JDK 已安装，SQL Developer 支持 Oracle JDK 8 版本或 11 版本[①]。

（2）从 Oracle 官网下载 SQL Developer 并安装，平台选择 Linux RPM。

（3）运行 sqldeveloper。首次运行时，按提示输入 JDK 路径，后续运行无须再输入。

以下为一个实际的安装示例，虚拟机通过 /vagrant 目录共享宿主机中的文件：

```
$ sudo rpm -Uvh /vagrant/jdk-8u281-linux-x64.rpm
$ sudo rpm -ivh /vagrant/sqldeveloper-20.4.0.379.2205-20.4.0-379.2205.noarch.rpm
$ ls -d /usr/java/jdk*
/usr/java/jdk1.8.0_281-amd64
$ sqldeveloper

Oracle SQL Developer
Copyright (c) 2005, 2020, Oracle and/or its affiliates. All rights reserved.

Default JDK not found
Type the full pathname of a JDK installation (or Ctrl-C to quit), the path will be stored
in /home/vagrant/.sqldeveloper/20.4.0/product.conf
/usr/java/jdk1.8.0_281-amd64
```

2.6.3　使用 SQL Developer 连接数据库

首先需要获取数据库监听的地址和需要连接的数据库服务，在数据库服务器上运行以下命

---

① https://www.oracle.com/java/technologies/javase/javase-jdk8-downloads.html

令，可知数据库服务为 orclpdb1，服务监听端口为 1521：

```
$ lsnrctl status
…
Listening Endpoints Summary…
  (DESCRIPTION=(ADDRESS=(PROTOCOL=ipc)(KEY=EXTPROC1)))
  (DESCRIPTION=(ADDRESS=(PROTOCOL=tcp)(HOST=0.0.0.0)(PORT=1521)))
…
Service "orclpdb1" has 1 instance(s).
  Instance "ORCLCDB", status READY, has 1 handler(s) for this service…
The command completed successfully
```

VirtualBox 创建的虚拟机通常会通过 NAT 方式连接到网络，这表示虚拟机可以连接到外部，但外部无法直接连入虚拟机。此时需要借助 VirtualBox 提供的端口转发功能。

在 VirtualBox 管理界面，可以查看端口转发规则，即客户机（子系统端口）与宿主机（主机端口）之间的端口映射。运行 gitbash，通过命令行也可以查看到相同的信息：

```
$ cd /c/Program\ Files/Oracle/VirtualBox/
$ ./VBoxManage showvminfo oracle-19c-vagrant | grep -i rule
NIC 1 Rule(0):   name = ssh, protocol = tcp, host ip = 127.0.0.1, host port = 2222, guest ip = , guest port = 22
NIC 1 Rule(1):   name = tcp1521, protocol = tcp, host ip = , host port = 1521, guest ip = , guest port = 1521
NIC 1 Rule(2):   name = tcp5500, protocol = tcp, host ip = , host port = 5500, guest ip = , guest port = 5500
```

在图 2-7 中，记录主机端口下显示的端口号，即 1521。在本例中，客户机与宿主机的端口恰好一样，但其实它们是可以不同的。例如假设有多个宿主机上安装了数据库服务，宿主机只有一个，而在一个端口地址只能启动一个监听。此时通过端口转发规则可以将宿主机上其他空闲端口映射到目标客户机上的数据库服务。

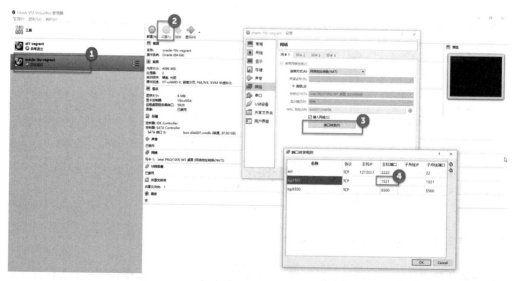

图 2-7　查看 VirtualBox 端口转发规则

最后在 SQL Developer 中新建一个数据库连接，填入用户名、口令及之前记录的主机端口和服务名，如图 2-8 所示。注意主机名输入 localhost 或 127.0.0.1 即可，这实际上是宿主机的 IP 地址，但通过端口转发可以最终连接到目标客户机的数据库服务。

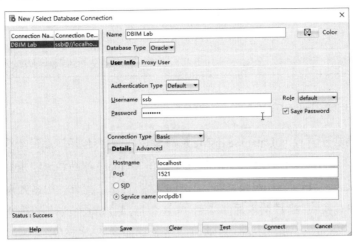

图 2-8　在 SQL Developer 中配置数据库连接

2.7　Database In-Memory 学习资源

2.7.1　文档资源

在 Oracle 帮助中心[①]包含了所有 Oracle 产品相关文档。在 Oracle 数据库的性能主题部分，可以找到 Database In-Memory 产品手册，这是学习 Database In-Memory 基本概念和日常备查必不可少的文档。

在 Oracle 官网 Database In-Memory 页面[②]资源部分，可以找到 Database In-Memory 相关技术白皮书。此页面还提供大量客户案例介绍，通过客户案例的研读，可以了解 Database In-Memory 适用的业务场景、解决的问题和使用效果。

表 2-2 列出了推荐阅读的 Database In-Memory 文档。

表 2-2　推荐阅读的 Database In-Memory 文档

文　　档	说　　明
Database In-Memory 产品手册[③]	系统介绍了基本概念、内存列式存储的配置与发布、查询优化、高可用特性等方面
Database In-Memory 产品介绍[④]	快速了解 Database In-Memory 主要特性和优势

---

① https://docs.oracle.com

② https://www.oracle.com/cn/database/technologies/in-memory.html

③ https://docs.oracle.com/en/database/oracle/oracle-database/21/inmem/database-memory-guide.pdf

④ http://www.oracle.com/technetwork/cn/database/options/database-in-memory-ds-2210927-zhs.pdf

文　　档	说　　明
何时使用 Database In-Memory[①]	Database In-Memory 的优势及应用场景，下载后请将后缀改为 pdf
Database In-Memory 快速入门指南[②]	从规划、配置到实施全过程的概要介绍
Database In-Memory 实施指南[③]	是快速入门指南的详解版，包括最佳实践和延伸阅读文档

2.7.2　Database In-Memory 官方博客

Oracle 官方博客中开设的 Database In-Memory 专栏[④]，内容包括产品最新特性、技术深入解读、活动与讲座预告等。文章作者大多是 Database In-Memory 产品经理和相关领域专家，是全面和深入了解 Database In-Memory 技术特性首选的学习资源。另外，很多知识点的介绍都会结合示例，兼顾理论和实践，解读清晰透彻。专栏中还包括部分精心策划的系列文章，其中 Getting started with Oracle Database In-Memory 系列[⑤]一共 5 篇，非常适合于初学者入门使用。

如图 2-9 所示，在此专栏还专门设计了资源汇总页面。其中包括了 Database In-Memory 相关的技术文档、视频、市场活动资料和动手实验材料等。

图 2-9　官方博客上的学习资源汇总页面

2.7.3　动手实验资源

Oracle LiveLabs[⑥]是由官方提供的动手实验资源，用户可按角色（DBA、应用开发者或

---

① https://www.oracle.com/technetwork/database/in-memory/overview/twp-dbim-usage-2441076.html
② https://www.oracle.com/technetwork/database/in-memory/learnmore/dbim-quickstart-guide-v5-6525689.pdf
③ https://www.oracle.com/technetwork/database/in-memory/learnmore/twp-oracle-dbim-implementation-3863029.pdf
④ https://blogs.oracle.com/in-memory
⑤ https://blogs.oracle.com/in-memory/getting-started-with-oracle-database-in-memory-part-i-installing-enabling
⑥ https://apexapps.oracle.com/pls/apex/dbpm/r/livelabs

DevOps 工程师)或按产品(Oracle 数据库或 Oracle 公有云)来选择实验。Oracle LiveLabs 中的实验均在 Oracle 公有云中运行,一部分实验只提供实验脚本,用户需要自己提供云账号;为方便大家学习,部分实验会为用户临时生成短期的公有云账号,Database In-Memory 基础实验即属于此类。运行实验需要 Oracle 账号,此账号可在 Oracle 官网[①]进行申请,同时实验就绪和到期时也会向此账号绑定的邮箱发送通知。

启用 Database In-Memory 实验的过程如下:

(1)在顶端搜索栏输入 Database In-Memory。

(2)确认搜索结果中包含 Database In-Memory Fundamentals 实验。

(3)单击 Database In-Memory Fundamentals 实验上的 Launch 按钮。

(4)在新页面中单击 Reserve Workshop on LiveLabs 按钮。

(5)输入 Oracle 账号进行登录。

(6)进入实验预约页面。如图 2-10 所示,指定实验开始时间,并提供 SSH 公钥。

(7)实验就绪后,在 My Reservations 页面查询实验主机公网 IP 地址。

(8)使用与之前提供的 SSH 公钥对应的私钥,以 opc 用户登录实验主机。

(9)按照实验指南开始 Database In-Memory 实验。

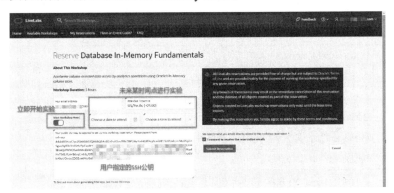

图 2-10 Oracle LiveLabs 实验预约页面

用户可通过此实验了解 Database In-Memory 基础的配置和启用操作、核心特性的演示以及与传统方式相比所带来的性能提升。此实验设计完备,并提供了大量精心制作的实验脚本,用户可以循序渐进对 Database In-Memory 关键特性逐一体验。此实验为用户预留了 6 小时的时间,在实验过期后可以再次预约。

2.7.4 My Oracle Support

My Oracle Support[②](简称 MOS)是 Oracle 官方服务支持网站,包含知识库、服务请求管理、补丁更新、认证列表和支持社区等功能,是 Oracle 与客户交互的重要渠道。

---

① https://profile.oracle.com/myprofile/account/create-account.jspx

② https://support.oracle.com

对于 Database In-Memory，My Oracle Support 的主要作用在三方面：补丁下载、认证列表查询和知识库。

Database In-Memory 建议数据库在主版本的基础上应用最新的补丁，这些补丁均可以在 MOS 网站下载。在 MOS 网站产品认证页面可以查询 Database In-Memory 所支持的操作系统版本，以 Oracle 19.0.0 为例，支持的主流操作系统包括：

（1）Linux x86-64：Red Hat Enterprise Linux 7 和 8、Oracle Linux 7 和 8、SLES12 和 15；

（2）IBM AIX：7.1 和 7.2；

（3）Microsoft Windows x64 (64-bit)：8.1、10、2012 R2、2016 和 2019。

在 MOS 知识库中，推荐阅读的文档如表 2-3 所示。

表 2-3　My Oracle Support 知识库推荐 Database In-Memory 文档

MOS 文档 ID	简　　介
1121043.1	为 Oracle EBS（电子商务套件）性能问题搜集诊断数据
1329441.1	Oracle 数据库数据仓库组件相关白皮书汇总
1903683.1	介绍 Database In-Memory 最为全面的 MOS 文档，包括了基本概念、特性概览、常见问题和推荐阅读等
1926052.1	如何确认 Database In-Memory 选件已使用
1929758.1	如何配置 Database In-Memory 和确认配置
1935305.1	介绍 12.1.0.2 支持的新特性：In-Memory 聚合
1937782.1	Database In-Memory 相关补丁
1947581.1	解释更新语句不会将表置入内存列式存储
1950831.1	执行计划和实际执行不符问题的处理
1954808.1	Database In-Memory 如何保证数据读取一致性
1956736.1	In-Memory 表在系统视图中不显示的问题
1965343.1	Database In-Memory 建议器
1965472.1	Database In-Memory 出现问题时的数据搜集
1986938.1	介绍如何利用 Enterprise Manager 中 In-Memory Central 页面管理 Database In-Memory
2025309.1	如何在 Oracle EBS（电子商务套件）中使用 Database In-Memory
2137389.1	如何发布 In-Memory Join Group
2211831.1	介绍 12.2.0.1 支持的新特性：自动数据优化
2232741.1	哪些情况下，数据库对象不会被发布到 In-Memory 内存列式存储
2242973.1	介绍 12.2.0.1 支持的新特性：内存表达式
2242985.1	介绍 12.2.0.1 支持的新特性：FastStart
2242995.1	介绍 12.2.0.1 支持的新特性：Join Group
2446648.1	如何启用 Automatic In-Memory 特性
2494116.1	介绍 18.1 新特性 Automatic In-Memory
2526168.1	RAC 中 INMEMORY DISTRIBUTE FOR SERVICE 不生效问题的解决
2606619.1	FastStart 工作原理与示例
2655885.1	在启用 Database In-Memory 后，SQL 查询变为并行问题的处理

2.7.5 视频学习资源

视频资源包括两个部分。第一个是 Database In-Memory Office Hour[1]，是由产品经理定期举办的专题讲座，每一期针对一个特定的主题进行深入探讨和解读，并可在线回答观众提问。由于讲座时间通常在北京时间凌晨，不能现场参加的观众可后期回看录制视频。

另一个视频学习资源是 YouTube。可以订阅 YouTube 上开设的 Oracle Database In-Memory 频道，其中包括产品演示、使用技巧和会议讲座。此外，也推荐由 Oracle 大学制作的 In-Memory Column Store Architecture Overview[2]，以及卡耐基梅隆大学数据库系统课程中的 Oracle In-Memory Databases 视频课。

---

[1] https://devgym.oracle.com/pls/apex/dg/office_hours/681

[2] https://www.youtube.com/watch?v=fMW2-TDheec

第3章

Database In-Memory 基本概念与架构

3.1 Oracle 数据库基本概念

3.1.1 Oracle 数据库版本简介

Oracle 数据库支持本地和公有云两种部署方式。相应地,不同的部署方式支持的数据库版本也不一样。

本地部署也称为用户内部部署,目前支持 4 个版本。

1. Oracle 数据库企业版

企业版是最经典、使用最为广泛的数据库版本,为任务关键型事务应用程序、查询密集型数据仓库以及混合负载应用提供高效、可靠和安全的数据管理。此外,数据库企业版可以搭配数据库选件和数据库管理包来扩展其功能。Oracle 提供了 RAC、ADG 和 Database In-Memory 等十多个数据库选件,可满足用户对性能、可扩展性、高可用性、安全性、可管理性、数据仓储和大数据等方面的特定需求。数据库管理包可实现单一平台、集中和自动化的数据库运维管理。

2. Oracle 数据库标准版 2

标准版 2 是从 Oracle 12.1.0.2 开始支持的版本,取代了之前的标准版和标准版 1。标准版 2 是一款适用于工作组和部门级应用的经济实惠、功能完备的数据库。和企业版不同,标准版 2 不能搭配数据库选件,也不能通过数据库管理包进行管理,对底层硬件的能力也有所限制,只能运行在最多 2 个 CPU 插槽的单个服务器上。

3. Oracle 数据库个人版

个人版只能用于单用户,适合于应用开发环境。它包括企业版的所有功能,以及除 RAC 和 RAC One Node 外的所有数据库选件。个人版只支持 Windows 和 Linux 两个平台,并且不能通过数据库管理包进行管理。

4. Oracle 数据库快捷版

快捷版是一个免费版本，适用于数据库功能体验和试用，目前只有 18.4 发行版支持快捷版。快捷版包含了分区、高级压缩、高级安全和 Database In-Memory 等选件，但不包括 RAC 和 ADG，对多租户只支持最多 3 个 PDB。和个人版一样，快捷版只支持 Windows 和 Linux 两个平台。由于主要用于试用，快捷版不提供支持和补丁，并且数据库只支持 2 CPU 线程、2GB 内存和 12GB 用户数据。

本地部署 4 个版本的主要差异可参见表 3-1，在这些版本中，企业版可搭配 Database In-Memory 选件，个人版和快捷版包含 Database In-Memory 选件。

表 3-1　Oracle 数据库本地部署版本

版本名称	简称	应用场景	数据库选件支持	支持通过数据库管理包管理	许可
Oracle 数据库企业版	EE	企业级 生产应用	可搭配所有选件	是	按 CPU 核
Oracle 数据库标准版 2	SE2	部门级 生产应用	无	否	按 CPU 槽位
Oracle 数据库个人版	PE	单用户 开发环境	除 RAC 和 RAC One Node 外的所有选件	否	按用户
Oracle 数据库快捷版	XE	功能体验与使用	分区、Database In-Memory 等少数选件	否	免费

Oracle 公有云也称为 Oracle 云基础设施（OCI）。由于采用了本地部署数据库相同的技术、架构和管理方式，并进一步实现了标准化、服务化和优化，因此可提供与用户自有数据中心部署相同甚至更高的安全性、可靠性和性能；并且供应、扩展和运行维护的过程更加简化和快速；用户只需为实际用量付费，无须任何前期资本投入，可极大地降低运营成本。

Oracle 公有云中部署的数据库包括 4 个版本，如表 3-2 所示，只有超强性能版提供 Database In-Memory。另外，由于云服务中的数据库数据是强制加密的，尽管透明数据加密属于高级安全选件的功能，但其已包含在所有数据库版本中。

表 3-2　Oracle 数据库云部署版本

版本名称	简称	描述	数据库云服务支持
数据库云服务标准版	DBCS SE	包括 Oracle 数据库标准版 2 软件	虚拟机数据库云服务 裸金属数据库云服务
数据库云服务企业版	DBCS EE	包括 Oracle 数据库企业版软件、数据脱敏与子集包、诊断包、调优包和真正应用测试选件	虚拟机数据库云服务 裸金属数据库云服务
数据库云服务高性能版	DBCS EE-HP	在数据库云服务企业版基础上增加了多租户、分区、高级压缩、高级安全、标签安全、数据库保险箱、空间与图、OLAP 和高级分析选件以及数据库生命周期和云管理包	虚拟机数据库云服务 裸金属数据库云服务
数据库云服务超强性能版	DBCS EE-EP	在数据库云服务高性能版基础上增加了 Database In-Memory、ADG 和 RAC 三个选件	虚拟机数据库云服务 裸金属数据库云服务 Exadata 云服务

3.1.2 Oracle 数据库企业版选件

正如在购车时可以加配天窗、行车记录仪和倒车影像等设备一样，对于数据库企业版，也可以通过搭配选件来扩展其功能，以满足用户对于高可用性、安全性、可管理性等多方面的企业级需求。

数据库企业版目前共有 15 个选件，其简要介绍可参见表 3-3。2019 年 12 月，Oracle 宣布 Machine Learning 和 Spatial and Graph 两个选件不再需要额外许可，开放给所有数据库版本使用，以支持其多模融合数据库的发展战略。

表 3-3　Oracle 数据库企业版选件

数据库选件名称	别　名	描　述	功　能　域
Multitenant	多租户	用于提供最高水平的数据库整合，而且无须更改任何现有应用程序	整合
Real Application Clusters	RAC, 真正应用集群	多实例共享缓存的集群数据库架构	高可用
Real Application Clusters One Node	RAC One Node	单实例运行的主备集群架构，可升级为 RAC	高可用
Active Data Guard	ADG, 活动数据卫士	数据保护、数据可用性和灾难恢复解决方案	高可用
Partitioning	分区	将大型数据库表和索引细分成更小的对象，从而在更细的粒度级别对它们进行管理	大型数据库，数据仓库，商务智能
Real Application Testing	RAT, 真正应用测试	数据库负载录制和重放，在测试环境中评估变更对应用的影响	可管理性
Advanced Compression	高级压缩	提供一组全面的压缩功能，帮助降低存储成本的同时提高性能	大型数据库，数据仓库，商务智能
Advanced Security	高级安全	提供加密和编辑两种重要的预防性安全控制，从源头保护敏感数据	安全
Label Security	标签安全	提供行级安全性，允许具有不同访问权限的用户查看他们能够查看的数据子集	安全
Database Vault	数据库保险箱	数据库权限分析、权限控制和敏感数据保护，提升安全与合规性	安全
OLAP	在线分析处理	数据库内置多维分析引擎，提供丰富的分析和出色的查询性能	大型数据库，数据仓库，商务智能
TimesTen Application-Tier Database Cache	TimesTen	内存优化的关系数据库，可独立使用或作为 Oracle 数据库的读写缓存	性能
Database In-Memory	In-Memory 数据库	通过列式内存优化了分析和混合负载应用，支持实时分析、商务智能和报表	性能
Machine Learning	机器学习，高级分析	支持 SQL 与 R 的数据库内机器学习算法	大型数据库，数据仓库，商务智能
Spatial and Graph	空间与图	通用属性图形数据库和图形分析特性，支持信息系统应用以及基于位置的服务应用	空间与图数据

在以上选件中，除 TimesTen 是独立产品外，其他选件均为 Oracle 数据库的一部分，并且大多数都不需要额外安装，即数据库安装完成后已包含在其中。从以下输出可知，Database In-Memory 中包含的组件 In-Memory Column Store 和 In-Memory Aggregation 均已安装：

```
SQL> select parameter, value as "INSTALLED" from v$option order by 1;

PARAMETER                               INSTALLED
--------------------------------------  ---------
Active Data Guard                       TRUE
Advanced Analytics                      TRUE
Advanced Compression                    TRUE
Data Mining                             TRUE
Data Redaction                          TRUE
In-Memory Aggregation                   TRUE
In-Memory Column Store                  TRUE
OLAP                                    TRUE
Oracle Data Guard                       TRUE
Oracle Database Vault                   FALSE
Oracle Label Security                   FALSE
Partitioning                            TRUE
Real Application Clusters               FALSE
Real Application Security               TRUE
Real Application Testing                TRUE
Snapshot time recovery                  TRUE
Spatial                                 TRUE
Transparent Data Encryption             TRUE
```

3.1.3 数据库管理包

Oracle Enterprise Manager（OEM）是 Oracle 内部部署的管理平台，可以对用户数据中心内部或公有云上的 IT 资产进行集中管理。OEM 支持非 Oracle 产品，但其优势仍体现在自身产品的管理上，包括 Oracle 数据库、中间件、应用和集成系统。

在所有数据库版本中，只有数据库企业版能通过 OEM 进行管理，这包括本地部署的企业版以及云中部署的企业版、高性能版和超强性能版。和数据库选件类似，用户可以搭配 OEM 数据库管理包来简化管理和实现功能扩展。OEM 提供 5 个数据库管理包，如表 3-4 所示。

表 3-4 Oracle Enterprise Manager 数据库管理包

名　称	别　名	描　述	依赖性
Diagnostics Pack	诊断包	通过内置于 Oracle 数据库内核中的性能诊断引擎发现性能问题根因，包括性能统计信息收集、实时和自动性能分析、比较和诊断，以及系统监视和通知功能	
Tuning Pack	调优包	Oracle 数据库环境的专业性能管理、实时 SQL 监视、自动 SQL 调优和建议、对象重新组织等功能，从而最终实现数据库调优流程自动化	诊断包
Database Lifecycle Management Pack	生命周期管理包	实现 Oracle 数据库生命周期管理所需过程简化和自动化，包括供应、修补、配置管理和持续变更管理、灾难保护等，并提供最佳建议和合规性报告	

名　　称	别　　名	描　　述	依　赖　性
Data Masking and Subsetting Pack	数据脱敏与子集包	依靠屏蔽和子集规则生成用于测试、开发和其他活动的生产数据副本，丢弃不必要的数据。同时满足开发测试与安全合规要求	
Cloud Management Pack for Oracle Database	数据库云管理包	帮助云管理员识别池化资源，配置基于角色的访问，定义服务目录和计费方案，允许用户请求数据库服务并按需使用	生命周期管理包

和 Database In-Memory 相关的管理包主要是诊断包和调优包，例如用于建立性能基线的 AWR（Automatic Workload Repository）报告属于诊断包，在工作负载捕获和分析时使用的 SQL Monitor 及用于内存对象建议的 Database In-Memory Advisor 都属于调优包。

3.1.4　数据库版本号

每一个数据库版本都有对应的版本号，平时所说的 11g、12c 或 19c 只是省略的叫法，只反映了数据库的主版本号，实际上完整的数据库版本号由 5 部分组成。在数据库升级、补丁下载、功能与兼容性确认时，就需要用到完整的数据库版本号。

数据库 11g 和 12c 采用的 5 段式版本号含义如图 3-1 左半部分所示。其中第 3 位与数据库无关，因此总是 0；第 4 位通常指补丁集版本，第 5 位通常指补丁集更新版本。有时为了简便也会省略末尾部分，例如 Database In-Memory 需要 12.1.0.2 以上才能支持，又如比较经典的版本 11.2.0.4 和 12.2.0.1。

为缩短新特性推出的周期，并给予用户更灵活的数据库升级与更新选择，2018 年后发行的数据库版本采用了新的、更为简单的 5 段式版本号命名规则，如图 3-1 右半部分所示。其中第 1 位是数据库版本首次发行年份；第 2 位指更新版本（Release Updates），以下简称 RU 或更新；第 3 位指更新修订版本（Release Update Revisions），以下简称 RUR 或修订；第 5 位预留或表示更新版本发布的日期。由于最后两位为预留，因此大部分时候只会用到前面 3 位，如 19.9.0。

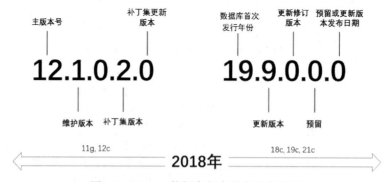

图 3-1　Oracle 数据库版本号定义与描述

3.1.5　数据库升级与更新

数据库升级与更新是数据库管理员日常工作的重要组成部分，但相比如何实施升级与更新，选择适合的数据库版本同样非常关键。

Oracle 数据库主发行版分为两类，即长期版本和创新版本。长期版本可避免频繁升级，并提供最高级别的稳定性和最长的错误修正支持时间，具有 5 年的标准支持期和随后 3 年的扩展支持期。在两个长期版本之间，Oracle 还会发布多个创新版本。创新版本包含了新特性和现有特性增强，这些变化也会融入下一个长期版本中。创新版本包括至少 2 年的标准支持，但没有扩展支持。如图 3-2 所示，目前最新的长期版本为 19c，而 21c 是最新的创新版本。

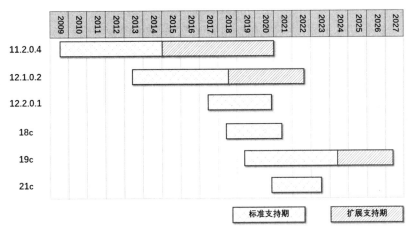

图 3-2　数据库发行版与支持生命周期

在标准支持期，用户可获得最全面的支持服务，包括与最新 Oracle 和第三方产品的认证、更新、修补和安全警示等。到了扩展维护期，主要的区别是对于新的第三方产品不一定会做认证。扩展维护到期后，将不再提供新的补丁和更新以及与 Oracle 和第三方产品的认证[①]。由于数据库只是应用运行环境中的一部分，其他还包括存储、主机、操作系统、备份和数据集成软件等，因此与这些产品的兼容性认证非常重要。这也意味着越到后期，数据库支持的范围和力度会越来越小。因此必须提前准备，选择合适的时间升级到新的主版本。

例如数据库版本为 12.1.0.2，也就是首个支持 Database In-Memory 的版本。按照 Oracle 支持文档[②]的建议，此版本应升级到目前最新的长期版本 19c。如图 3-2 所示，在 19c 发布时，12.1.0.2 已进入扩展支持期，到其扩展支持期结束还有 3 年多的时间可用来准备升级。

升级完成后，下一个重要事项是更新。更新是指对软件中的缺陷和安全漏洞进行修补，从而使软件运行更稳定。更新实施策略方面，分为主动和被动两种。被动策略是指针对突发问题临时采取应对措施。这种策略可能会给业务带来巨大的风险和损失，故障处理的压力和难度也会更大。而主动策略则是按照更新和修订发布的规律，积极评估和分析，并加以实施的策略。对于 Oracle 数据库而言，更新和修订就是之前提到的 RU 和 RUR。

从一个主版本到下一个主版本称为升级，主版本之后的修补和完善则称为更新，更新通过

---

① http://www.oracle.com/us/support/library/lifetime-support-technology-069183.pdf

② https://support.oracle.com/epmos/faces/DocContentDisplay?id=742060.1

部署 RU 或 RUR 实现，RU 和 RUR 都属于补丁类型。

RU 在每个季度的首月发布。RU 对客户最有可能遇到的错误进行修复，包括安全相关的修复以及高优先级的非安全修复。RU 是累积的，当前 RU 包含了之前所有 RU 中包含的修复。在 RU 发布后的 6 个月内，会有两个针对此 RU 的修订，即 RUR。RUR 也是每季度发布，其中包括了对其关联 RU 中缺陷的修复和最新的安全修复。另外，针对初始 RU 不会发布 RUR，例如 19.1.0 和 21.1.0。初始 RU 仅用于 Oracle 公有云，下一个季度才会发布本地部署版本。

按照 Oracle 官方建议①，数据库应尽可能保持最新的 RU 或 RUR 以保证系统稳定，以避免触发已知缺陷或安全问题。如图 3-3 所示，用户可以选择两种更新方式。第一种是按照 RU 发布周期，例如从 19.3.0 向右一直更新到当前最新的 RU 19.13.0。另一种方式是当用户希望稳定到某一 RU 版本后，按照 RUR 发布周期更新。例如用户当前版本为 19.3.0，然后更新到 19.3.1，直到 19.3.2。

当前日期：2021年3月31日

尚未发布

	2019				2020				2021			
	1月	4月	7月	10月	1月	4月	7月	10月	1月	4月	7月	10月
RU	19.2.0	19.3.0	19.4.0	19.5.0	19.6.0	19.7.0	19.8.0	19.9.0	19.10.0	19.11.0	19.12.0	19.13.0
RUR			19.3.1	19.4.1	19.5.1	19.6.1	19.7.1	19.8.1	19.9.1	19.10.1	19.11.1	19.12.1
RUR				19.3.2	19.4.2	19.5.2	19.6.2	19.7.2	19.8.2	19.9.2	19.10.1	19.11.2
RU									21.1.0	21.2.0	21.3.0	21.4.0
RUR											21.2.1	21.3.1

更新路径（按RU）

更新路径（按RUR）

新特性版本

图 3-3 Oracle 数据库更新（RU）与修订（RUR）发布

RU 中包含了优化器、功能、安全和回归四类修补，而 RUR 中仅包含后两类修补。换句话说，即使应用了最新的 RUR，同样可能遇到最新 RU 中已经解决的问题。因此长期来看，仍然建议按照 RU 发布周期来对数据库软件进行更新。从 RUR 更新到 RU 是支持的，只需目标版本号的第 1 位和源版本的第 1 位相同，并且目标版本号的第 2 位与第 3 位之和大于或等于源版本的第 2 位与第 3 位之和即可。例如从 19.5.2 更新到 19.8.0 是支持的，更新到 19.6.0 则不支持。另外，从图 3-3 也可以知道，如果 3 位版本号的后两位之和相同，表示更新是在同一季度发布的；如果不同，则较大的和表示较新的更新。

对于数据库更新和补丁的下载，Oracle 支持网站提供了一个数据库更新下载助手②，可以非常容易地找到所需的更新版本。例如之前安装的测试环境数据库版本为 19.3.0，如果希望更新到当前最新的 RU，可以如图 3-4 所示进行操作，第①步先选择希望下载 RU，第②步选择主版

① https://support.oracle.com/epmos/faces/DocContentDisplay?id=2285040.1

② https://support.oracle.com/epmos/faces/DocContentDisplay?id=2118136.2

本，第③步就会显示出所有的 RU。在第①步紧挨着所选项目的下面一项表示 RUR，最上面一项表示主版本。

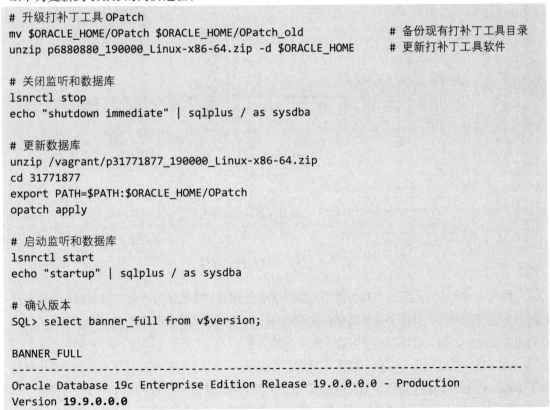

图 3-4　Oracle 数据库更新下载助手

在具体操作上，RU 和 RUR 更新过程没有区别。在第 2 章通过 Vagrant 创建的数据库实验环境中，数据库的版本为 19.3.0。假设已经下载了 19.9.0 RU 以及打补丁工具 OPatch 的最新补丁，以下为更新到 19.9.0 的简要过程：

```
# 升级打补丁工具OPatch
mv $ORACLE_HOME/OPatch $ORACLE_HOME/OPatch_old          # 备份现有打补丁工具目录
unzip p6880880_190000_Linux-x86-64.zip -d $ORACLE_HOME  # 更新打补丁工具软件

# 关闭监听和数据库
lsnrctl stop
echo "shutdown immediate" | sqlplus / as sysdba

# 更新数据库
unzip /vagrant/p31771877_190000_Linux-x86-64.zip
cd 31771877
export PATH=$PATH:$ORACLE_HOME/OPatch
opatch apply

# 启动监听和数据库
lsnrctl start
echo "startup" | sqlplus / as sysdba

# 确认版本
SQL> select banner_full from v$version;

BANNER_FULL
--------------------------------------------------------------------------------
Oracle Database 19c Enterprise Edition Release 19.0.0.0.0 - Production
Version 19.9.0.0.0
```

综合来看，对于 Database In-Memory 而言，应首先升级到较新的主版本如 19c，以延长支持周期，获取最新功能，然后应定期更新，以保证系统稳定，并避免触发已知缺陷或安全问题。

3.2 Database In-Memory 体系架构

3.2.1 双格式存储

传统的 Oracle 数据库中，数据在磁盘和内存中都是以行格式存储的。事务生成的新纪录对应于数据库表中的一行，每一行数据又由多列组成，每一列代表记录的一个属性。由于这些属性都组织在一起，因此行格式非常适合于快速访问一条记录中的单个或多个属性，换句话说，非常适合于在线事务处理类的应用。

相反，列格式则是将记录的属性单独存放，即每一列只存放记录某一个属性的所有值。由于记录的每个属性都集中存放，因此列式结构非常适合于读取绝大部分数据中少数几个属性用于统计和汇总等操作，也就是通常所说的分析类应用。但列式数据库有一个明显的问题，当插入、更新和删除记录时，由于属性是分散存放的，所有涉及列的结构都必须改动。因此，列式存储通常用于数据更改相对较少的数据仓库和数据集市系统。

图 3-5 展示了数据库表按行和列两种不同的组织形式，以及各自适合的操作。可以看到，行格式的订单表顺时针旋转 90 度，然后水平翻转即可转换为列格式。

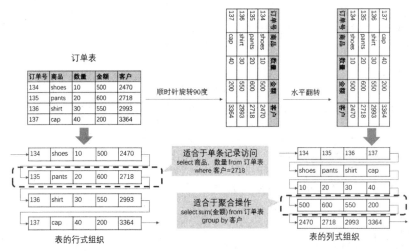

图 3-5 行式存储与列式存储

但现实世界中的应用并不能简单划分为纯事务处理类或纯数据分析类。例如财务系统，平时以事务处理为主，月结和年结时存在大量统计分析操作；又如电商系统，实际也包括了很多分析类的操作，如实时推荐、实时定价和实时预警等。

为同时应对这两种不同类型的工作负载，Oracle 在 12cR1 的第一个补丁集 12.1.0.2 中新增了 Database In-Memory 选件，从传统的仅支持行式存储的数据库转变为同时支持行列存储的双格式数据库。用户不再为选择行式还是列式陷入两难境地。如图 3-6 所示，对于事务处理类负载，Oracle 将在传统的行格式存储 Buffer Cache 中进行处理，而对于分析类负载，则在列格式的 In-Memory Column Store（内存列式存储）中进行处理。Oracle 会自动判断工作负载类型并选择适合的格式，应用无须做任何修改。唯一需要做的就是为列式存储增加内存，并将指定的对象加载到内存列式存储中。

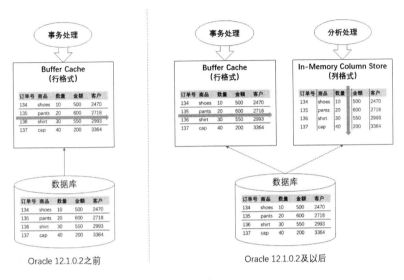

图 3-6　Oracle 数据库双格式架构

3.2.2　Oracle 数据库内存结构与管理

数据库实例启动后，Oracle 为其分配内存并启动后台进程。Oracle 数据库内存中最重要的部分为 SGA（System Global Area）和 PGA（Program Global Area）。SGA 由一组共享的内存结构组成，包含一个数据库实例所使用的数据和控制信息，例如 Buffer Cache 中缓存的数据块和 Shared Pool 中共享的 SQL 区域。PGA 则是非共享的，每一个服务进程和后台进程都会分配专属的 PGA，包含其私有的数据和控制信息。所有进程的 PGA 构成实例 PGA。SGA 和 PGA 在数据库实例中的内存结构如图 3-7 所示。

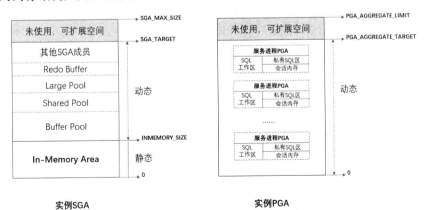

图 3-7　Oracle 数据库实例内存结构与管理

通过设置不同的初始化参数，Oracle 内存管理分为自动、纯手动和半自动 3 种方式。

（1）自动内存管理通过设置 MEMORY_TARGET 指定 SGA 和 PGA 合用的内存空间，Oracle 按需自动调节 SGA 和 PGA 的大小。自动内存管理还可以通过可选参数 MEMORY_MAX_TARGET 设置内存的上限。

（2）纯手动需要逐个设置 SGA 或 PGA 中各成员的内存大小，后续再做调整。

（3）半自动方式内存管理包括自动的 SGA 管理和自动的 PGA 管理，即通过为 SGA 和 PGA 设置一个总的内存容量，然后系统按需自动调节 SGA 或 PGA 内部各成员的内存大小。半自动方式兼顾了控制度和灵活性，是推荐的内存管理方式。如图 3-7 所示，自动的 SGA 管理通过 SGA_TARGET 和 SGA_MAX_SIZE 指定实例 SGA 的总容量和上限；自动的 PGA 管理则通过参数 PGA_AGGREGATE_TARGET 和 PGA_AGGREGATE_LIMIT 指定实例 PGA 的总容量和上限；而 SGA 和 PGA 内部各成员内存大小可按需自动调节。

Oracle 推出 Database In-Memory 后，SGA 中新增了一个可选的成员 In-Memory Area，用来存放列式数据。In-Memory Area 的大小通过初始化参数 INMEMORY_SIZE 设定。但 In-Memory Area 的内存大小是静态的，由 Database In-Memory 独享，不受自动或半自动内存管理控制。也就是说，当 Database In-Memory 内存不够时，它不会从 SGA 其他成员借用内存，反之亦然。因此当启用 Database In-Memory 功能时，需要考虑为数据库服务器添加额外的内存，否则可能会因为内存不足影响数据库的运行。

表 3-5 为启用 Database In-Memory 后的 SGA 内存设置建议。运行在 RAC 环境下的 Database In-Memory，系统会额外分配锁和元数据信息，因此增加了放大系数。

表 3-5　启用 Database In-Memory 后的 SGA 内存设置

数据库类型	启用 Database In-Memory 后的 SGA_TARGET
单实例数据库	原 SGA_TARGET + INMEMORY_SIZE
RAC 数据库	(原 SGA_TARGET + INMEMORY_SIZE) × 1.1

PGA 设置对于 Database In-Memory 同样重要，因为 Database In-Memory 查询通常会执行大量的联结、聚合和排序操作。如果 PGA 内存不够，这些操作将转移到位于磁盘的临时表空间中运行，系统性能将受到严重影响。

通过以下 SQL，可以查询内存管理相关的初始化参数。其中 ISSYS_MODIFIABLE 为 IMMEDIATE 表示修改后立即生效，若为 FALSE 则表示重启实例后生效。ISBASIC 为 TRUE 表示其为基础初始化参数：

```
SQL>
SELECT
    name,
    display_value,
    issys_modifiable,
    isbasic
FROM
    v$parameter
WHERE
```

```
name IN
('sga_target', 'sga_max_size', 'pga_aggregate_target', 'pga_aggregate_limit',
       'memory_target', 'memory_max_target', 'inmemory_size')
ORDER BY
   name;

NAME                      DISPLAY_VALUE         ISSYS_MODIFIABLE      ISBASIC
------------------------- --------------------- --------------------- ----------
inmemory_size             0                     IMMEDIATE             FALSE
memory_max_target         0                     FALSE                 FALSE
memory_target             0                     IMMEDIATE             FALSE
pga_aggregate_limit       2G                    IMMEDIATE             FALSE
pga_aggregate_target      768M                  IMMEDIATE             TRUE
sga_max_size              2304M                 FALSE                 FALSE
sga_target                2304M                 IMMEDIATE             TRUE

7 rows selected.
```

对于现有系统，可通过 AWR 报告来获取初始 SGA 和 PGA 设置的建议。

3.2.3 Database In-Memory 架构

Database In-Memory 架构如图 3-8 所示，其中最核心的部分是 In-Memory Area，它的大小由初始化参数 INMEMORY_SIZE 设置。In-Memory Area 被自动划分为列式数据池（Column Store）和元数据池，它们的大小由系统自动设定，用户无法更改。列式数据池中的对象均按列格式组织，和磁盘上以行格式组织的数据库数据文件及内存中以行格式组织的 Buffer Cache 一起形成双格式存储。

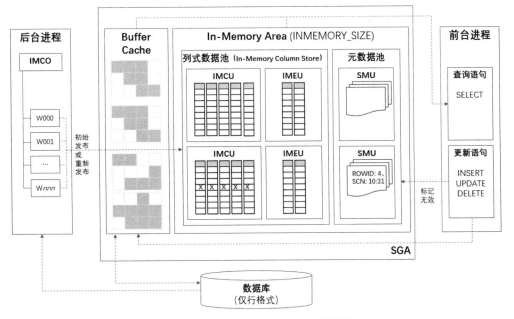

图 3-8　Database In-Memory 架构图

列式数据池包含内存压缩单元（IMCU）和内存表达式单元（IMEU）两部分。IMCU 存放单个数据库对象部分或所有列的数据。IMEU 是 IMCU 的逻辑扩展，与 IMCU 一一对应。例如，若某数据库表由 4 个 IMCU 组成，则系统也会为此表相应分配 4 个 IMEU。IMEU 中存放了由 IMCU 中的一列或多列衍生出来的虚拟列或表达式结果。假设 IMCU 有两列为"价格"和"折扣"，可以将表达式"价格×（1－折扣）"的结果存放在 IMEU 中，从而通过预计算提升查询性能。

和 Buffer Cache 可读可写不同，IMCU 是只读的。那么数据库如何保证两者的数据一致性呢？这实际上是通过 SMU 来实现的。SMU（Snapshot Metadata Unit）与 IMCU 一一对应，包含了其关联 IMCU 的元数据（如 IMCU 对应的数据库对象、行和数据块数量等）和事务信息。当用户通过更新语句修改 IMCU 中的某行数据时，由于 IMCU 是只读的，因此改动只能在 Buffer Cache 中进行，也就是采用和没有 Database In-Memory 之前一样的传统方式。但此时，SMU 中会记录被修改行的 ROWID 以及修改时的 SCN（逻辑时间戳），也就是说，此行被标记为无效（因为不是最新的数据）。后续当用户查询 IMCU 中数据时，如果所涉及行被标记为无效并且查询所指定的 SCN 比 SMU 中记录的 SCN 较新时，查询语句不会采用 IMCU 中的数据，而是转到 Buffer Cache 中读取最新的数据。当然，其他没有被标记为无效的行仍然在 IMCU 中读取，最终的结果集通过在 IMCU 和 Buffer Cache 中的数据合并形成。

如果更新的数据越来越多，SMU 中被标记为无效的行将不断增加，这种跨两处读取再合并的查询操作效率必然下降。此时就需要借助于后台进程 IMCO 来解决这个问题。IMCO 又称为内存协调进程，其主要任务是负责内存列式数据的发布，即将行式数据转换为列式数据，压缩后存入 IMCU 中。IMCO 负责列式数据的初始发布和后续的重新发布。初始发布是指数据库实例启动时或数据库对象被初次访问时将其发布到列式存储中。重新发布又分为基于阈值的重新发布和涓流式重新发布。基于阈值的重新发布是指当 IMCU 中无效数据超过内部设置的陈旧度阈值时，IMCO 将启动重新发布流程，将无效的数据更新。涓流式重新发布是对基于阈值重新发布的补充，是指 IMCU 中存在无效数据但没有达到陈旧度阈值时的数据重新发布。IMCO 每两分钟唤醒一次，系统自动判断是否执行涓流式重新发布。正是基于阈值和涓流式这两种重新发布机制，使 IMCU 中的数据可以保持"新鲜度"，后续查询可以完全或尽可能在列式存储中进行，从而保证查询效率。

在发布和重新发布过程中，IMCO 只负责判断和调度，实际的发布任务由空间管理工作进程（图 3-8 中的 W000 到 W*nnn*）完成。对于用户而言，数据陈旧度阈值和是否真正进行涓流式重新发布均属于内部设置，但用户可以指定空间管理工作进程的数量，以加快数据发布的速度。

3.2.4　内存压缩单元架构

内存压缩单元（IMCU）是列式存储中最重要的部分，包含了以列格式存储的实际用户数据。如图 3-9 所示，一个表由一个到多个 IMCU 组成，而 IMCU 又包含一个或多个 CU（Compression Unit），每一个 CU 对应发布到内存中的一列。

CU 由两部分组成。CU 主体中存放以指定压缩级别存放的列数据。CU 头中存放元数据信息,包括本地字典和存储索引。本地字典中存放了此 CU 中不同列数据对应的字典编码,由于每一个 CU 中的数据不同,因此字典编码也不尽相同。存储索引中存放此 CU 中列数据的最小和最大值。后续会详细介绍 Database In-Memory 如何利用字典编码和存储索引来提升数据扫描的性能。每一个 IMCU 中还存放了一个特殊的 CU:ROWID,用来与行式存储中的数据行对应。例如,用户指定了部分列发布到 IMCU 中,如果用户需查询某行的所有列的数据,Database In-Memory 可以利用 ROWID 从 Buffer Cache 中查询对应的未发布到 IMCU 中其他列的数据,然后和 IMCU 中查询到的数据拼接到一起返回给用户。

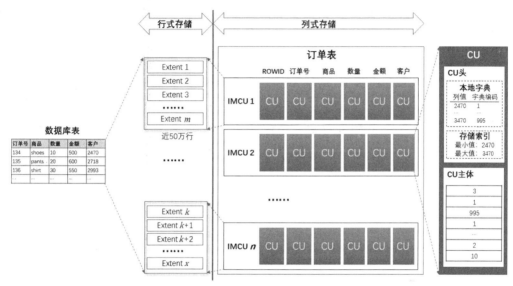

图 3-9 IMCU 结构及与行式存储的映射

　　IMCU 中的列数据并未进行排序,与从磁盘读取的顺序保持一致。为便于并行处理和保持空间和性能的平衡,每一个 IMCU 存储的行数尽可能保持一致,即 50 万行左右。因此,对于同一个数据库对象而言,IMCU 的大小是相近的。但对于不同的数据库对象,IMCU 则不尽相同,这取决于发布到 IMCU 中列的数量和每列的宽度。

　　在数据库的逻辑存储结构中,表是由 Extent 组成的,而 Extent 由数据块组成,默认的数据库块大小为 8KB。IMCU 正是通过 Extent 实现与行式存储的映射,每一个 IMCU 对应连续存放的 Extent,或者反过来说,一个 Extent 只能对应一个 IMCU。第 4 章会通过示例进一步说明这种映射关系。

3.3 Exadata 作为最佳 Database In-Memory 平台

　　Exadata 是 Oracle 于 2008 年推出的软硬件工程整合系统,中文官方名称为 Exadata 数据库云平台,俗称数据库一体机。从最初与 HP 合作的 V1 系统到 2019 年发布的 X8 系统,Exadata 历经了 9 代的发展演进,已经是非常成熟的产品。

Database In-Memory 可运行于标准的 X86 或 IBM 等小型机系统，但最佳运行平台则非 Exadata 莫属。由于 Exadata 与 Oracle 数据库软件的深度集成，以及较高的计算、存储和网络设计标准，因此在 Exadata 上运行 Database In-Memory 可以提供更高的性能和可扩展性。实际上，Exadata 正是 Database In-Memory 的开发平台，Database In-Memory 的问题都是首先在 Exadata 上发现并解决的[①]。Exadata 也是 Oracle 数据库测试、HA 最佳实践验证、集成和支持的主要平台。因此，Exadata 作为数据库最佳平台，这一点也同样适用于 Database In-Memory。

以下 Database In-Memory 特性仅在 Exadata 数据库云平台上提供。

（1）In-Memory 复制也称为 In-Memory 容错，适用于 RAC 集群环境。可以对内存列式存储中的数据在另一个和所有 RAC 示例上提供副本。此功能可提升系统整体的容错性，与并行执行结合还可以提升查询的性能。In-Memory 复制将在第 8 章详细介绍。

（2）ADG 上的 Database In-Memory 支持。Oracle ADG（Active Data Guard）是 Oracle 最高可用性架构中的核心组件，可用来搭建数据库灾备系统，在生产数据库系统发生故障时切换到备数据库。在系统正常时，由于备数据库允许读取，因此可以将报表查询、备份等只读负载转移到备数据库中执行，从而减轻生产数据库的压力。只有在 Exadata 上，才允许在备数据库上启用 Database In-Memory。因此可以进一步将实时分析工作负载转移到备数据库执行，从而提升了灾备系统建设的投资回报，提升了整个灾备系统的处理能力。ADG 与 Database In-Memory 的结合将在第 8 章详细介绍。

（3）闪存缓存上的 In-Memory 列格式支持。Database In-Memory 使用纯列格式存放数据，但最初只在内存中支持。在 DDL 语句中，INMEMORY 关键字即表示与内存中纯列格式相关的属性。到 Exadata 系统软件 12.2 版本，在 Exadata 的闪存中也支持纯列格式，为了区分，此项特性被命名为 CELLMEMORY。闪存的性能虽不如内存，但与普通硬盘相比仍有很大的提升。而且 Exadata 在存储层拥有大量闪存，以 Exadata X8M-2 高容量 1/4 标准配置为例，所有计算节点上的内存为 768GB，而所有存储节点上的闪存容量为 76.8TB。在 Exadata 闪存上支持列格式，可以将 In-Memory 列压缩、SIMD 向量处理和存储索引等一系列性能优化应用于更多的数据，同时拓展了 Exadata 和 Database In-Memory 的应用范围和使用场景。Exadata 对 In-Memory 列格式的支持将在第 10 章详细介绍。

Automatic In-Memory 和 In-Memory 外部表最初也是专属于 Exadata 的功能，在 2021 年 8 月去除了此限制。

Exadata 与 Database In-Memory 的配合可谓相得益彰。例如，Exadata 是极佳的数据库整合平台，与 Database In-Memory 的结合可以更有效地利用硬件资源，提供所有数据的整合能力。RAC 环境下的 Database In-Memory 可实现内存容量和处理能力的横向扩展，Exadata 在设计上综合考虑了 CPU、I/O 和网络等各个层面，从而保证在横向扩展时不会产生瓶颈，并且可以高效执行分布式环境下的并行内存中查询。Exadata 的内联网络早期采用 40Gb/s 的 Infiniband，从 X8M 开始使用 100Gb/s 的 RoCE 网络，均比传统网络高出数倍。互联协议采用 DMA（Direct

---

① https://www.oracle.com/technetwork/database/in-memory/learnmore/twp-dbim-exadata-2556211.pdf

Memory Access），极大降低了 CPU 开销，并保证了数据的高速移动，以及内存列式存储中数据的高效复制。数据发布的速度取决于存储的性能，Exadata 提供卓越的 I/O 性能可加快数据的发布过程，从而使应用尽快享用 Database In-Memory 的好处。此外，Exadata 均衡的设计保证了数据在重新发布时不会对业务应用造成影响。

Exadata 与 Database In-Memory 的结合使得应用可以在内存、闪存和磁盘所有存储层存放和移动数据，全面匹配业务在性能和成本上的需求。Exadata 本身也提供一系列软件优化特性，如混合列压缩、智能闪存缓存和智能扫描等。混合列压缩与 Database In-Memory 中的纯列压缩针对不同特征的数据，均能节省空间和提升性能。智能扫描与 Database In-Memory 中的操作下推使用相同的原理，都可以减少数据的传输 I/O，都是将操作推向数据。这些特性并不冲突，互相配合可以全面满足不同层面应用的需求，也使得 Exadata 和 Database In-Memory 的适应面更广。

3.4 何时使用 Database In-Memory

3.4.1 Database In-Memory 适用场景

按照 Oracle 数据仓库概念和设计指南[①]中的定义，传统的 Oracle 企业数据仓库参考架构分为 3 层。

（1）临时数据层也称为数据准备区，在数据进入数据仓库之前充当数据操作的临时存储区域。此层从一个或多个数据源获取数据，并对其进行清洗、丰富、转换以保证数据的完整性和一致性，最后进入到基础数据层，转换为数据仓库所需的模式。可以利用批量 ETL 工具（如 Oracle Data Integrator）或实时 ETL 工具（如 Oracle GoldenGate）将数据源的数据传输到临时数据层，如果是实时传输，临时数据层也可以部分或完全替代 ODS（Operational Data Store）的功能。

（2）基础数据层是企业数据仓库的核心，该层以尽可能低的粒度级别记录数据，因此也称为原子数据层。该层的主要任务是负责信息管理，包含当前数据与所有的历史数据。数据通常只是插入，而不会进行删除和修改。为减少数据冗余和提升存储效率，该层以规范化的第三范式（3NF）存储数据。基础数据层在设计上要保持业务和查询中立，并具备一定的灵活性，以保证其不会因外部业务的频繁变化导致不必要的数据重组。

（3）访问与性能层，第三范式模型适合于存放全生命周期数据，能够不受变化的业务需求所影响，但并不适合于数据访问。实际上，业务部门采用的查询工具多种多样，而且不断有新的工具产生，访问的模型也各不相同。访问和性能层的基本目的就是具备在物理或逻辑上将数据重组为不同表示形式的能力，如立方体和星型模型。另一方面，此层通过采用索引、分区、物化视图、并行计算等技术来提升数据分析的性能。

对于企业级 OLTP 系统和企业数据仓库，Database In-Memory 都可以被用来提供实时分析，

---

① https://docs.oracle.com/database/121/TDPDW/part2.htm

如图 3-10 所示。企业级 OLTP 系统通常是混合负载系统，同时包括在线事务处理和定期分析报表需求，如 ERP、财务系统、人力资源系统。如果需要实时报表，传统的方式通常采用 ODS 来实现。使用 Database In-Memory 可以避免搭建额外的 ODS 系统，直接在 OLTP 系统上提供实时报表功能。可以删除专为分析建立的索引，这可以带来一系列的好处，包括节省空间、提速 DML 操作、提高缓存命中率和减少 I/O。对于企业数据仓库，Database In-Memory 在基础数据层提供基于第三范式的时间敏感的分析，在访问与性能层，Database In-Memory 具备快速扫描与过滤能力，以及 In-Memory 联结和 In-Memory 聚合等技术，可以很好地与星形模型相结合。物化视图等预计算对象可以直接加载到内存列式存储中查询。Database In-Memory 也可能替换掉物化视图，同时满足性能和简化架构。无论是在哪一层，分析型索引都可以被删除以释放空间和减少 DBA 调优和管理工作量。对于索引无法优化的不可预知分析，Database In-Memory 也可以提供很好的性能，并且没有索引所需的持久化成本。

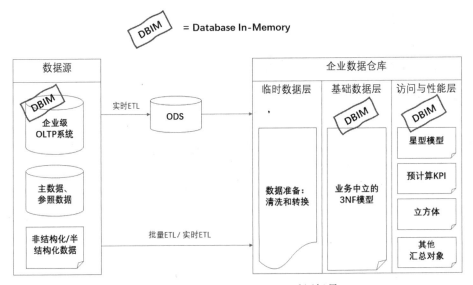

图 3-10　Database In-Memory 适用场景

3.4.2　分析型查询与应用的考虑

Database In-Memory 适合于混合负载的企业级 OLTP 系统和企业数据仓库两大场景，本质上都是针对分析型查询提供实时报表与分析能力。那么什么是分析型查询？分析型查询具备什么样的特征？按照 Oracle 官方白皮书 Oracle Database In-Memory 快速入门指南[①]的说法，分析型查询"扫描大量数据、访问有限数量的列并使用聚合和过滤条件返回少量聚合值。花费大部分时间扫描和过滤数据的查询会获得最大的好处"。再具体一点，以下分析型查询从 Database In-Memory 中受益最多。

（1）扫描大量的行，并通过谓词过滤掉大部分的数据。

---

① https://www.oracle.com/a/otn/docs/dbim-quickstart-guide.pdf

（2）带联结、聚合操作的查询，并且联结条件可以过滤掉大部分的数据。

这两类查询都有一个共同点，首先是处理的数据较多，其次是通过更具选择性的谓词或联结过滤掉大部分数据。也就是说，查询访问的总数据量与查询实际处理的数据量的比值越大，Database In-Memory 的潜在好处就越大。例如对于这两类查询，通常在 WHERE 条件中都具有>、<、=和 IN 等操作符，而等值查询的选择性更强，因此受益也最多。查询返回的行和列越少，从 Database In-Memory 中获得的好处越大。例如对于聚合函数，由于是通过多行计算得到一行，因此也非常适合于 Database In-Memory。

从监控的角度来说，如果执行计划中大部分的时间用于扫描与过滤数据，或者大部分时间用于数据联结操作，都可以考虑使用 Database In-Memory 来进行优化。Database In-Memory 可以缓解 I/O 瓶颈，但如果瓶颈在 CPU，也就是大部分的时间都用于计算，则 Database In-Memory 的作用有限。

理想情况下，Database In-Memory 可以对数据一次性执行扫描、过滤和聚合，也就是边扫描、边过滤，边建立报表框架和执行聚合操作。第 6 章将对 Database In-Memory 集成性能优化功能中的纯列格式、SIMD、压缩、存储索引、In-Memory 联结和 In-Memory 聚合进行详细介绍。

为了最大化 Database In-Memory 的好处，可以参考 Oracle 官方白皮书[①]中提供的 4 条原则。这些原则并不是新的，但对于 Database In-Memory 同样重要，遵循这些原则可以使 Database In-Memory 发挥最大的效益。

（1）在数据库而非在应用中处理数据。如果将大量数据取回到应用中处理，不仅数据往复的成本很高，同时也会导致 I/O 瓶颈。数据库服务器的处理能力通常很强，结合操作下推功能，在数据库中进行处理效率更高。操作下推将在第 6 章详细介绍。

（2）批量而非逐行处理数据。也就是说，应尽量减少与数据库之间的往复。例如，在执行 DML 操作时，应使用批量提交而非逐行提交。Oracle 也提供数组绑定（Array Binding）API，可以将多个 SQL 语句一次性发往数据库执行。由于内存的性能和一系列软件优化，Database In-Memory 可以提供超强的处理能力，因此为数据库提供足够的数据以在每次调用时进行处理就非常重要。

（3）提供准确的优化器统计信息。准确的统计信息是最优执行计划和最快执行路径的基础，因此应选择合适的时间间隔，定期手工或自动执行统计信息搜集任务。优化器统计信息将在第 5 章详细介绍。

（4）尽可能使用并行。当查询瓶颈位于 CPU 而非 I/O，并且 CPU 资源富余时，使用并行是非常有效的性能提升手段。特别是在 RAC 环境中，需要启动自动并行（Auto DOP）以充分利用 CPU 的能力。Database In-Memory 与 RAC 的结合将在第 8 章详细介绍。

---

① https://www.oracle.com/technetwork/database/in-memory/overview/twp-dbim-usage-2441076.html

第4章

Database In-Memory 配置与发布

4.1 配置与发布基本流程

要使分析型或混合负载应用受益于 Database In-Memory，必须首先启用 Database In-Memory，然后将查询语句涉及的数据库对象发布到列式存储中。配置和发布的主要步骤如下。

（1）设置内存列式存储大小，为数据库实例启用内存列式存储。
（2）为需要发布到内存列式存储中的对象设置 INMEMORY 属性。
（3）可选步骤，设定发布对象的数据压缩级别，可使用默认压缩级别或单独指定。
（4）可选步骤，设定发布对象的发布优先级，可使用默认发布优先级或单独指定。
（5）自动或手工将数据库对象发布到内存列式存储。

经过以上配置后，数据库查询优化器就可以感知到内存列式存储的存在，并自动将符合条件的分析型查询定向到内存列式存储中，应用无须修改即可得到性能提升并保证行式和列式数据间的一致性。

4.2 启用 Database In-Memory

4.2.1 In-Memory Area 初始设置

启用 Database In-Memory 的目的是在 SGA 中划分出一块内存区域单独供内存列式存储使用，这片区域就是第 3 章介绍的 In-Memory Area，它的大小通过数据库初始化参数 INMEMORY_SIZE 指定，最小值为 100MB。

在设置前，需要先确认 INMEMORY_SIZE 初始化参数尚未设置，并且 COMPATIBLE 设置为 12.1.0（首个支持 Database In-Memory 的版本）以上：

```
SQL> SHOW PARAMETER inmemory_size
```

```
NAME                                     TYPE         VALUE
---------------------------------------- ------------ ------------------------------
inmemory_size                            big integer  0

SQL> SHOW PARAMETER compatible

NAME                                     TYPE         VALUE
---------------------------------------- ------------ ------------------------------
compatible                               string       19.0.0
```

然后即可设置 INMEMORY_SIZE，设置完成后，需要重启数据库实例后才能生效。以下为设置 In-Memory Area 的示例。注意，在最后启动实例时会显示 In-Memory Area 的信息：

```
-- 设置 In-Memory Area 为 1G
SQL> ALTER SYSTEM SET inmemory_size = 1G SCOPE = SPFILE;

System altered.

-- 关闭数据库
SQL> SHUTDOWN IMMEDIATE
Database closed.
Database dismounted.
ORACLE instance shut down.

-- 启动数据库
SQL> STARTUP
ORACLE instance started.

Total System Global Area 2415918560 bytes
Fixed Size                  9137632 bytes
Variable Size             335544320 bytes
Database Buffers          989855744 bytes
Redo Buffers                7639040 bytes
In-Memory Area           1073741824 bytes
Database mounted.
Database opened.

-- 确认设置生效
SQL> SHOW PARAMETER inmemory_size

NAME                                     TYPE         VALUE
---------------------------------------- ------------ ------------------------------
inmemory_size                            big integer  1G
```

启用成功后，在视图 V$INMEMORY_AREA 中可以看到，1GB 的 In-Memory Area 被分割为约 700MB 的列式数据池和约 300MB 的元数据池。这种分配比例由系统自动设置，用户无法指定。另外，由于尚无对象发布到内存列式存储中，因此 USED_BYTES 为 0：

```
SQL> SELECT * FROM v$inmemory_area;
POOL          ALLOC_BYTES USED_BYTES POPULATE_STATUS      CON_ID
------------- ----------- ---------- -------------------- ----------
1MB POOL        737148928          0 DONE                          3
```

```
64KB POOL        318767104             0 DONE                                 3
```

4.2.2 多租户下的 In-Memory Area 设置

对于多租户数据库，INMEMORY_SIZE 可以在 CDB 和 PDB 两个层面设置。CDB 一级的设置决定了 In-Memory Area 的总容量，此容量是可以保证的，因为已经实际从 SGA 中划分出来；PDB 一级的设置决定了此 PDB 可分配的内存容量上限，并且所有 PDB 共享 CDB 一级设置的总容量。默认情况下，PDB 一级的 INMEMORY_SIZE 设置与 CDB 一级的相同。

如图 4-1 所示，如果仅在 CDB 一级设置，则 PDB 的设置继承自 CDB。也可以进一步在 PDB 一级进行细粒度控制，例如所有 PDB 的设置之和不超过 CDB 一级设置的总容量，表示不会出现资源超用的情况，而 PDB3 的设置为 0 表示禁用内存列式存储。

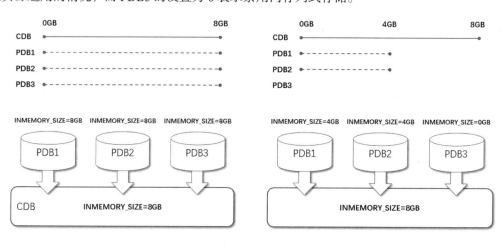

图 4-1　多租户下的 In-Memory Area 设置

新建一个名为 orclpdb2 的 PDB，其 INMEMORY_SIZE 已经被设置成与 CDB 相同的值；在 PDB 一级可以修改 INMEMORY_SIZE 设定，但不能超过 CDB 一级设置的值，代码如下：

```
SQL> ALTER SYSTEM SET db_create_file_dest = "/opt/oracle/oradata/ORCLCDB";
System altered.

SQL> CREATE PLUGGABLE DATABASE orclpdb2 ADMIN USER pdbadmin IDENTIFIED BY Welcome1;
Pluggable database created.

SQL> ALTER PLUGGABLE DATABASE orclpdb2 OPEN;
Pluggable database altered.

SQL> ALTER SESSION SET CONTAINER = orclpdb2;
Session altered.

SQL> SHOW PARAMETER inmemory_size;
NAME                                 TYPE        VALUE
------------------------------------ ----------- ------------------------------
```

```
inmemory_size                        big integer 1G

SQL> ALTER SYSTEM SET inmemory_size = 800M;
System altered.

SQL> ALTER SYSTEM SET inmemory_size = 1G;
System altered.

SQL> ALTER SYSTEM SET inmemory_size = 2G;
ALTER SYSTEM SET inmemory_size = 2G
*
ERROR at line 1:
ORA-02097: parameter cannot be modified because specified value is invalid
ORA-02095: specified initialization parameter cannot be modified
```

4.2.3 调整 In-Memory Area 大小

在 Oracle 12.2 之前，数据库实例的 In-Memory Area 设置只能静态调整，也就是说，无论容量扩大或缩小，数据库实例必须重启才能生效。从 Oracle 12.2 开始，In-Memory Area 支持动态扩容，每次增加的容量必须在 128MB 以上，缩小容量仍然只能静态调整。对于 PDB，由于其 In-Memory Area 设置继承自 CDB，因此可以动态扩大或缩小，只要不超过 CDB 一级的容量即可。

假设当前 In-Memory Area 大小为 1GB（1024M），可以将其动态增加到 1152MB。如果调整为 1150MB，会因为增量不足 128MB 而失败。

```
SQL> SHOW PARAMETER inmemory_size

NAME                                 TYPE        VALUE
------------------------------------ ----------- ------------------------------
inmemory_size                        big integer 1G

SQL> ALTER SYSTEM SET inmemory_size = 1150M SCOPE = MEMORY;
ALTER SYSTEM SET inmemory_size = 1150M SCOPE = MEMORY
*
ERROR at line 1:
ORA-02097: parameter cannot be modified because specified value is invalid
ORA-02095: specified initialization parameter cannot be modified

SQL> ALTER SYSTEM SET INMEMORY_SIZE = 1152M SCOPE = MEMORY;
System altered.

SQL> SHOW PARAMETER inmemory_size
NAME                                 TYPE        VALUE
------------------------------------ ----------- ------------------------------
inmemory_size                        big integer 1152M
```

缩小容量不支持动态调整，仍需要使用传统的静态方式：

```
SQL> ALTER SYSTEM SET inmemory_size = 1G SCOPE = MEMORY;
ALTER SYSTEM SET inmemory_size = 1G SCOPE = MEMORY
```

```
        *
ERROR at line 1:
ORA-02097: parameter cannot be modified because specified value is invalid
ORA-02095: specified initialization parameter cannot be modified

SQL> ALTER SYSTEM SET inmemory_size = 1G SCOPE = SPFILE;
System altered.
```

注意，以上内存动态调整的规则是针对实例，也就是 CDB 层面的。对于 PDB 内的 In-Memory Area 设置，增大和缩小均支持动态调整。

4.2.4 Base Level 特性

Oracle 在 2020 年 4 月宣布 Database In-Memory 支持 Base Level 特性，后续又向前移植到 Oracle 19.8 RU 中。此特性允许用户无须许可即可使用最多 16GB 内存列式存储。对于单实例数据库，此容量限制是指 CDB 一级不超过 16GB；对于 RAC 数据库，是指每一个数据库实例不超过 16GB。

除容量限制外，Base Level 特性只能使用默认的压缩级别，不支持 Automatic In-Memory 和 Exadata 上的 CellMemory 特性。另外，Base Level 特性属于数据库企业版，在 SE2 版本中并不提供。

只需将 INMEMORY_FORCE 初始化参数设置为 BASE_LEVEL，即可启用 Base Level 特性。启用 Base Level 特性的简要过程如下：

```
SQL> CONNECT / AS SYSDBA

SQL> SHOW PARAMETER inmemory_force

NAME                                 TYPE        VALUE
------------------------------------ ----------- ------------------------------
inmemory_force                       string      DEFAULT

SQL> ALTER SYSTEM SET inmemory_force = base_level SCOPE = SPFILE;
SQL> ALTER SYSTEM SET INMEMORY_SIZE = 1G SCOPE = SPFILE;

SQL> SHUTDOWN IMMEDIATE
SQL> STARTUP

SQL> SHOW PARAMETER inmemory_force

NAME                                 TYPE        VALUE
------------------------------------ ----------- ------------------------------
inmemory_force                       string      BASE_LEVEL
```

启用后，INMEMORY_SIZE 只能设置为小于或等于 16GB 的值。如果超过 16GB，数据库在启动时会报错如下：

```
$ oerr ora 64901
64901, 00000, "INMEMORY_SIZE may not exceed 16 GB with Database In-Memory Base Level"
```

```
// *Document: YES
// *Cause: INMEMORY_SIZE cannot be greater than 16 GB because the Database In-Memory
//         Base Level was enabled.
// *Action: Set inmemory_size to less than 16 GB or disable Database In-Memory Base Level
//          with INMEMORY_FORCE parameter.
```

4.2.5 常见错误与处理

1. 在 PDB 中首次启用

INMEMORY_SIZE 指定的内存区域是 SGA 的一部分，而 SGA 是数据库实例的属性，对于多租户数据库而言，实例一级属性需要在 CDB 中设置。因此只有首先在 CDB 中启用，后续各 PDB 才能继承或单独设置自己的 INMEMORY_SIZE。

如果 CDB 一级尚未设置，直接在 PDB 中设置是不允许的：

```
SQL> SHOW CON_NAME
CON_NAME
------------------------------
CDB$ROOT

SQL> SHOW PARAMETER inmemory_size
NAME                                 TYPE        VALUE
------------------------------------ ----------- ------------------------------
inmemory_size                        big integer 0

SQL> ALTER SESSION SET CONTAINER = orclpdb1;
Session altered.

SQL> ALTER SYSTEM SET inmemory_size = 1 G;
ALTER SYSTEM SET inmemory_size = 1 G
*
ERROR at line 1:
ORA-02097: parameter cannot be modified because specified value is invalid
ORA-02095: specified initialization parameter cannot be modified
```

2. 内存不足导致实例无法启动

一方面，In-Memory Area 需要设置足够大，以便容纳内存列式存储的所有对象。另一方面，SAG 设置也要足够大，才能保证有足够的内存分配给 In-Memory Area。但在设置 INMEMORY_SIZE 时，系统只是将其记录在参数文件中，直到重新启动实例时才进行实际的分配，也就是说，只有在设置完成后的下一次实例启动时，才能确定设置是否成功。

例如，将 INMEMORY_SIZE 设置为 100GB，远远超过当前 SGA 的大小，因此实例无法启动：

```
SQL> SHOW PARAMETER inmemory_size

NAME                                 TYPE        VALUE
------------------------------------ ----------- ------------------------------
inmemory_size                        big integer 0

SQL> ALTER SYSTEM SET inmemory_size = 100G SCOPE = SPFILE;
```

```
System altered.

SQL> SHUTDOWN IMMEDIATE
Database closed.
Database dismounted.
ORACLE instance shut down.

SQL> STARTUP
ORA-00821: Specified value of sga_target 2304M is too small, needs to be at least 102720M
ORA-01078: failure in processing system parameters
```

因此，为保证有足够的内存分配给 In-Memory Area，可能需要先调整 SGA 的大小，进而可能需要增加底层操作系统内存。

出现以上错误后，为保证实例能正常启动，需要修改启动参数文件，恢复正确的配置。过程如下：

```
# 基于二进制的启动参数文件 spfile, 创建可编辑的启动参数文件 pfile
SQL> CREATE PFILE FROM SPFILE;
File created.

SQL> EXIT

# 查看 pfile 文件中 inmemory_size 设置，当前为 100GB
$ cd $ORACLE_HOME/dbs
$ grep inmemory_size initORCLCDB.ora
*.inmemory_size=107374182400

# 修改 pfile 文件中 inmemory_size 为 1G
$ sed -i 's/inmemory_size=107374182400/inmemory_size=1073741824/' initORCLCDB.ora
$ grep inmemory_size initORCLCDB.ora
*.inmemory_size=1073741824

# 基于修改正确的 pfile 恢复 spfile, 然后使用 spfile 启动成功
SQL> CREATE SPFILE FROM PFILE;
File created.

SQL> STARTUP
ORACLE instance started.

Total System Global Area 2415918560 bytes
Fixed Size                  9137632 bytes
Variable Size             335544320 bytes
Database Buffers          989855744 bytes
Redo Buffers                7639040 bytes
In-Memory Area           1073741824 bytes
Database mounted.
Database opened.
```

4.3 指定 Database In-Memory 对象

4.3.1 哪些对象可以发布到内存列式存储

只有开启了 INMEMORY 属性的对象，才具备发布到内存列式存储的资格，实际的发布还需要另外的流程。默认情况下，INMEMORY 属性设置为关闭。

可以开启 INMEMORY 属性的数据库对象包括：表空间、表（内部表或外部表）、表分区、表子分区和物化视图。

不能开启 INMEMORY 属性的数据库对象包括：索引、索引组织（IOT）表和哈希集群。

对于开启了 INMEMORY 属性的表，以下类型的列不支持发布到内存列式存储中：具有 LONG 或 LONG RAW 数据类型的列、扩展数据类型列、带外列（varrays 和嵌套表列）。

INMEMORY 属性可以继承。对于分区表，如果在表一级开启了 INMEMORY 属性，则所有的分区将继承表一级的设置，也可以只为部分分区开启 INMEMORY 属性。例如，某表按月分为 12 个分区，可以只为最近 3 个月的分区开启 INMEMORY 属性。同样，表中的列也会继承在表一级的 INMEMORY 设置。如果希望节省内存，也可以关闭部分列的 INMEMORY 设置，从而避免将它们发布到内存列式存储中。但是，无法通过 WHERE 条件将部分行数据发布到列式内存中，因为 Database In-Memory 并非缓存。例如，对于非分区表或表分区的某一列，要么不发布此列，要么发布此列的所有行的数据，发布此列的部分行是无法实现的。

最后强调一点，开启 INMEMORY 属性并不一定会将对象置入内存列式存储，但关闭 INMEMORY 会将对象从内存列式存储中移除。

4.3.2 开启和关闭 INMEMORY 属性

必须通过 DDL 语句开启和关闭对象的 INMEMORY 属性，如 CREATE TABLE 和 ALTER TABLE。以下为对于已有表开启和关闭 INMEMORY 属性的示例，在视图中可以查询到相应的状态为 ENABLED 或 DISABLED：

```
SQL> SELECT inmemory FROM user_tables WHERE table_name = 'LINEORDER';
INMEMORY
--------
DISABLED

SQL> ALTER TABLE lineorder INMEMORY;
Table altered.

SQL> SELECT inmemory FROM user_tables WHERE table_name = 'LINEORDER';
INMEMORY
--------
ENABLED

SQL> ALTER TABLE lineorder NO INMEMORY;
Table altered.
```

在表已经开启 INMEMORY 属性的前提下，可以进而关闭指定列的属性：

```
-- 排除指定的列
SQL> ALTER TABLE customer INMEMORY NO INMEMORY(c_address);
Table altered.

SQL> SELECT column_name, inmemory_compression FROM v$im_column_level
WHERE table_name = 'CUSTOMER' AND column_name = 'C_ADDRESS';
COLUMN_NAME             INMEMORY_COMPRESSION
-------------------     --------------------------
C_ADDRESS               NO INMEMORY

-- 增加指定的列
SQL> ALTER TABLE customer INMEMORY INMEMORY(c_address);
Table altered.

SQL> SELECT column_name, inmemory_compression FROM v$im_column_level
WHERE table_name = 'CUSTOMER' AND column_name = 'C_ADDRESS';
COLUMN_NAME             INMEMORY_COMPRESSION
-------------------     --------------------------
C_ADDRESS               DEFAULT
```

开启物化视图 INMEMORY 属性的语法与数据库表完全一样，例如：

```
SQL> ALTER TABLE sh.fweek_pscat_sales_mv INMEMORY;
```

表分区的 INMEMORY 属性既可以直接在分区一级设置，也可以先在表一级设置，然后再在分区一级更改。例如：

```
SQL> ALTER TABLE sh.sales MODIFY PARTITION sales_q1_1998 NO INMEMORY;
Table altered.

SQL> ALTER TABLE sh.sales INMEMORY;
Table altered.

SQL> ALTER TABLE sh.sales MODIFY PARTITION sales_q2_1998 NO INMEMORY;
Table altered.

SQL> SELECT table_name, partition_name, inmemory FROM user_tab_partitions WHERE
table_name = 'SALES';

TABLE_NAME              PARTITION_NAME          INMEMORY
-------------------     --------------------    --------
SALES                   SALES_Q1_1998           ENABLED
SALES                   SALES_Q2_1998           DISABLED
SALES                   SALES_Q3_1998           ENABLED
...
```

INMEMORY 是段（segment）对象的属性，数据库表、表分区和物化视图都属于段对象。但表空间并非段对象，只是存放段对象的容器。修改表空间的 INMEMORY 属性，之后新建的对象会继承此设置，但已有对象不受影响。这些对象也可以单独设置 INMEMORY 属性以覆盖表空间一级的设置：

```
SQL> CREATE TABLE T1(a INT);
```

```
Table created.

SQL> ALTER TABLESPACE users DEFAULT INMEMORY;
Tablespace altered.

-- 新建对象会继承表空间的设置
SQL> CREATE TABLE T2(a INT);
Table created.

SQL> SELECT table_name, inmemory FROM user_tables WHERE table_name IN ('T1', 'T2');
TABLE_NAME      INMEMORY
-----------     --------
T1              DISABLED
T2              ENABLED

-- 已有对象不受影响
SQL> ALTER TABLESPACE users DEFAULT NO INMEMORY;
Tablespace altered.

SQL> SELECT table_name, inmemory FROM user_tables WHERE table_name IN ('T1', 'T2');
TABLE_NAME      INMEMORY
-----------     --------
T1              DISABLED
T2              ENABLED
```

4.4 发布 INMEMORY 对象

发布是指将行格式存储的数据库对象先转换为列格式，然后经过压缩存入内存列式存储的过程。只有将查询相关对象发布后，分析查询才能实现查询速度的提升。

开启了 INMEMORY 属性的对象成为了发布的候选人，数据库并不会马上发布它们。可以通过设置发布优先级、查询或调用 PL/SQL 触发发布。本章只介绍单实例的发布，RAC 的发布有所不同，会在第 8 章详细介绍。另外，21c 的自动内存管理功能可实现完全自动化的对象发布和清除，将在第 9 章详细介绍。

此外，考虑到内存浪费和性能提升有限，即使已经开启 INMEMORY 属性，64KB 及以下的对象也不会被发布。以下为对应的隐含参数设置：

```
Parameter                             Instance Value
---------------------------------     --------------
_inmemory_small_segment_threshold     65536
```

4.4.1 确认对象已发布

可以通过系统视图监控发布状态和查看发布的结果。其中最重要的两个视图为 V$INMEMORY_AREA 和 V$IM_SEGMENTS。

V$IM_SEGMENTS 包含了发布到内存列式存储（列式数据池）中每一个对象的状态。例如，可以查询到对象是正在发布还是发布完成，在磁盘上的大小和在内存中占用的容量，从而可以计算出压缩比。

```
SQL>
SELECT
    segment_name,
    inmemory_size / 1024              AS "IN_MEM_SIZE(KB)",
    bytes / 1024                      AS "ON_DISK_SIZE(KB)",
    round(bytes / inmemory_size, 2)   AS "COMPRESSION_RATIO",
    populate_status
FROM
    v$im_segments;

SEGMENT_NAME   IN_MEM_SIZE(KB)   ON_DISK_SIZE(KB)   COMPRESSION_RATIO   POPULATE_STATUS
------------   ---------------   ----------------   -----------------   ---------------
DATE_DIM                 1,280                344                 .27   COMPLETED
SUPPLIER                 2,304              1,952                 .85   COMPLETED
CUSTOMER                17,664             30,176                1.71   COMPLETED
LINEORDER              579,136          1,376,128                2.38   STARTED
PART                    21,952            105,256                4.79   COMPLETED
```

V$INMEMORY_AREA 包含了 In-Memory Area 的总容量以及列式数据池和元数据池的内存分配情况。例如，以下输出显示了 SSB 示例中 5 个表发布完成后的内存分配与使用情况。

```
SQL> SELECT * FROM v$inmemory_area;

POOL         ALLOC_BYTES   USED_BYTES   POPULATE_STATUS   CON_ID
---------    -----------   ----------   ---------------   ------
1MB POOL      1466957824    631242752   DONE                   3
64KB POOL      612368384      6029312   DONE                   3
```

如果由于内存不足，对象只有一部分发布到内存中，称为不完整发布。不完整发布的对象不能有效被 Database In-Memory 利用，应避免出现这种情况。

对于单实例数据库，如果视图 V$IM_SEGMENTS 的 BYTES_NOT_POPULATED 列不为 0，表示此对象为不完整发布。此时应进一步查看 POPULATE_STATUS，如果状态为 COMPLETED，表示数据库物理内存足够，可能的原因是 PDB 一级的 INMEMORY_SIZE 设置太小；如果状态为 OUT OF MEMORY，则表示内存不足，需要调高 CDB 一级 INMEMORY_SIZE 的设置。V$INMEMORY_AREA 也有 POPULATE_STATUS 列，但其取值只能是 DONE（发布完成）或 POPULATING（正在发布），并不能反映出不完整发布的现象。

```
SQL> SELECT segment_name, bytes_not_populated, populate_status FROM v$im_segments;

SEGMENT_NAME   BYTES_NOT_POPULATED   POPULATE_STATUS
------------   -------------------   ---------------
DATE_DIM                         0   COMPLETED
SUPPLIER                         0   COMPLETED
CUSTOMER                         0   COMPLETED
LINEORDER                615391232   OUT OF MEMORY
PART                             0   COMPLETED
```

```
SQL> SELECT * FROM v$inmemory_area;

POOL          ALLOC_BYTES USED_BYTES POPULATE_STATUS      CON_ID
------------- ----------- ---------- ---------------- ----------
1MB POOL       1500512256  631242752 DONE                      3
64KB POOL       629145600    6029312 DONE                      3
```

第二种确认方式是通过执行计划。如果在执行计划的输出中出现了 TABLE ACCESS INMEMORY FULL 关键字，则说明已经使用了内存列式存储，但此时对象也可能为不完整发布。

```
SQL> SELECT COUNT(*) FROM lineorder;

  COUNT(*)
----------
  11997996

SQL> SELECT * FROM TABLE(dbms_xplan.display_cursor());

PLAN_TABLE_OUTPUT
---------------------------------------------
SQL_ID  0g5kz6486txsb, child number 0
---------------------------------------------
SELECT COUNT(*) FROM lineorder

Plan hash value: 2267213921

-----------------------------------------------------------------------------
| Id  | Operation                  | Name     | Rows  | Cost (%CPU)| Time     |
-----------------------------------------------------------------------------
|   0 | SELECT STATEMENT           |          |       |  1767 (100)|          |
|   1 |  SORT AGGREGATE            |          |     1 |            |          |
|   2 |   TABLE ACCESS INMEMORY FULL| LINEORDER|   11M|  1767   (2)| 00:00:01 |
-----------------------------------------------------------------------------

14 rows selected.
```

4.4.2 发布优先级与自动发布

通过设置对象的发布优先级，可以控制对象发布的顺序，以及实现在数据库实例启动时自动发布对象。如表 4-1 所示，发布优先级设为非 NONE 的对象，都将在数据库启动时自动发布，发布的顺序取决于它们的优先级设置。

表 4-1 数据库对象发布优先级

发布优先级	描 述
NONE	默认的发布优先级。数据库启动时不会自动发布，需要通过全表扫描或 PL/SQL 过程触发发布
LOW	数据库启动时自动发布，优先级最低
MEDIUM	数据库启动时自动发布，优先级仅高于 LOW
HIGH	数据库启动时自动发布，优先级仅低于 CRITICAL
CRITICAL	数据库启动时自动发布，优先级最高

在表空间一级可以设置发布优先级，未来在表空间中新建的对象将继承此属性。列不能单

独设置发布优先级，只能继承其父对象，即表、表分区和物化视图的设置。在视图中可以查看这些设置：

```
-- 查看表空间的发布优先级设置
SQL> SELECT def_inmemory_priority FROM dba_tablespaces WHERE tablespace_name = 'USERS';

DEF_INME
--------

-- 普通表的发布优先级设置
SQL> ALTER TABLE lineorder INMEMORY PRIORITY HIGH;
Table altered.

SQL> SELECT inmemory_priority FROM user_tables WHERE table_name = 'LINEORDER';

INMEMORY_PRIORITY
-----------------
HIGH

SQL> ALTER TABLE lineorder INMEMORY PRIORITY NONE;
Table altered.

-- 分区表的发布优先级设置
SQL> SELECT table_name, inmemory_priority FROM dba_tables WHERE table_name = 'SALES';

TABLE_NAME      INMEMORY_PRIORITY
------------    -----------------
SALES

SQL> ALTER TABLE sh.sales INMEMORY PRIORITY LOW;
Table altered.

SQL> SELECT table_name, partition_name, inmemory_priority FROM dba_tab_partitions
WHERE partition_name = 'SALES_Q1_1998';

TABLE_NAME      PARTITION_NAME        INMEMORY_PRIORITY
------------    ------------------    -----------------
SALES           SALES_Q1_1998         LOW
```

如果当前优先级为 NONE（最低优先级）或未设置，修改为其他任一优先级会立刻触发发布。但如果对象已发布，修改优先级为 NONE 并不会将对象从内存中清除。例如：

```
SQL> SELECT inmemory_priority FROM user_tables WHERE table_name = 'LINEORDER';

INMEMORY_PRIORITY
-----------------
NONE

SQL> ALTER TABLE lineorder INMEMORY PRIORITY LOW;
```

```
Table altered.

SQL> SELECT populate_status FROM v$im_segments WHERE segment_name = 'LINEORDER';

POPULATE_STATUS
---------------
STARTED
```

4.4.3 通过查询手工发布

如果查询语句引发全表扫描，对象将发布到内存列式存储。Oracle 建议使用 FULL 提示来强制实现全表扫描。以下是 Oracle 建议的两种命令格式，可以任选其一：
1. SELECT /*+ FULL(表名) NO_PARALLEL(表名) */ COUNT(*) FROM 表名;
2. SELECT /*+ FULL(t) NO_PARALLEL(t) */ COUNT(*) FROM 表名 t;

在下例中，强制全表扫描引发了发布，输出中 POPULATE_STATUS 变为 POPULATING 可以证实这一点：

```
$ cd ~/dbimbook/chap04
$ ./dbim_ima_usage.sh
POOL        ALLOC_BYTES  USED_BYTES  POPULATE_STATUS   CON_ID
----------  -----------  ----------  ----------------  ------
1MB POOL    1500512256    40894464   DONE                   3
64KB POOL    629145600      983040   DONE                   3

SQL> SELECT /*+ FULL(t) NO_PARALLEL(t) */ COUNT(*) FROM lineorder t;

  COUNT(*)
----------
  11997996


$ ./dbim_ima_usage.sh
POOL        ALLOC_BYTES  USED_BYTES  POPULATE_STATUS   CON_ID
----------  -----------  ----------  ----------------  ------
1MB POOL    1500512256   199229440   POPULATING             3
64KB POOL    629145600     2621440   POPULATING             3
```

需要注意，单纯查询所有行的数据并不一定会导致发布，例如以下的查询语句。特别是当表具有索引时，优化器可能选择通过索引查询数据而非全表扫描。因此仍然建议使用 FULL 提示来确保全表扫描，继而引发发布。

```
SQL> SELECT COUNT(*) FROM lineorder;
```

4.4.4 通过 PL/SQL 子程序手工发布

在 Database In-Memory 发展过程中，Oracle 先后提供了 3 个 PL/SQL 子程序来实现数据库对象的手工发布，如表 4-2 所示。

表 4-2 支持发布的 PL/SQL 过程

子程序名	类型	PL/SQL 包	运行方式	范围	数据量
POPULATE	过程	DBMS_INMEMORY	后台	指定单个对象	全量
REPOPULATE	过程	DBMS_INMEMORY	后台	指定单个对象	增量
POPULATE_WAIT	函数	DBMS_INMEMORY_ADMIN	前台	大于或等于指定优先级的对象	全量

POPULATE 过程可以实现表和分区的手工发布,实现的功能与前面介绍的全表扫描查询是一样的。以下代码演示了表和表分区的手工发布,以及调用过程的两种不同方式:

```
SQL> EXEC DBMS_INMEMORY.POPULATE('SSB', 'LINEORDER');
PL/SQL procedure successfully completed.

SQL>
BEGIN
    DBMS_INMEMORY.POPULATE('SH', 'SALES', 'SALES_Q1_1998');
END;
/
```

只有开启了 INMEMORY 属性的对象,才能通过 PL/SQL 过程进行发布,否则报错如下:

```
SQL> ALTER TABLE lineorder NO INMEMORY;
Table altered.

SQL> EXEC DBMS_INMEMORY.POPULATE('SSB', 'LINEORDER');
BEGIN DBMS_INMEMORY.POPULATE('SSB', 'LINEORDER'); END;
*
ERROR at line 1:
ORA-64399: In-Memory population or repopulation cannot be run for this segment.
ORA-06512: at "SYS.DBMS_INMEMORY", line 22
ORA-06512: at "SYS.DBMS_INMEMORY", line 265
ORA-06512: at line 1
```

最先支持的 POPULATE 过程只能实现后台发布,或称为异步发布。也就是说,此过程并不会等到对象完成发布后才返回,用户需要其他的手段来确认对象是否发布完成。但对于一些依赖 Database In-Memory 的分析性应用来说,确认对象已发布完成是这些应用正常运行的先决条件。为了弥补这一缺憾,Oracle 在 19c 版本中新增了基于前台的同步发布手段,即 POPULATE_WAIT 函数。

如表 4-3 所示,POPULATE_WAIT 函数支持 4 个参数,均为入参,并带默认值。可以看出,POPULATE_WAIT 只能发布指定发布优先级及以上的对象,这点和 POPULATE 过程针对单个对象操作不同。由于 POPULATE_WAIT 函数是前台方式执行的,因此会一直阻塞等待,直到所有对象发布完成后才返回。如果希望跟踪发布进度,也可以让 POPULATE_WAIT 函数在发布完指定比例的对象或超时后返回,然后余下的工作会继续在后台完成。当然,后续还可以再次调用 POPULATE_WAIT 来跟踪余下的进度,直到全部完成。

表 4-3 POPULATE_WAIT 函数参数

参数名	类型	默认值	描述
priority	VARCHAR2	'LOW'	大于或等于此发布优先级的 INMEMORY 对象将会被发布

续表

参 数 名	类 型	默 认 值	描 述
percentage	NUMBER	100	认为发布已完成的比例，达到比例后函数返回
timeout	NUMBER	99999999	超时返回的时间，单位为秒
force	VARCHAR2	FALSE	将内存列式存储中大于或等于指定优先级的对象移除，并重新发布

POPULATE_WAIT 的一种使用场景为：在启动时先进入管理员模式，然后调用 POPULATE_WAIT 函数，确认全部对象发布完成后，退出管理员模式并进入用户模式。这样可以保证在启动用户应用程序前，所有的 INMEMORY 对象均已就绪，从而避免所谓的"失控查询"现象。

在下面的示例中，系统以 RESTRICT 模式启动并进入管理员模式，循环调用设置了超时的 POPULATE_WAIT 函数，直到最终全部发布成功。在超时返回时，可以显示进度或进行其他一些处理。最后从管理员模式切换到用户模式，启动需要依赖内存列式存储的应用程序。

```
STARTUP RESTRICT
ALTER PLUGGABLE DATABASE ORCLPDB1 OPEN RESTRICTED;
ALTER SESSION SET CONTAINER=orclpdb1;
SET SERVEROUTPUT ON
DECLARE
    pop_status NUMBER;
BEGIN
    LOOP
        SELECT
            dbms_inmemory_admin.populate_wait(priority => 'NONE', percentage => 100, timeout => 60)
        INTO pop_status
        FROM
            dual;

        dbms_output.put_line('POP STATUS IS('
                            || pop_status
                            || ')');

        IF pop_status = -1 THEN
            dbms_output.put_line('Not ready, Please Wait…');
            CONTINUE;
        END IF;

        IF pop_status = 0 THEN
            dbms_output.put_line('Populate completed, You can start application now!');
        END IF;

        EXIT;
    END LOOP;
END;
/
```

```
ALTER SESSION SET CONTAINER=CDB$ROOT;

ALTER SYSTEM DISABLE RESTRICTED SESSION;
```

以下为程序运行的部分输出信息：

```
_INMEMORY.POPULATE_WAIT
timeout: 60
now: 05-MAR-21 12.18.19.334722000 PM +08:00
timeout time: 05-MAR-21 12.19.19.310835 PM
percentage: 100
priority: NONE
instances: 1
DBA_SEGMENTS:
populate:
…
POP STATUS IS (-1)
Not ready, Please Wait…
…
sleep(5)
GV$IM_SEGMENTS:
INST_ID: 1
OWNER: SSB
SEGMENT_NAME: CUSTOMER
PARTITION_NAME:
BYTES: 30900224
BYTES_NOT_POPULATED: 0
BYTES-BNP: 30900224
POPULATE_STATUS: COMPLETED
INST_ID: 1
OWNER: SSB
SEGMENT_NAME: LINEORDER
PARTITION_NAME:
BYTES: 1409155072
BYTES_NOT_POPULATED: 999759872
BYTES-BNP: 409395200
POPULATE_STATUS: STARTED
SUM(BYTES): 1440055296
SUM(BYTES_NOT_POPULATED): 999759872
SUM(BYTES-BNP): 440295424
TOTAL BYTES: 1440055296
%COMPLETE: 30.00
…
%COMPLETE: 100.00
SUM(BYTES): 1440055296
SUM(BYTES_NOT_POPULATED): 0
SUM(BYTES-BNP): 1440055296
POPULATION FULLY COMPLETED
POP STATUS IS(0)
Populate completed, You can start application now!
```

从以上程序输出可知，POPULATE_WAIT 函数在后台每 5 秒检查一次并输出状态信息，因此返回时的超时或完成比例并不一定精确地等于设定值。例如设定的超时为 32 秒，返回时超时可能为 35 秒。另外，完成比例并非指对象数量的比例，而是指所有对象存储容量的比例。最后，POPULATE_WAIT 的返回值有多个取值，除表示成功、超时外，还可能是内存列式存储未设置，内存不足或没有符合条件的对象。

  POPULATE 过程实现 INMEMORY 对象的初始发布，而 REPOPULATE 过程实现对象的重新发布，也就是当变更数据达到一定程度时刷新 IMCU 数据。因此，REPOPULATE 最主要的功能是实现增量数据刷新。例如 IMCU 1 包含第 1 行到 500 000 行数据，IMCU 2 包含第 500 001 行到 1 000 000 行数据，如果应用修改了第 100 行数据，则 REPOPULATE 只会重新发布 IMCU 1。REPOPULATE 也可以通过指定参数重新发布所有数据，当对象尚未发布到内存中时，REPOPULATE 过程的作用和 POPULATE 相同。

  POPULATE 过程只能后台发布对象，POPULATE_WAIT 函数虽然可以前台发布，但只能发布大于或等于指定发布优先级的对象。如果希望前台发布单个对象，可以自行编写代码实现。下面的过程通过循环查询动态性能视图 V$IM_SEGMENTS 中的 POPULATE_STATUS 字段实现了单个对象的前台发布：

```sql
CREATE OR REPLACE PROCEDURE popwait(
    schema_name IN VARCHAR2 DEFAULT 'SSB',
    table_name IN VARCHAR2 DEFAULT 'LINEORDER',
    timeout IN NUMBER DEFAULT 120
)AS
    ok  NUMBER:= 0;
    ts  NUMBER:= 0;
    t1  NUMBER:= dbms_utility.get_time;
BEGIN
    DBMS_INMEMORY.POPULATE(schema_name, table_name);

    LOOP
        dbms_session.sleep(1);
        ts := ts + 1;

        SELECT
                count(*)
        INTO ok
        FROM
            v$im_segments
        WHERE
            segment_name = table_name and populate_status = 'COMPLETED';

        EXIT WHEN ok = 1 OR ts > timeout;
    END LOOP;

    dbms_output.put_line('Pop time for table(' || upper(table_name) || '): '
                        ||(dbms_utility.get_time - t1) / 100
                        || ' seconds');

END popwait;
```

以下为使用此过程前台发布 LINEORDER 表的示例：

```
SQL>
SET SERVEROUTPUT ON
ALTER TABLE lineorder NO INMEMORY;
ALTER TABLE lineorder INMEMORY;
EXEC popwait('SSB', 'LINEORDER');

Pop time for table (LINEORDER): 41.06 seconds

PL/SQL procedure successfully completed.
```

4.4.5 通过初始化参数控制发布

初始化参数 INMEMORY_FORCE 可以控制是否将对象发布到内存列式存储中。INMEMORY_FORCE 的取值和说明参见表 4-4。

表 4-4 初始化参数 INMEMORY_FORCE 的取值与说明

INMEMORY_FORCE 值	说　明
DEFAULT	默认值。表示设置了 INMEMORY 属性的对象可以发布到内存列式存储中
OFF	关闭 Database In-Memory 功能。对象不会发布到内存列式存储中，即使对象设置了 INMEMORY 属性
BASE_LEVEL	启用 Database In-Memory Base Level 特性，通常用于对 Database In-Memory 功能的试用
CELLMEMORY_LEVEL	支持在内部（非公有云）部署的 Exadata 系统中的闪存存储中使用内存列式格式

4.5 重新发布

发布，或称为初始发布，是指在内存列式存储中创建 IMCU。重新发布则是当数据发生改变时刷新 IMCU，因此重新发布只针对已发布的对象。为何 IMCU 需要刷新？这是由于 IMCU 是只读的，数据更新并不能立即反映到 IMCU 中。如果不刷新，随着变化数据量的增加，数据库不得不频繁地从 Buffer Cache 甚至磁盘中读取最新的数据，从而导致性能下降。

Database In-Memory 提供两种重新发布机制，即基于阈值的重新发布和涓流式重新发布。当 IMCU 中的数据变化量超过内部设定的陈旧度阈值时，系统会进行基于阈值的重新发布。涓流式发布则是数据变化量未达到陈旧度阈值时，定期进行的小批量发布。IMCO 进程每 2 分钟唤醒一次，如果发现有变化的数据，会自动判断是否进行发布。这两种重新发布方式都属于内部机制，用户唯一可调整的只有发布进程的数量，用以控制发布速度。

除发布进程数量外，影响重新发布的因素还包括以下几点。

（1）数据修改的频度。数据修改越频繁，需要从 Buffer Cache 或磁盘读取数据的可能性就越大，性能受影响的几率也越大。

（2）数据修改操作的类型。删除和更新操作的性能开销比插入操作大。因为插入操作通常会产生新的数据块甚至新的 IMCU。

（3）压缩级别。压缩级别越高，重新发布的开销越大。

（4）被修改数据的位置。数据修改位置比较集中时，开销比修改位置分散时小，因为受影响的数据块或 IMCU 会较少。

IMCU 重新发布采用了一种称为双缓冲（double buffering）的技术。如图 4-2 所示，最初在老版本的 IMCU 已经包含了部分"陈旧"数据，这些数据对应行的 ROWID 被记录在 SMU 中的日志中。当后台进程决定进行重新发布时，会建立新版本的 IMCU。新版本 IMCU 中的数据是老版本 IMCU（没有发生改变的行）和 Buffer Cache 数据（发生改变行的最新版本）的合并。在重新发布期间，如果有应用对数据进行修改，实际的数据更新只会在 Buffer Cache 中进行。同时这些修改会被记录在 SMU 中，留待下一次重新发布时处理。

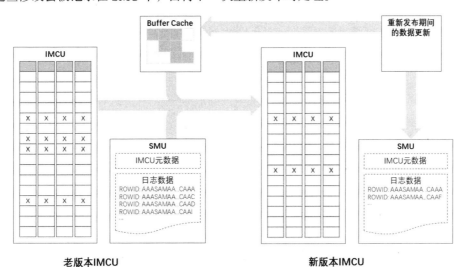

图 4-2　IMCU 重新发布时使用的双缓冲技术

由于在重新发布期间，新版本 IMCU 处于不可用状态。为保证性能，老版本 IMCU 会保留一段时间（默认为 2 分钟），以供需要访问之前版本数据的查询使用。这也是其被称为双缓冲的原因。

通过一个示例来了解一下重新发布的过程。首先创建一个 100 万行的表，然后将其前台发布到内存列式存储中：

```
CREATE TABLE t1 INMEMORY AS
     SELECT LEVEL AS ID, 'AAAAAAAA' AS VALUE
     FROM dual CONNECT BY 1 = 1 AND LEVEL <= 1000000;

SELECT dbms_inmemory_admin.populate_wait(PRIORITY => 'NONE') FROM dual;
```

更新 15 000 行数据，也就是 1.5% 的数据。通过查询 In-Memory 视图，可以发现无效（陈旧）行数据也是 15 000 条，涉及 40 个数据块：

```
SQL> UPDATE t1 SET VALUE = 'BBBBBBBB' WHERE ROWNUM <= 15000;
SQL> COMMIT;
```

```
SQL>
SELECT
    total_rows,
    invalid_rows,
    invalid_blocks,
    prepopulated,
    repopulated,
    trickle_repopulated
FROM
    v$im_smu_head    a,
    v$im_header      b
WHERE
    a.objd = b.objd
AND a.total_rows = b.num_rows;

TOTAL_ROWS INVALID_ROWS INVALID_BLOCKS PREPOPULATED REPOPULATED TRICKLE_REPOPULATED
---------- ------------ -------------- ------------ ----------- -------------------
    525293        15000             40            0           0                   0
    474707            0              0            0           0                   0
```

执行查询语句,从执行计划中可以看出:尽管扫描是在内存中进行的,但其中 15 000 行数据需要到 Buffer Cache 中读取,而这将影响性能。

```
---------------------------------------------------------------------------------
| Id | Operation                    | Name | Rows | Bytes | Cost (%CPU)| Time    |
---------------------------------------------------------------------------------
|  0 | SELECT STATEMENT             |      |      |       |   53 (100) |         |
|  1 |  SORT AGGREGATE              |      |    1 |     9 |            |         |
|* 2 |   TABLE ACCESS INMEMORY FULL | T1   |    2 |    18 |   53  (48) | 00:00:01|
---------------------------------------------------------------------------------
...
NAME                                                              VALUE
---------------------------------------------------------------- -------
IM scan blocks cache                                                 40
IM scan rows cache no delta                                       15000
IM scan rows excluded                                             15000
```

经过一段时间,系统启动重新发布。重新查询,无效数据行数和无效数据块数变为 0。

```
TOTAL_ROWS INVALID_ROWS INVALID_BLOCKS PREPOPULATED REPOPULATED TRICKLE_REPOPULATED
---------- ------------ -------------- ------------ ----------- -------------------
    429327            0              0            0           1                   1
    570673            0              0            0           0                   0
```

经过多次测试发现。重新发布完结的时间跨度从 2 分钟到 6 分钟不等。这也说明,尽管涓流式重新发布每 2 分钟唤醒一次,但未必每次都会真正执行,可能还需要考虑是否有应用负载访问等因素。另外,如果实验中将修改的行数增加,则重新发布会更早发生。从系统隐含参数中,可以找到相关阈值设置。作为内部设置,不要去修改这些参数,但可以大致推断出重新发布综合了时间、数据块和数据行修改比例等因素。

```
_inmemory_repopulate_threshold_blocks_percent          10
_inmemory_repopulate_threshold_mintime_factor          5
_inmemory_repopulate_threshold_rows_percent            5
```

4.6 发布进程与发布速度

发布优先级只能影响发布的顺序，并不能控制发布的速度。发布速度在硬件层面受机器性能的影响，如存储性能、处理器数量和能力；在软件层面则受 Oracle 服务器进程的控制。

在第 3 章已经提到，内存列式数据的发布与重新发布是由 IMCO（内存协调进程）来控制的。具体而言，IMCO 负责内存列式数据的初始发布、基于阈值的重新发布、涓流式重新发布。在操作系统中可以查看到此进程：

```
$ ps -ef|grep imco|grep -v grep
oracle    2402    1  0 08:06 ?        00:00:00 ora_imco_ORCLCDB
```

IMCO 只负责调度和协调，具体发布的工作由 W*nnn*（空间管理工作者）进程代替 IMCO 来完成。在发布过程中，W*nnn* 进程负责创建 IMCU、SMU 和 IMEU。在重新发布过程中，W*nnn* 进程基于现有 IMCU 和 SMU 中的日志记录创建新版本的 IMCU，并临时保留老版本的 IMCU，即前面介绍的双缓冲技术。对于某一个对象，多个 W*nnn* 进程并行处理，每一个进程负责此对象的部分数据块。但只有发布完成一个对象后，才会接下来处理下一个对象。也就是说，对于多个对象，发布是顺序进行的。

数据库会提前生成一组 W*nnn* 进程，等待 IMCO 的调遣。在操作系统中可以看到这一组进程：

```
$ ps -ef|grep w00|grep -v grep
oracle    2351    1  0 08:06 ?        00:00:00 ora_w000_ORCLCDB
oracle    2355    1  0 08:06 ?        00:00:00 ora_w001_ORCLCDB
oracle    2413    1  0 08:06 ?        00:00:00 ora_w002_ORCLCDB
oracle    2522    1  0 08:06 ?        00:00:00 ora_w003_ORCLCDB
oracle    2665    1  0 08:06 ?        00:00:00 ora_w004_ORCLCDB
oracle    2831    1  0 08:07 ?        00:00:00 ora_w005_ORCLCDB
oracle    2835    1  0 08:07 ?        00:00:00 ora_w006_ORCLCDB
oracle    2839    1  0 08:07 ?        00:00:00 ora_w007_ORCLCDB
```

但并非所有的 W*nnn* 进程都会参与数据的初始发布和重新发布，实际参与进程数量的最大值由初始化参数 INMEMORY_MAX_POPULATE_SERVERS 控制，此参数默认值为 CPU_COUNT 的一半，并且只能在 CDB 一级设置。如果将此参数设为 0，则会禁止发布：

```
SQL> SHOW PARAMETER inmemory_max_populate_servers
NAME                                 TYPE        VALUE
------------------------------------ ----------- ------------------------------
inmemory_max_populate_servers        integer     4

SQL> SHOW PARAMETER cpu_count
NAME                                 TYPE        VALUE
------------------------------------ ----------- ------------------------------
```

```
cpu_count                                          integer        8
```

在初始发布时,可以通过 top 命令查看实际参与发布的 W*nnn* 进程。如图 4-3 所示,可以看到总共有两个 W*nnn* 进程参与了发布,它们的 CPU 使用率均在 90%左右,这也说明内存列式数据发布是 CPU 密集型的操作。

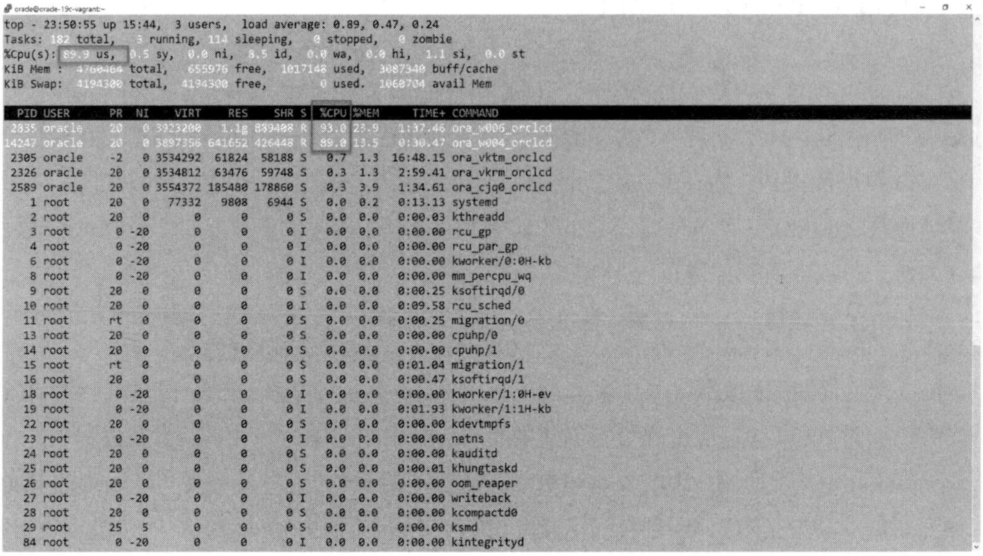

图 4-3 监控内存对象的发布

更多的进程带来更快的发布,更少的进程则给系统预留出更多的处理器资源。注意不要将此参数设置过高,例如等于或非常接近于 CPU_COUNT,否则在发布时系统将没有可用资源用于运行其他任务。

下面通过一个实验来了解不同的发布进程数对发布速度的影响。这里选取了一个 CPU_COUNT 为 8 的数据库,然后将发布进程的数量从 1 增加到 8,总共运行 8 轮。测试结果如图 4-4 所示。

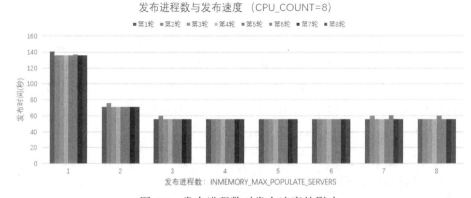

图 4-4 发布进程数对发布速度的影响

可以看到,每一轮测试的结果都非常接近。发布进程数从 3 开始,发布时间基本保持稳定。

因此，系统默认的设置（即 CPU_COUNT 的一半，本例为 4）是合适的。既可以保证最佳的发布速度，也可以预留足够的资源给其他任务。

另外，本测试发布所有的 SSB 样例表，并将 LINEORDER 表的数据增加了一倍（23 995 992 行）。总共的磁盘数据量为 2822MB，对应发布进程数为 4 时的发布速度为 50MB/s。

本实验的测试脚本如下：

```
$ cat dbim_pop_wait_simple.sql
SET TIMING ON

VARIABLE pop_status NUMBER

BEGIN
    SELECT
        dbms_inmemory_admin.populate_wait(priority => 'NONE', percentage => 100)
    INTO :pop_status
    FROM
        dual;
END;
/

PRINT pop_status

SET TIMING OFF

$ cat dbim_pop_speed_test.sh
for i in {1..8}; do
        CPU_COUNT=8
        for i in $( seq 1 $CPU_COUNT ); do
                echo "# pop_processes: $i"
                sqlplus -S / as sysdba <<-EOF
                        alter system set INMEMORY_MAX_POPULATE_SERVERS = $i;
                        @../userlogin.sql
                        !./dbim_alter_table_off.sh > /dev/null
                        !./dbim_alter_table_on.sh > /dev/null
                        @dbim_pop_wait_simple.sql
                        !./dbim_alter_table_off.sh > /dev/null
                EOF
        done
done
```

两种重新发布类型中，基于阈值的重新发布是 Oracle 内部机制，用户无法参与；涓流式重新发布则可以通过初始化参数 INMEMORY_TRICKLE_REPOPULATE_SERVERS_PERCENT 控制。此参数限制了参与涓流式重新发布的最大进程数。

此值表示 INMEMORY_MAX_POPULATE_SERVERS 初始化参数的百分比。默认值 1 在大部分情况下都是适合的。可以设为 0 以临时禁止涓流式重新发布。但不能设置为 50 或以上，因为另外一半的进程需要用于其他类型的发布和重新发布任务。

```
SQL> COL name_col_plus_show_param FOR a45
SQL> COL value_col_plus_show_param FOR a6
SQL> SHOW PARAMETER repop
NAME                                          TYPE         VALUE
--------------------------------------------- ------------ ------
inmemory_trickle_repopulate_servers_percent   integer      1
```

4.7 指定内存压缩级别

行格式存储的数据库对象,在转换为列格式后,还可以根据性能和空间的需要,选择不同的压缩级别。列式存储非常适合压缩,因为用于数据分析的列通常重复值和空值较多,因此压缩比通常比较高,例如性别、年龄、省份和城市等。除了节省空间外,更重要的是可以提升查询的性能,因为扫描和过滤操作可以在更小的数据集上操作。对于某些压缩级别,一些查询操作甚至可以直接针对压缩格式的数据进行扫描和过滤,数据库只在必要时对数据进行解压。

可选的压缩级别如表 4-5 所示。不同的压缩级别在空间节省和查询性能两方面各有侧重,FOR QUERY 倾向于查询性能,FOR CAPACITY 倾向于空间节省。其中 FOR QUERY LOW 是默认的压缩级别,具有最好的查询性能。FOR QUERY LOW、FOR QUERY HIGH 和 FOR DML 三种压缩级别支持直接在压缩格式数据上进行扫描和过滤。

表 4-5 内存列式存储压缩级别

压缩级别	无需解压即可查询	描述
NO MEMCOMPRESS	是	无任何压缩
FOR DML	是	压缩比最低,但 DML 操作性能最好
FOR QUERY LOW	是	查询性能最好,压缩比高于 FOR DML。为默认的压缩级别
FOR QUERY HIGH	否	查询性能好,但不如 FOR QUERY LOW。压缩比高于 FOR QUERY LOW
FOR CAPACITY LOW	否	综合考虑性能和压缩比,但倾向于后者。压缩比高于 FOR QUERY HIGH 但低于 FOR CAPACITY HIGH。使用了 OZIP 专利算法
FOR CAPACITY HIGH	否	压缩比最高
AUTO	取决于实际压缩级别	最高级别自动内存管理,21c 新特性

表或表分区中的列默认继承其父对象的压缩级别设置,也可以为列单独设置不同的压缩级别;表分区中的各个分区也可以设置不同的压缩级别。以下为各类设置命令的示例:

```
SQL> ALTER TABLE lineorder INMEMORY;
Table altered.

-- 查看默认压缩级别
SQL> SELECT inmemory_compression FROM user_tables WHERE table_name = 'LINEORDER';

INMEMORY_COMPRESS
-----------------
FOR QUERY LOW

-- 修改表的压缩级别
```

```
SQL> ALTER TABLE lineorder INMEMORY MEMCOMPRESS FOR QUERY HIGH;
Table altered.
```

-- 修改部分列的压缩级别
```
SQL> ALTER TABLE lineorder INMEMORY MEMCOMPRESS FOR CAPACITY HIGH (lo_tax) INMEMORY NO
MEMCOMPRESS (lo_shipmode);
Table altered.
```

-- 在视图中查看设置的压缩级别
```
SQL> SELECT column_name, inmemory_compression FROM v$im_column_level WHERE table_name
= 'LINEORDER';

COLUMN_NAME              INMEMORY_COMPRESSION
------------------       --------------------------
...
LO_SUPPLYCOST            DEFAULT
LO_TAX                   FOR CAPACITY HIGH
LO_COMMITDATE            DEFAULT
LO_SHIPMODE              NO MEMCOMPRESS

17 rows selected.

SQL> ALTER TABLE sales INMEMORY;
Table altered.
```

-- 查询分区表的压缩级别，返回为空表示默认压缩级别
```
SQL> SELECT inmemory_compression FROM user_tables WHERE table_name = 'SALES';

INMEMORY_COMPRESS
-----------------
```

-- 修改分区表中某分区的压缩级别
```
SQL> ALTER TABLE sales MODIFY PARTITION sales_q4_2001 INMEMORY MEMCOMPRESS FOR CAPACITY
HIGH;
Table altered.
```

-- 查询指定分区的压缩级别
```
SQL> SELECT segment_name, partition_name, inmemory_compression FROM user_segments
WHERE segment_name = 'SALES';

SEGMENT_NA  PARTITION_NAME    INMEMORY_COMPRESS
----------  ----------------  -----------------
SALES       SALES_Q1_1998     FOR QUERY LOW
...
SALES       SALES_Q4_2000     FOR QUERY LOW
SALES       SALES_Q4_2001     FOR CAPACITY HIGH

16 rows selected.
```

不同的压缩级别除了空间效率不同，还会影响到数据发布时间和查询时间。以下通过一个

实验来了解它们之间的差异。针对 SSB 样例的 5 个表，在不同的压缩级别下测试其压缩比、发布时间和查询时间。表 4-6 为测试 10 轮后的结果。

表 4-6 不同压缩比下空间和性能测试结果

压缩级别	压缩比	平均发布时间/s	平均查询时间/（百分之一秒）
NO MEMCOMPRESS	1.25	26.131	77.8
FOR DML	1.25	25.886	68.7
FOR QUERY LOW（默认）	2.38	46.294	28
FOR QUERY HIGH	3.55	47.164	34.6
FOR CAPACITY LOW	4.06	51.157	36.1
FOR CAPACITY HIGH	5.12	46.09	40.3
AUTO	N/A	N/A	N/A

从以上实验结果可以得出以下结论。

（1）默认压缩级别 FOR CAPACITY LOW 具有最好的查询性能，压缩比处于中位。

（2）FOR CAPACITY HIGH 压缩级别具有最好的空间效率。

（3）FOR DML 和 NO MEMCOMPRESS 压缩比一样，说明在本实验中并没有进行压缩。事实上，FOR DML 只有在列中的所有值相同时才进行压缩[①]。

（4）和未使用压缩相比，使用压缩后查询时间更快，说明压缩可以提升性能。

（5）压缩级别增加，查询时间和发布时间也随之增加。

在选择压缩级别时，首要的考虑因素是查询性能，这也是用户选用 Database In-Memory 的重要原因。因此，默认的 FOR QUERY LOW 压缩级别是绝大多数情况下的最佳选择。当内存不足以容纳所有对象时，可以考虑扩展内存或增加压缩级别，以保证对象的完整发布。如果希望加快对象的发布速度，可以增加发布进程数量，或使用第 8 章介绍的 FastStart 技术。

4.8 内存列式存储与行式存储映射

第 3 章已经介绍到，内存列式存储由 IMCU 和 SMU 组成。IMCU 是只读、压缩的存储单元，包含一到多列的数据。一个数据库对象对应于一个或多个 IMCU，一个 IMCU 只包含来自一个数据库对象的数据。下面通过一个例子说明内存列式存储与行式存储之间的映射关系。

首先建立一个 200 万行的表 T2。此表包含 4 列，其中第 2、3 列为随机数和随机字符串：

```
CREATE TABLE T2 AS SELECT
    LEVEL AS ID,
    round(dbms_random.VALUE(1,1000)) AS v1,
    dbms_random.STRING('L', 2) AS v2,
    'ABCDEFGH' AS v3
FROM dual
```

[①] https://blogs.oracle.com/in-memory/database-in-memory-compression

```
CONNECT BY 1=1 AND LEVEL <= 2000000;
```

在探索表 T2 实际的行式存储结构之前，先来了解一下 Oracle 数据库存储结构的基本概念。如图 4-5 所示，Oracle 数据库存储结构包括逻辑和物理两个层面。逻辑层面，数据库被划分为逻辑存储单元，即表空间。表空间是 Segment 对象的容器，像表、表分区和索引等都属于 Segment 对象。Segment 包含一组为数据库对象分配的 Extent，而 Extent 由一到多个连续的数据库块组成，默认的数据库块大小为 8KB。物理层面，表空间对应一到多个数据文件。

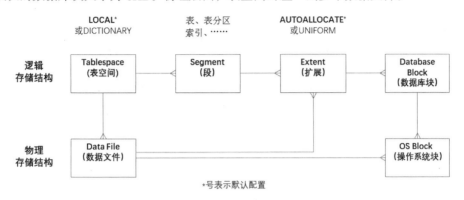

图 4-5　Oracle 数据库存储结构

当 Segment 对象中的数据不断增加时，数据库通过分配额外的 Extent 来容纳新的数据。Oracle 通过逻辑空间管理在表空间中分配 Extent 和跟踪使用情况。Oracle 提供 LOCAL 和 DICTIONARY 两种逻辑空间管理方式，分别用位图和数据字典来管理 Extent。LOCAL 是默认和推荐的方式。在 LOCAL 方式下，Extent 的大小可以是固定（UNIFORM）或变化（AUTOALLOCATE）的。AUTOALLOCATE 是默认和推荐的配置，适用于表空间中对象大小不一且用户希望简化空间管理任务的情形。当表空间指定了 AUTOALLOCATE 属性时，Extent 的大小最初为 64KB，然后随着对象容量的增长会依次变为 1MB、8MB 和 64MB。

基于以上基本概念，接下来可以查看 T2 表实际的存储分配。首先可以查出 T2 表所在的表空间为 USERS，表空间的空间管理方式为 LOCAL，Extent 大小为自动管理。同时可以从系统表中查询出各种方式的描述：

```
SQL> SELECT tablespace_name FROM user_segments WHERE segment_type = 'TABLE'
AND segment_name = 'T2';

TABLESPACE_NAME
---------------
USERS

SQL>    SELECT    tablespace_name,    extent_management,    allocation_type,
segment_space_management FROM dba_tablespaces WHERE tablespace_name ='USERS';

TABLESPACE                     EXTENT          ALLOCATION SEGMENT SPACE
NAME                           MANAGEMENT      TYPE       MANAGEMENT
------------------------------ --------------- ---------- --------------
```

```
USERS                                LOCAL      SYSTEM    AUTO

SQL> SELECT column_name, comments FROM all_col_comments
WHERE table_name ='DBA_TABLESPACES' AND column_name IN
   ('EXTENT_MANAGEMENT','ALLOCATION_TYPE','SEGMENT_SPACE_MANAGEMENT');

COLUMN_NAME                 COMMENTS
--------------------------- ----------------------------------------------------------
EXTENT_MANAGEMENT           Extent management tracking: "DICTIONARY" or "LOCAL"
ALLOCATION_TYPE             Type of extent allocation in effect for this tablespace
SEGMENT_SPACE_MANAGEMENT    Segment space management tracking: "AUTO" or "MANUAL"
```

然后，可以查询出 T2 表由 74 个 Extent 组成，其中前 16 个 Extent 的大小为 64KB（8 个数据库块），余下 58 个 Extent 的大小为 1MB（1024KB，128 个数据库块）。如果数据不断增加，Extent 的大小会变为 8MB 甚至 64MB。

```
SQL> SHOW PARAMETER db_block_size

NAME                                 TYPE        VALUE
------------------------------------ ----------- ------------------------------
db_block_size                        integer     8192

SQL> SELECT COUNT(*) AS num_extents, SUM(BYTES)/1024 AS "size(kb)", SUM(blocks)
FROM user_extents WHERE segment_name = 'T2';

NUM_EXTENTS   SIZE(KB)  SUM(BLOCKS)
----------- ---------- -----------
         74      60416        7552

SQL> SELECT extent_id, bytes, blocks FROM user_extents WHERE segment_name = 'T2';

 EXTENT_ID      BYTES     BLOCKS
---------- ---------- ----------
         0      65536          8
         1      65536          8
         2      65536          8
...
        13      65536          8
        14      65536          8
        15      65536          8
        16    1048576        128
        17    1048576        128
        18    1048576        128
        19    1048576        128
...
        69    1048576        128
        70    1048576        128
        71    1048576        128
        72    1048576        128
```

```
           73    1048576           128

74 rows selected.
SQL> SELECT COUNT(*), bytes/1024 AS "size(kb)", BYTES/1024/8 AS blocks FROM user_extents
WHERE segment_name = 'T2' GROUP BY BYTES ORDER BY BYTES;

  COUNT(*)   SIZE(KB)    BLOCKS
---------- ---------- ----------
        16         64          8
        58       1024        128
```

接下来，为表 T2 启用 INMEMORY 属性并发布。发布完成后查看状态，可知 T2 表在内存中总容量为 33 600KB，包括实际的列式数据（IMCU）和元数据（SMU）：

```
SQL> ALTER TABLE T2 INMEMORY;
SQL> SELECT /*+ FULL(p) NO_PARALLEL(p) */ COUNT(*) FROM t2 p;

$ ./dbim_get_popstatus.sh
         SEGMENT        IN MEM      ON DISK   COMPRESSION   BYTES NOT   POPULATED
OWNER    NAME           SIZE(KB)    SIZE(KB)  RATIO         POPULATED   STATUS
-------  -------------  ----------  --------  ------------  ----------  ----------
SSB      T2             33,600      58,424    1.74                  0   COMPLETED


$ ./dbim_ima_usage.sh
POOL          ALLOC(KB)    USED(KB)   POPULATE_STATUS          CON_ID
----------   ----------   ---------   ----------------      ---------
1MB POOL        1072128       32768   DONE                         3
64KB POOL        446464         832   DONE                         3
```

数据库对象在内存列式存储中由一到多个 IMCU 组成，IMCU 中只能包括一个数据库对象的数据，IMCU 的存储分配单元为 1MB。通过查看视图 V$IM_HEADER，可知表 T2 发布到内存列式存储后，由 4 个 IMCU 组成，每个 IMCU 对应行式存储中 Extent 的数量依次为 31、13、16 和 14，看上去似乎不太均衡。

```
SQL>
SELECT imcu_addr, allocated_len/1024/1024 "alloc(mb)", round(used_len/1024/1024,2)
"used(mb)", num_disk_extents FROM v$im_header WHERE table_objn IN
(SELECT object_id FROM user_objects WHERE object_name = 'T2') AND num_rows != 0;

IMCU_ADDR            ALLOC(MB)    USED(MB)   NUM_DISK_EXTENTS
----------------    ----------   ---------   ----------------
0000000087AFFF70             9        8.04                 31
00000000873FFF70             7        6.52                 13
0000000086AFFF70             9        8.29                 16
00000000863FFF70             7         6.7                 14
```

实际上，每一个 IMCU 中行数的期望值和实际值均在 50 万左右。这可以让 IMCU 的大小比较均衡，也便于对发布、扫描和过滤等操作实施并行处理。因此，每一个 IMCU 的发布时间也是接近的，这里使用的单位是毫秒。

```
SQL> SELECT num_rows, num_target_rows, num_blocks, num_target_blocks,
 time_to_populate FROM v$im_header WHERE table_objn IN
(SELECT object_id FROM user_objects WHERE object_name = 'T2') AND num_rows != 0;

 NUM_ROWS NUM_TARGET_ROWS NUM_BLOCKS NUM_TARGET_BLOCKS TIME_TO_POPULATE
---------- --------------- ---------- ----------------- ----------------
    560658          500280       2008              1792              786
    455735          500280       1638              1792              639
    542561          500280       2016              1792              853
    441046          500280       1641              1792              694
```

系统视图 DBA_EXTENTS 记录了每一个 Extent 起始数据库块的地址和随后数据库块的数量，据此可以计算出 Extent 的起始和终止地址。在视图 V$IM_TBS_EXT_MAP 中则包含了 IMCU 与 Extent 起始地址之间的映射。利用这些信息，以下的 PL/SQL 代码可以显示内存列式存储和行式存储之间的映射关系。

```
$ cat imcs_rowstore_mapping.sql
SET PAGES 999
SET SERVEROUTPUT ON
COL imcu_addr HEADING "imcu_addr" FOR 9999999999999999
COMPUTE NUMBER LABEL "#extents in imcu" OF extent_id ON imcu_addr
BREAK ON imcu_addr SKIP 1 DUPLICATES
DECLARE
    v_rc SYS_REFCURSOR;
    v_objd NUMBER;
    v_objname VARCHAR2(32) := UPPER('&tabname');
BEGIN
    SELECT object_id INTO v_objd FROM user_objects WHERE object_name = v_objname;

    OPEN v_rc FOR
    SELECT  A.imcu_addr,  b.extent_id,  b.blocks,  A.start_dba,  b.ext_start_dba,
b.ext_end_dba  FROM(
    (SELECT imcu_addr, start_dba, v_objd dataobj, COUNT(1) FROM v$im_tbs_ext_map WHERE
dataobj = v_objd GROUP BY imcu_addr, start_dba) A JOIN
    (SELECT extent_id+1 AS extent_id,blocks
     ,dbms_utility.make_data_block_address(relative_fno,block_id) ext_start_dba
     ,dbms_utility.make_data_block_address(relative_fno,block_id+blocks-1)
ext_end_dba
     ,v_objd dataobj
        FROM dba_extents WHERE OWNER = USER AND segment_name = v_objname) b
   ON  (A.dataobj = b.dataobj AND  (A.start_dba BETWEEN  b.ext_start_dba  AND
b.ext_end_dba)))
    ORDER BY 4;

    dbms_sql.return_result(v_rc);

END;
```

执行以上代码，从输出中可以查看到每一个 IMCU 中所对应的 Extent，基本上是按照 Extent 的顺序映射的。每一个 IMCU 中所包含的 Extent 数量与之前的信息也是一致的。另外，第一个

IMCU 中包含的 Extent 较多,余下 3 个 IMCU 中包含的 IMCU 数量接近,这是由于最初 16 个 Extent 的大小仅为 8KB,实际上每一个 IMCU 的大小和行数都应该是差不多的。

```
$ sqlplus ssb@orclpdb1 @imcs_rowstore_mapping.sql

Enter value for tabname: t2
old   4:       v_objname varchar2(32) := upper('&tabname');
new   4:       v_objname varchar2(32) := upper('t2');

PL/SQL procedure successfully completed.

ResultSet #1

      IMCU_ADDR   EXTENT_ID    BLOCKS   START_DBA  EXT_START_DBA  EXT_END_DBA
--------------- ----------- ---------- ---------- -------------- -----------
     2276458352           1          8   50561939       50561936     50561943
     2276458352           2          8   50561944       50561944     50561951
     2276458352           3          8   50561953       50561952     50561959
     2276458352           4          8   50561960       50561960     50561967
     2276458352           5          8   50561969       50561968     50561975
     2276458352           6          8   50561976       50561976     50561983
     2276458352           7          8   50561985       50561984     50561991
     2276458352           8          8   50561992       50561992     50561999
     2276458352           9          8   50562001       50562000     50562007
     2276458352          10          8   50562008       50562008     50562015
     2276458352          11          8   50562017       50562016     50562023
     2276458352          12          8   50562024       50562024     50562031
     2276458352          13          8   50562033       50562032     50562039
     2276458352          14          8   50562040       50562040     50562047
     2276458352          15          8   50564609       50564608     50564615
     2276458352          16          8   50564616       50564616     50564623
     2276458352          17        128   50575106       50575104     50575231
     2276458352          18        128   50575234       50575232     50575359
     2276458352          19        128   50575490       50575488     50575615
     2276458352          20        128   50575618       50575616     50575743
     2276458352          21        128   50575746       50575744     50575871
     2276458352          22        128   50575874       50575872     50575999
     2276458352          23        128   50576002       50576000     50576127
     2276458352          24        128   50576130       50576128     50576255
     2276458352          25        128   50576258       50576256     50576383
     2276458352          26        128   50576386       50576384     50576511
     2276458352          27        128   50576514       50576512     50576639
     2276458352          28        128   50576642       50576640     50576767
     2276458352          29        128   50576770       50576768     50576895
     2276458352          30        128   50576898       50576896     50577023
     2276458352          31        128   50577026       50577024     50577151
***************             ----------
#EXTENTS IN IMCU         31

     2269118320          47        128   50578050       50578048     50578175
     2269118320          32        128   50578946       50578944     50579071
```

```
                 2269118320         33           128      50579074       50579072       50579199
                 2269118320         34           128      50579202       50579200       50579327
                 2269118320         35           128      50579330       50579328       50579455
                 2269118320         36           128      50579458       50579456       50579583
                 2269118320         37           128      50579586       50579584       50579711
                 2269118320         38           128      50579714       50579712       50579839
                 2269118320         39           128      50579842       50579840       50579967
                 2269118320         40           128      50579970       50579968       50580095
                 2269118320         41           128      50580098       50580096       50580223
                 2269118320         42           128      50580226       50580224       50580351
                 2269118320         43           128      50580354       50580352       50580479
****************               ----------
#EXTENTS IN IMCU                   13

                 2259681136         44           128      50580482       50580480       50580607
                 2259681136         45           128      50580610       50580608       50580735
                 2259681136         46           128      50580738       50580736       50580863
                 2259681136         48           128      50580866       50580864       50580991
                 2259681136         49           128      50580994       50580992       50581119
                 2259681136         50           128      50581122       50581120       50581247
                 2259681136         51           128      50581250       50581248       50581375
                 2259681136         52           128      50581378       50581376       50581503
                 2259681136         53           128      50581506       50581504       50581631
                 2259681136         54           128      50581634       50581632       50581759
                 2259681136         55           128      50581762       50581760       50581887
                 2259681136         56           128      50581890       50581888       50582015
                 2259681136         57           128      50582018       50582016       50582143
                 2259681136         58           128      50582146       50582144       50582271
                 2259681136         59           128      50582274       50582272       50582399
                 2259681136         60           128      50582402       50582400       50582527
****************               ----------
#EXTENTS IN IMCU                   16

                 2252341104         61           128      50582530       50582528       50582655
                 2252341104         62           128      50582658       50582656       50582783
                 2252341104         63           128      50582786       50582784       50582911
                 2252341104         64           128      50582914       50582912       50583039
                 2252341104         65           128      50583042       50583040       50583167
                 2252341104         66           128      50583170       50583168       50583295
                 2252341104         67           128      50583298       50583296       50583423
                 2252341104         68           128      50583426       50583424       50583551
                 2252341104         69           128      50583554       50583552       50583679
                 2252341104         70           128      50583682       50583680       50583807
                 2252341104         71           128      50583810       50583808       50583935
                 2252341104         72           128      50583938       50583936       50584063
                 2252341104         73           128      50584066       50584064       50584191
                 2252341104         74           128      50584194       50584192       50584319
****************               ----------
#EXTENTS IN IMCU                   14

74 rows selected.
```

在以上实验基础上，将 T2 表的内存压缩级别修改为 FOR CAPACITY HIGH。可以发现，尽管压缩比从 1.74 提高到了 4.83，但与行式存储中 Extent 的映射关系并没有发生变化，每个 IMCU 包括 Extent 的数量仍然为 31、13、16 和 14，每个 IMCU 的目标行数也仍然为 50 万左右，只不过每个 IMCU 的大小降至 3MB 左右。

```
SQL> SELECT SUM(BYTES)/1024 AS "total_disk_size(kb)", SUM(inmemory_size)/1024 AS
 "total im size(kb)", round(SUM(BYTES)/SUM(inmemory_size))
 AS overall_compression_ratio FROM v$im_segments;

TOTAL_DISK_SIZE(KB) TOTAL_IM_SIZE(KB) OVERALL_COMPRESSION_RATIO
------------------ ----------------- -------------------------
             58424             12096                      4.83

SQL>
SELECT imcu_addr, allocated_len/1024/1024 "alloc(mb)", round(used_len/1024/1024,2)
"used(mb)", num_disk_extents FROM v$im_header WHERE table_objn IN
(SELECT object_id FROM user_objects WHERE object_name = 'T2') AND num_rows != 0;

IMCU_ADDR           ALLOC(MB)    USED(MB)   NUM_DISK_EXTENTS
----------------   ----------   ----------  ----------------
0000000092BFFFB8            3        2.77                 31
00000000928FFFB8            3        2.44                 13
00000000925FFFB8            3        2.82                 16
00000000923FFFB8            2        1.71                 14

SQL> SELECT num_rows, num_target_rows, num_blocks, num_target_blocks,
 time_to_populate FROM v$im_header WHERE table_objn IN
(SELECT object_id FROM user_objects WHERE object_name = 'T2') AND num_rows != 0;

NUM_ROWS NUM_TARGET_ROWS NUM_BLOCKS NUM_TARGET_BLOCKS TIME_TO_POPULATE
-------- --------------- ---------- ----------------- ----------------
  560658          500280       2008              1792              853
  455735          500280       1638              1792              686
  542561          500280       2016              1792              806
  441046          500280       1641              1792              758
```

通过以上两个实验，可以得出以下结论。

（1）数据库对象发布到内存列式存储中后，数据按照其在行式存储中自然存储的顺序映射为一到多个 IMCU。

（2）行式存储中的数据以 Extent 为最小单元映射到 IMCU 中。一个 IMCU 包含一到多个 Extent，一个 Extent 只能属于一个 IMCU。

（3）对于一个对象，每个 IMCU 的大小都是接近的，每个 IMCU 的目标行数为 50 万左右。

（4）调低或调高内存压缩级别，IMCU 的大小也随之增减。

4.9 移除数据库对象

当对象已发布到内存列式存储后，以下 3 种情况下会将对象从内存列式存储中移除。

（1）通过 ALTER TABLE 命令关闭 INMEMORY 属性（NO INMEMORY）。

（2）删除或移动对象。

（3）设置自动数据优化（ADO）策略。

自动数据优化基于热图统计信息将对象发布到内存或从内存移除，让最适合的对象驻留在宝贵的内存中，此特性将在第 9 章介绍。

关闭 INMEMORY 属性除了将对象从内存中清除外，还会将发布优先级、压缩级别等属性清空并恢复为默认值。例如：

```
$ ./dbim_get_table_attr.sh
                                          INMEMORY   INMEMORY
SEGMENT_NAME     SEGMENT_TYPE    INMEMORY  PRIORITY   COMPRESSION
---------------  --------------  --------  ---------  --------------
CUSTOMER         TABLE           ENABLED   LOW        FOR QUERY LOW
DATE_DIM         TABLE           ENABLED   LOW        FOR QUERY LOW
LINEORDER        TABLE           ENABLED   LOW        FOR QUERY LOW
PART             TABLE           ENABLED   LOW        FOR QUERY LOW
SUPPLIER         TABLE           ENABLED   LOW        FOR QUERY LOW

SQL> ALTER TABLE lineorder NO INMEMORY;
SQL> ALTER TABLE part NO INMEMORY;
SQL> ALTER TABLE customer NO INMEMORY;
SQL> ALTER TABLE supplier NO INMEMORY;
SQL> ALTER TABLE date_dim NO INMEMORY;

$ ./dbim_get_table_attr.sh
                                          INMEMORY   INMEMORY
SEGMENT_NAME     SEGMENT_TYPE    INMEMORY  PRIORITY   COMPRESSION
---------------  --------------  --------  ---------  --------------
CUSTOMER         TABLE           DISABLED
DATE_DIM         TABLE           DISABLED
LINEORDER        TABLE           DISABLED
PART             TABLE           DISABLED
SUPPLIER         TABLE           DISABLED
```

需要注意，修改除发布优先级之外的所有 INMEMORY 属性（如压缩级别）也会将对象从内存列式存储中移除。

移动对象或改变对象的物理结构也会将对象从内存中移除，但表更名操作不会。在下例中，为表启用或禁止行级压缩后，此表将从内存列式存储中移除：

```
-- 确认表已发布
SQL> SELECT inmemory_size FROM v$im_segments WHERE segment_name = 'T2';

INMEMORY_SIZE
-------------
     17629184

-- 为表启用行级压缩
SQL> ALTER TABLE T2 MOVE ONLINE TABLESPACE USERS COMPRESS;

Table altered.

-- 由于表的物理结构改变，因此被清除出内存列式存储
```

```
SQL> SELECT inmemory_size FROM v$im_segments WHERE segment_name = 'T2';

no rows selected

-- 通过全表扫描重新发布对象
SQL> SELECT /*+ FULL(p) NO_PARALLEL(p) */ COUNT(*) FROM t2 p;

  COUNT(*)
----------
   2000000

-- 确认表已发布
SQL> SELECT inmemory_size FROM v$im_segments WHERE segment_name = 'T2';

INMEMORY_SIZE
-------------
     18481152

-- 为表禁用行级压缩
SQL> ALTER TABLE T2 MOVE ONLINE TABLESPACE USERS NOCOMPRESS;

Table altered.

-- 由于表的物理结构改变,因此被清除出内存列式存储
SQL> SELECT inmemory_size FROM v$im_segments WHERE segment_name = 'T2';

no rows selected
```

4.10 禁用 Database In-Memory

如果希望在 PDB 一级禁用,可以在 PDB 中将 INMEMORY_SIZE 设置为 0,立即生效,无须重启:

```
SQL> SHOW CON_NAME;

CON_NAME
------------------------------
ORCLPDB1

SQL> ALTER SYSTEM SET inmemory_size = 0 CONTAINER = CURRENT;
System altered.
```

如果希望在数据库实例层面禁用,可以在 CDB 中将 INMEMORY_SIZE 设置为 0,然后重启数据库即可生效:

```
SQL> ALTER SESSION SET CONTAINER = cdb$root;
Session altered.

SQL> ALTER SYSTEM SET inmemory_size = 0 SCOPE = SPFILE;
System altered.
```

```
SQL> SHUTDOWN IMMEDIATE
Database closed.
Database dismounted.
ORACLE instance shut down.

SQL> STARTUP
```

　　将初始化参数 INMEMORY_QUERY 设为 DISABLE 可以在数据库系统或会话层面禁用 Database In-Memory 查询，即使对象已经发布到内存列式存储中。这非常适合于测试，例如比较使用或禁用 Database In-Memory 时对性能的影响。此参数默认值为 ENABLE。在下述代码中，在会话层面临时禁用 Database In-Memory 后，执行计划也由内存全表扫描变为了普通全表扫描。

```
SQL> SET AUTOTRACE ON
SQL> SELECT COUNT(*) FROM lineorder;

  COUNT(*)
----------
  11997996

Execution Plan
----------------------------------------------------------
Plan hash value: 2267213921

-----------------------------------------------------------------------------------
| Id  | Operation                   | Name     | Rows  | Cost (%CPU)| Time     |
-----------------------------------------------------------------------------------
|   0 | SELECT STATEMENT            |          |     1 |  1767   (2)| 00:00:01 |
|   1 |  SORT AGGREGATE             |          |     1 |            |          |
|   2 |   TABLE ACCESS INMEMORY FULL| LINEORDER|   11M |  1767   (2)| 00:00:01 |
-----------------------------------------------------------------------------------

SQL> ALTER SESSION SET inmemory_query = DISABLE;
Session altered.

SQL> SELECT COUNT(*) FROM lineorder;

  COUNT(*)
----------
  11997996

Execution Plan
----------------------------------------------------------
Plan hash value: 2267213921

-----------------------------------------------------------------------------------
| Id  | Operation          | Name     | Rows  | Cost (%CPU)| Time     |
-----------------------------------------------------------------------------------
|   0 | SELECT STATEMENT   |          |     1 | 46691   (1)| 00:00:02 |
|   1 |  SORT AGGREGATE    |          |     1 |            |          |
|   2 |   TABLE ACCESS FULL| LINEORDER|   11M | 46691   (1)| 00:00:02 |
```

数据库提示是在 SQL 语句注释中实现的特殊指令，可以影响优化器对执行计划的选择。使用数据库提示 INMEMORY 和 NO_INMEMORY 可以在 SQL 语句一级启用和禁用 Database In-Memory 查询。在 INMEMORY_QUERY 禁用前提下，INMEMORY 提示仍可以启用 Database In-Memory 查询：

```
SQL> SELECT /*+ NO_INMEMORY */ COUNT(*) FROM lineorder;

  COUNT(*)
----------
  11997996

Execution Plan
----------------------------------------------------------
Plan hash value: 2267213921

--------------------------------------------------------------------
| Id | Operation          | Name     | Rows | Cost (%CPU)| Time     |
--------------------------------------------------------------------
|  0 | SELECT STATEMENT   |          |    1 | 46691   (1)| 00:00:02 |
|  1 |  SORT AGGREGATE    |          |    1 |            |          |
|  2 |   TABLE ACCESS FULL| LINEORDER|  11M | 46691   (1)| 00:00:02 |
--------------------------------------------------------------------

SQL> ALTER SESSION SET inmemory_query = DISABLE;
Session altered.

SQL> SELECT /*+ INMEMORY */ COUNT(*) FROM lineorder;

  COUNT(*)
----------
  11997996

Execution Plan
----------------------------------------------------------
Plan hash value: 2267213921

-----------------------------------------------------------------------------
| Id | Operation                   | Name     | Rows | Cost (%CPU)| Time     |
-----------------------------------------------------------------------------
|  0 | SELECT STATEMENT            |          |    1 |  1767   (2)| 00:00:01 |
|  1 |  SORT AGGREGATE             |          |    1 |            |          |
|  2 |   TABLE ACCESS INMEMORY FULL| LINEORDER|  11M |  1767   (2)| 00:00:01 |
-----------------------------------------------------------------------------
```

第 5 章

Database In-Memory 管理工具

5.1 SQL Developer

SQL Developer 可以看作是具有图形界面的 SQL Plus，但其不仅仅适用于开发者，也可以执行一些常用的数据库管理任务。SQL Developer 实际上结合了 SQL Plus 和 Enterprise Manager 两者的特点，比 SQL Plus 直观，功能更丰富，同时安装比 Enterprise Manager 简单，资源消耗更少。实际上这 3 种工具可以相互配合和补充，开发和管理人员可以根据任务需要选择最适合的工具。

在 SQL Developer 中，直接和 Database In-Memory 相关的只有一个 In-Memory 窗格，如图 5-1 所示。可以选择对象后，右击 Edit 菜单打开。其中可以编辑表和列的各种 INMEMORY 属性。

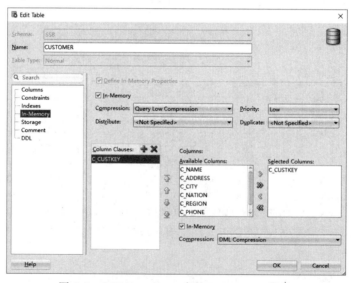

图 5-1　SQL Developer 中的 In-Memory 窗格

此外，SQL Developer 中的以下功能也有助于针对 Database In-Memory 的管理和开发任务。

（1）SQL 工作表（SQL Worksheet）可以输入和执行 SQL、PL/SQL 和 SQL Plus 脚本。由于是图形界面，因此复制、粘贴和保存都会比 SQL Plus 方便，并可以在多个数据库间切换执行。SQL 工作表具有强大的编辑功能，可以利用 Ctrl+F7 组合键对代码进行美化，Ctrl+Space 组合键实现代码自动完成。可以将对象拖拽到其中生成简单的 DML 语句，或通过查询构建器（Query Builder）生成复杂的 SQL 语句。如图 5-2 所示，将 SSB 示例中的 5 个表拖曳到 SQL 工作表中的 Query Builder 区域，然后通过连线、勾选和下方输入条件，即可生成右侧复杂的 SQL，实际上就是 SSB 示例 SQL 4_3。

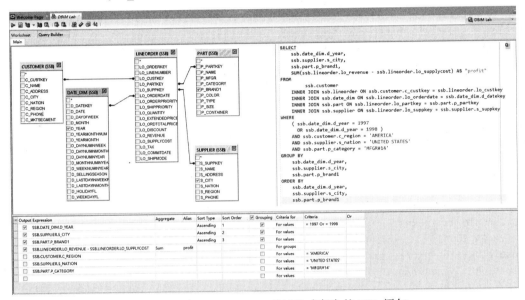

图 5-2　通过 Query Builder 快速生成复杂的 SQL 语句

此外，SQL Developer 提供了 Snippets 工具，可以插入常用的代码片段到 SQL 语句中。这些片段包括 SQL 函数、优化器提示等，用户也可以定制自己常用的代码片段。

（2）版本控制。SQL Developer 实现了与最流行的版本控制工具 Git 的集成。

（3）DBA 管理工具可以查看和修改初始化参数，生成 AWR 和 ASH 报告，对比 AWR 报告及实时 SQL 监控。

（4）Data Modeler 通过反向工程了解数据库对象的属性和相互关系，或进行逻辑模型设计，最终转换为物理模型。

更多 SQL Developer 的功能可参见 Oracle 官方文档[①]。

5.2　Oracle Enterprise Manager

Oracle Enterprise Manager 是管理 Oracle 数据库资产的一体化、自动化的图形管理工具，其

---

① https://docs.oracle.com/en/database/oracle/sql-developer/index.html

深度交互式分析有助于数据库管理员更快地发现、修复和验证问题。除数据库外，Enterprise Manager 还可以管理 Oracle 应用、中间件、硬件和集成系统，适用于本地数据中心和云环境。

5.2.1　In-Memory Central

Enterprise Manager 在 12c 及以上版本提供 In-Memory Central 页面管理 Database In-Memory，目前最新版本为 13.4。安装 Enterprise Manager 最简单快速的方式是通过 Oracle 公有云上的应用市场[①]。安装完成后，登录 Enterprise Manager 可以查看其端口、访问连接和各组件信息：

```
$ emctl status oms -details
Oracle Enterprise Manager Cloud Control 13c Release 4
Copyright (c) 1996, 2020 Oracle Corporation.  All rights reserved.
Enter Enterprise Manager Root (SYSMAN) Password :
Console Server Host        : oms1
HTTP Console Port          : 7788
HTTPS Console Port         : 7799
HTTP Upload Port           : 4889
HTTPS Upload Port          : 4900
EM Instance Home           : /u01/app/oracle/em/gc_inst_134/em/EMGC_OMS1
OMS Log Directory Location : /u01/app/oracle/em/gc_inst_134/em/EMGC_OMS1/sysman/log
OMS is not configured with SLB or virtual hostname
Agent Upload is locked.
OMS Console is locked.
Active CA ID: 1
Console URL: https://oms1:7799/em
Upload URL: https://oms1:4900/empbs/upload

WLS Domain Information
Domain Name            : GCDomain
Admin Server Host      : oms1
Admin Server HTTPS Port: 7101
Admin Server is RUNNING

Oracle Management Server Information
Managed Server Instance Name: EMGC_OMS1
Oracle Management Server Instance Host: oms1
WebTier is Up
Oracle Management Server is Up
JVMD Engine is Up

BI Publisher Server Information
BI Publisher Managed Server Name: BIP
BI Publisher Server is Up

BI Publisher HTTP Managed Server Port  : 9701
BI Publisher HTTPS Managed Server Port : 9801
```

---

① https://console.us-ashburn-1.oraclecloud.com/marketplace/application/70051865

```
BI Publisher HTTP OHS Port                 : 9788
BI Publisher HTTPS OHS Port                : 9899
BI Publisher is locked.
BI Publisher Server named 'BIP' running at URL:
 https://oms1:9899/xmlpserver/servlet/home
BI Publisher Server Logs:
 /u01/app/oracle/em/gc_inst_134/user_projects/domains/GCDomain/servers/BIP/logs/
BI Publisher Log:
 /u01/app/oracle/em/gc_inst_134/user_projects/domains/GCDomain/servers/BIP/logs/bipu
blisher/bipublisher.log
```

输出中的 Console URL 为 Enterprise Manager 控制台的访问地址。通过浏览器打开此地址，以用户名 sysman 登录。如图 5-3 所示，选择目标数据库后，在 Administration 下拉菜单中选择 In-Memory Central，即可进入 In-Memory Central 管理主页面。

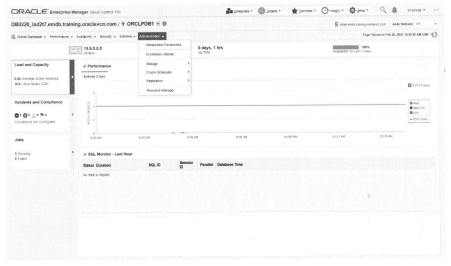

图 5-3　In-Memory Central 菜单位置

如图 5-4 所示，In-Memory Central 管理主页面左上方显示了 Database In-Memory 的重要配置信息，包括 SGA 和内存列式存储的大小、初始化参数设置、性能概要和 In-Memory 建议器入口。

左侧中部为 In-Memory 对象的概要信息，包括磁盘和内存占用空间、压缩比、发布比例等。下方为 In-Memory 对象的详细信息，包括每一个对象的类型、磁盘和内存占用空间、是否完整发布、压缩比、发布优先级和压缩级别等。右上方为对象的热图信息，红色表示对象在内存列式存储中被重度使用，这正是我们期望的情况。蓝色表示对象未被有效利用，内存紧张时，可以考虑将其从内存中移除。在红色和蓝色之间，暖色调表示对象使用频繁，而冷色调表示对象很少被访问。灰色表示尚无访问数据。

如图 5-5 所示，在对象编辑页面，In-Memory Column Store 标签页中可以针对单个对象设置各种 In-Memory 属性，如是否启用 In-Memory、压缩级别、内存发布优先级、数据分布、服务和复制策略等。在基于表的设置基础上，还可以单独针对列进行设置，目前只能设置是否启用

In-Memory 和压缩级别两项。在各项属性选择完成后，如果并不想马上应用，可以单击 Show SQL 按钮生成 SQL 命令，保存好留待以后执行。

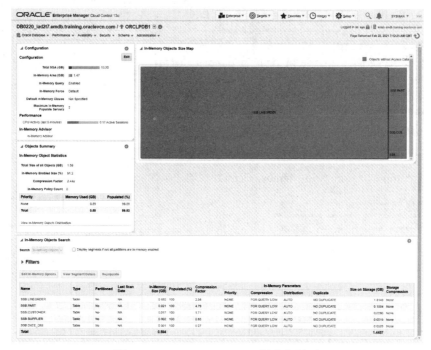

图 5-4　In-Memory Central 管理主页面

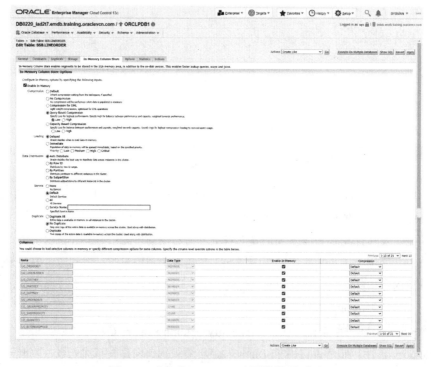

图 5-5　对象的 In-Memory 属性设置页面

5.2.2 SQL 性能分析器

SQL 性能分析器（SQL Performance Analyzer，SPA）属于 Oracle 数据库选件 Real Application Testing。SPA 主要用于评估系统改变对于 SQL 的性能影响，以充分预估和消除系统变更带来的风险。这些改变包括数据库、操作系统和硬件升级或配置改变，数据库参数调整，数据库结构变化，优化器统计信息变化等。

使用 SPA 评估变更对性能的影响，主要包括以下步骤。

（1）从源系统捕捉需要分析的工作负载，并存入 SQL 调优集。源系统通常为生产系统，SQL 调优集是 SPA 的输入。

（2）建立测试系统，并导入 SQL 调优集。虽然也可以在生产系统导入 SQL 调优集，但通常都会建立额外的测试系统以避免对生产的影响。测试系统环境应尽可能与源系统保持一致。

（3）建立系统改变前的测试档案。即建立基线，可测试执行或生成解释性的执行计划。

（4）实施系统改变。Enterprise Manager 也提供多种方式模拟系统的改变。

（5）建立系统改变后的测试档案。

（6）可测试执行或生成解释性的执行计划，建议与系统改变前所使用的方法保持一致。对改变前后的测试档案进行比较，生成 SPA 报告。

通过一个示例了解 SPA 在 Enterprise Manager 中的使用过程，将比较开启 Database In-Memory 前后性能的差异。以下为简要步骤。

（1）创建 SQL 调优集。菜单位于 Performance 标签页下的 SQL>SQL Tuning Sets。SQL 调优集可由用户定制，或基于游标缓存和 AWR 创建。本例使用游标缓存实时捕获最近执行的 13 个 SSB 示例 SQL，如图 5-6 所示。后续可以在 SQL 调优集添加或删除 SQL。

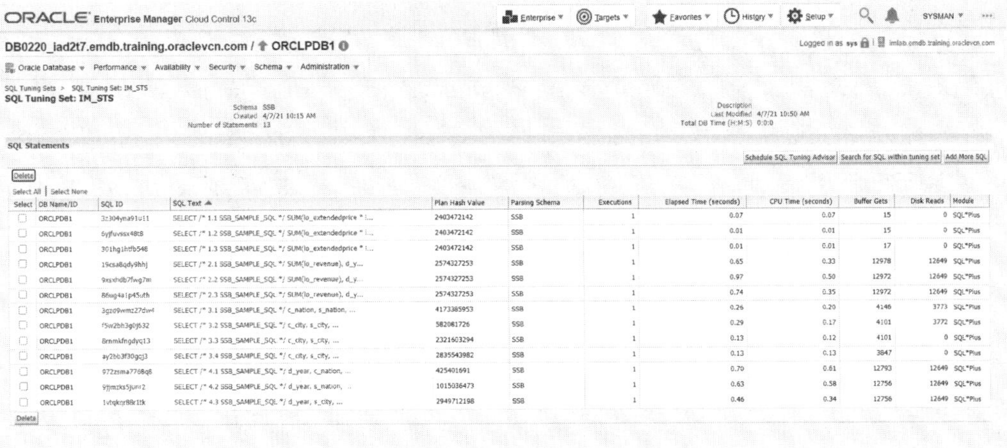

图 5-6　SQL 调优集

（2）创建 SPA 工作负载。菜单位于 Performance 标签页下的 SQL>SQL Performance Analyzer Home。如图 5-7 所示，Enterprise Manager 提供了多种工作负载模板，如初始化参数的改变、优化器统计信息的改变、模拟 Exadata 等。

图 5-7　SQL 性能分析器工作负载

本例选择 Parameter Change，并指定初始化参数 INMEMORY_QUERY，如图 5-8 所示。也就是说，在 Database In-Memory 已启用前提下，通过此参数控制 SQL 执行时是否使用内存列式存储。

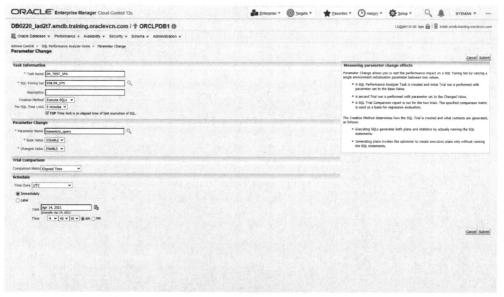

图 5-8　使用 SPA 测试初始化参数改变带来的性能影响

（3）生成 SPA 报告。Enterprise Manager 会在应用变更前后分别执行一次 SPA 工作负载，然后对执行情况进行比较后生成 SPA 报告。从图 5-9 中的报告可知，启用 Database In-Memory 后，SSB 示例中的 13 个 SQL 执行时间由 11 秒下降为不到 1 秒，由此可以证明 Database In-Memory 对性能的明显提升。

图 5-9　SQL 性能分析（SPA）报告

5.3　命令行管理工具

5.3.1　SQL Plus

SQL Plus 包含在每一个 Oracle 数据库安装中，是交互式和批量查询数据库的命令行工具，也是 DBA 和开发者最常用的工具。在 SQL Plus 中可以完成以下的工作。

（1）执行查询，对查询结果进行格式化、计算、存储和打印。

（2）查看对象的定义。

（3）开发和运行过程、函数和批处理脚本。

（4）执行数据库管理任务。

Oracle 提供免费、轻量级的数据库工具 Oracle Instant Client[①]，可以很方便地安装到容器和客户端中，可以安装开发运维的需求单独或组合安装这些工具包。例如 oracle-instantclient-tools 包中包含了 SQL Loader 和 Oracle Data Pump，oracle-instantclient-sqlplus 包中仅包含 SQL Plus，大小只有 706KB。

```
# yum search instantclient
```

---

① https://www.oracle.com/database/technologies/instant-client.html

```
oracle-instantclient-basic.x86_64 : Oracle Instant Client Basic package
oracle-instantclient-basiclite.x86_64 : Oracle Instant Client Light package
oracle-instantclient-devel.x86_64 : Development header files for Oracle Instant Client.
oracle-instantclient-jdbc.x86_64 : Supplemental JDBC features for the Oracle Instant
Client
oracle-instantclient-odbc.x86_64 : Oracle Instant Client ODBC
oracle-instantclient-release-el7.x86_64 : Oracle Instant Client yum repository
configuration
oracle-instantclient-sqlplus.x86_64 : Oracle Instant Client SQL*Plus package
oracle-instantclient-tools.x86_64 : Tools for Oracle Instant Client

# yum info oracle-instantclient-sqlplus.x86_64
Name        : oracle-instantclient-sqlplus
Arch        : x86_64
Version     : 21.1.0.0.0
Release     : 1
Size        : 706 k
Repo        : ol7_oracle_instantclient21/x86_64
Summary     : Oracle Instant Client SQL*Plus package
URL         : https://www.oracle.com/
License     : Oracle
Description : The Oracle Instant Client SQL*Plus package provides the SQL*Plus command
line tool for executing SQL and PL/SQL statements in Oracle
            : Database.

# repoquery -l oracle-instantclient-sqlplus.x86_64
/usr/bin/sqlplus
/usr/bin/sqlplus64
/usr/lib/oracle
/usr/lib/oracle/21
/usr/lib/oracle/21/client64
/usr/lib/oracle/21/client64/bin
/usr/lib/oracle/21/client64/bin/sqlplus
/usr/lib/oracle/21/client64/lib
/usr/lib/oracle/21/client64/lib/glogin.sql
/usr/lib/oracle/21/client64/lib/libsqlplus.so
/usr/lib/oracle/21/client64/lib/libsqlplusic.so
/usr/share/oracle
/usr/share/oracle/21
/usr/share/oracle/21/client64
/usr/share/oracle/21/client64/doc
/usr/share/oracle/21/client64/doc/SQLPLUS_LICENSE
/usr/share/oracle/21/client64/doc/SQLPLUS_README
```

SQL Plus 除支持标准 SQL 和扩展的 PL/SQL 外，还可运行自身的 SQL Plus 命令，例如：

```
SQL> help index

Enter Help [topic] for help.

 @              COPY            PASSWORD              SHOW
 @@             DEFINE          PAUSE                 SHUTDOWN
```

/	DEL	PRINT	SPOOL
ACCEPT	DESCRIBE	PROMPT	SQLPLUS
APPEND	DISCONNECT	QUIT	START
ARCHIVE LOG	EDIT	RECOVER	STARTUP
ATTRIBUTE	EXECUTE	REMARK	STORE
BREAK	EXIT	REPFOOTER	TIMING
BTITLE	GET	REPHEADER	TTITLE
CHANGE	HELP	RESERVED WORDS (SQL)	UNDEFINE
CLEAR	HISTORY	RESERVED WORDS (PL/SQL)	VARIABLE
COLUMN	HOST	RUN	WHENEVER OSERROR
COMPUTE	INPUT	SAVE	WHENEVER SQLERROR
CONNECT	LIST	SET	XQUERY

作为工具，SQL Plus 的使用并不复杂，实际上比较难的是 SQL 和 PL/SQL 的编程，SQL 也一直是最流行的编程语言之一[①]。由于 SQL 编程的话题涉及面太广，以下只对本书示例代码中针对 Database In-Memory 测试的常用任务和小技巧进行说明。

第一个常用任务是通过 SQL Plus 执行脚本。如果是交互式执行，可以通过@和@@命令，例如：

```
SQL> !ls
1_1.sql 1_2.sql 1_3.sql 2_1.sql 2_2.sql 2_3.sql 3_1.sql 3_2.sql 3_3.sql 3_4.sql
4_1.sql 4_2.sql 4_3.sql runall.sh

SQL> @1_1

   REVENUE
----------
8.9428E+11
```

如果是批处理方式执行，可以使用以下 4 种方式：

```
# 第 1 种方式，执行脚本文件中命令。后缀 sql 可以省略
$ sqlplus ssb@orclpdb1 @SQL 脚本文件名

# 第 2 种方式，通过 here document
$ sqlplus ssb/<ssb 用户口令>@orclpdb1 <<-EOF
    select 1 from dual;
EOF
# 第 3 种方式，通过 here string
$ sqlplus ssb/<ssb 用户口令>@orclpdb1 <<< 'select 1 from dual;'

# 第 4 方式，通过管道
echo 'select 1 from dual;' | sqlplus ssb/<ssb 用户口令>@orclpdb1
```

第二个常用任务是统计任务的执行时间，例如，想评估使用 Database In-Memory 后，查询时间是否明显减少。

交互式执行可使用 SET TIMING ON 命令，例如：

```
SQL> SET TIMING ON
```

[①] https://insights.stackoverflow.com/survey/2020

```
SQL> EXEC dbms_session.sleep(5);

PL/SQL procedure successfully completed.
Elapsed: 00:00:05.07
```

如果是编写脚本，可以在任务前后调用 DBMS_UTILITY PL/SQL 包中的 GET_TIME 函数，然后将两个值相减。此函数返回当前系统时间，单位为 1/100 秒。

```
SET SERVEROUTPUT ON
DECLARE
  ptime      NUMBER;
BEGIN
  ptime := dbms_utility.get_time;
  dbms_session.sleep(5);  -- 需要统计执行时间的任务
  ptime := (dbms_utility.get_time - ptime ) / 100;
  dbms_output.put_line('Elapsed Seconds: ' || ptime);
END;
```

第三个常用任务是自动生成大批量测试数据。第一种方式是通过循环加随机函数，例如：

```
CREATE TABLE t1 (
    id     NUMBER,
    nvar   NUMBER,
    cvar   VARCHAR2(8)
);

BEGIN
    FOR i IN 1..&&rowcnt LOOP
        INSERT INTO t1 VALUES (
            i,
            floor(dbms_random.value(1, 10000000)),
            dbms_random.string('p', 8)
        );

    END LOOP;

    COMMIT;
END;
/
```

另一种方式是利用 Oracle SQL 的层级查询功能：

```
CREATE TABLE t1
    AS
        SELECT
            level                                         AS id,
            floor(dbms_random.value(1, 10000000))         AS v1,
            dbms_random.string('p', 8)                    AS v2
        FROM
            dual
        CONNECT BY level <= &&rowcnt;
```

最后一个技巧是计算 SQL 语句所使用的资源或和 In-Memory 相关的统计指标。在动态性能视图 V$SESSTAT 中存放了当前会话的所有 SQL 执行统计信息，由于其是累加值，因此需要执行两次并取其差值。这可以通过 SQL Plus 的 COLUMN 命令来实现，NEW_VALUE 会自动创建一个替代变量，并将查询的返回值存入此变量中。例如以下脚本计算执行 SSB 示例 SQL 4_3 所消耗的 CPU：

```
COLUMN value NOPRINT NEW_VALUE start_cpu

SELECT value
FROM v$sesstat s, v$statname n
WHERE sid = SYS_CONTEXT('USERENV','SID')
AND s.statistic# = n.statistic#
AND n.name IN ('CPU used by this session')
/

@4_3

SELECT value - &start_cpu cpu_consumed
FROM v$sesstat s, v$statname n
WHERE sid = SYS_CONTEXT('USERENV','SID')
AND s.statistic# = n.statistic#
AND n.name IN ('CPU used by this session')
/

CPU_CONSUMED
------------
         765
```

5.3.2 SQLcl

SQLcl 是 SQL Developer 的命令行形式。和 SQL Developer 一样，SQLcl 也是基于 Java 编写的免费程序。不同的是，SQLcl 无须 JDK，只需要 JVM 即可运行。尽管 SQL Plus 是最常使用的数据库交互命令行程序，但 SQLcl 独有的行内编辑、命令自动完成、命令历史查看与回调、命令别名和 JavaScript 支持等功能是对 SQL Plus 的有益补充，可以很方便地完成某些特定任务。

SQLcl 是 SQL Developer 的一部分，由于默认的数据库安装已经包含了 SQL Developer，这也意味着 SQLcl 在默认情况下已经安装。如果希望安装最新版本的 SQLcl，可以从 Oracle 官方网站[①]或下载。下载后解压即可完成安装。然后需要将新版本程序的路径设置在 PATH 环境变量中，否则仍将调用老版本的 SQLcl 程序：

```
$ sql -V
SQLcl: Release 19.1.0.0 Production

$ curl -O https://download.oracle.com/otn_software/java/sqldeveloper/sqlcl-latest.zip
```

---

① https://download.oracle.com/otn_software/java/sqldeveloper/sqlcl-latest.zip

```
$ unzip sqlcl-latest.zip -d ~

$ export PATH=~/sqlcl/bin:$PATH
$ sql -V
SQLcl: Release 21.2.0.0 Production Build: 21.2.0.169.1529
```

SQLcl 的可执行程序为 sql，连接数据库的语法与 sqlplus 相同。连接成功后，help 可以显示 SQLcl 命令帮助，其中带星号的命令（新版本改为下画线）是 SQLcl 独有的命令。

```
$ sql ssb@orclpdb1
SQL> help
ALIAS*
APEX*
BRIDGE*
CD*
CTAS*
DDL*
FIND*
FORMAT*
HISTORY*
INFORMATION*
LIQUIBASE*
LOAD*
NET*
OERR*
REPEAT*
REST*
SCRIPT*
SODA*
SSHTUNNEL*
TNSPING*
VAULT*
WHICH*
...
```

SQLcl 支持丰富的结果集展示形式，默认的 ansiconsole 格式可根据列宽自动格式化数据，增强了结果集的可读性。其他支持的格式包括 CSV、JSON、XML 和 SQL Loader 等，甚至可以直接转换为 insert 语句：

```
SQL> HELP SET SQLFORMAT
SET SQLFORMAT
  SET SQLFORMAT { csv,html,xml,json,ansiconsole,insert,loader,fixed,default}
...

SQL> SHOW SQLFORMAT
SQL Format : Default

SQL> SET SQLFORMAT INSERT
SQL> SELECT * FROM hr.regions WHERE ROWNUM <= 2;
REM INSERTING into HR.REGIONS
SET DEFINE OFF;
```

```
Insert into HR.REGIONS (REGION_ID,REGION_NAME) values (1,'Europe');
Insert into HR.REGIONS (REGION_ID,REGION_NAME) values (2,'Americas');

SQL> SET SQLFORMAT JSON
SQL> SELECT * FROM hr.regions WHERE ROWNUM <= 2;
{"results":[{"columns":[{"name":"REGION_ID","type":"NUMBER"},{"name":"REGION_NAME",
"type":"VARCHAR2"}],"items":
[
{"region_id":1,"region_name":"Europe"}
,{"region_id":2,"region_name":"Americas"}
]}]}

SQL> SET SQLFORMAT ANSICONSOLE
SQL> SELECT * FROM hr.regions WHERE ROWNUM <= 2;
  REGION_ID REGION_NAME
          1 Europe
          2 Americas

SQL> SET SQLFORMAT DEFAULT
SQL Format Cleared
```

info 命令可以显示表的概要信息，如记录数，是否启用 INMEMORY。info+命令额外显示了每一列的最小值、最大值和唯一值的数量：

```
SQL> INFO+ lineorder;
TABLE: LINEORDER
        LAST ANALYZED:2021-03-03 14:00:07.0
        ROWS           :11997996
        SAMPLE SIZE    :11997996
        INMEMORY       :DISABLED
        COMMENTS       :

Columns
  NAME              DATA TYPE      NULL  DEFAULT  LOW_VALUE  HIGH_VALUE  NUM_DISTINCT
  LO_ORDERKEY       NUMBER         Yes            1          12000000    3019008
  LO_LINENUMBER     NUMBER         Yes            1          7           7
  LO_CUSTKEY        NUMBER         Yes            1          59999       40336
  LO_PARTKEY        NUMBER         Yes            1          200000      201968
  LO_SUPPKEY        NUMBER         Yes            1          4000        4000
  LO_ORDERDATE      NUMBER         Yes            19920101   19980802    2406
  LO_ORDERPRIORITY  CHAR(15 BYTE)  Yes                                   5
  LO_SHIPPRIORITY   CHAR(1 BYTE)   Yes                                   1
  LO_QUANTITY       NUMBER         Yes            1          50          50
  LO_EXTENDEDPRICE  NUMBER         Yes            90100      10494950    952256
  LO_ORDTOTALPRICE  NUMBER         Yes            85771      55528516    2848000
  LO_DISCOUNT       NUMBER         Yes            0          10          11
  LO_REVENUE        NUMBER         Yes            81360      10494950    4565504
  LO_SUPPLYCOST     NUMBER         Yes            54060      125939      12939
  LO_TAX            NUMBER         Yes            0          8           9
  LO_COMMITDATE     NUMBER         Yes            19920131   19981031    2466
  LO_SHIPMODE       CHAR(10 BYTE)  Yes                                   7
```

第 2 章使用 SQL Loader 工具 sqlldr 将 SSB 示例数据导入到 Oracle 中。sqlldr 导入需要定义控制文件，语法比较复杂。实际上，SQLcl 的 load 命令可以更简单地完成此任务。无须控制文件，可以直接将 dbgen 生成的数据文件直接导入数据库。首先通过预览数据文件，了解各字段的分隔符、左右定界符、是否有列头等信息。然后使用 set loadformat 定义对应的格式信息，然后即可执行数据文件导入。如果需要，也可以通过 set load 设置提交频率、日期时间格式等：

```
-- 了解数据文件格式
SQL> ! head -2 /home/oracle/ssb-dbgen/date.tbl
19920101|January 1, 1992|Thursday|January|1992|199201|Jan1992|5|1|1|1|1|…
19920102|January 2, 1992|Friday|January|1992|199201|Jan1992|6|2|2|1|1|…

-- 设置对应格式
SQL> SET LOADFORMAT COLUMN_NAMES OFF delimiter | enclosure_left OFF enclosure_right OFF

-- 显示设置的格式
SQL> SHOW LOADFORMAT
format csv

column_names off
delimiter |
enclosure_left off
enclosure_right off
encoding UTF8
row_limit off
row_terminator default
skip_rows 0
skip_after_names

-- 查看数据加载设置
SQL> SHOW LOAD
batch_rows 50
batches_per_commit 10
commit on
date_format
errors 50
method insert
timestamp_format
timestamptz_format
locale
truncate off

-- 创建空的数据加载目标表
SQL> CREATE TABLE date_dim_bak AS SELECT * FROM date_dim WHERE 1=2;
Table DATE_DIM_BAK created.

-- 导入数据到目标表
SQL> LOAD date_dim_bak /home/oracle/ssb-dbgen/date.tbl
format csv

column_names off
```

```
delimiter |
enclosure_left off
enclosure_right off
encoding UTF8
row_limit off
row_terminator default
skip_rows 0
skip_after_names

--Number of rows processed: 2,556
--Number of rows in error: 0
--Last row processed in final committed batch: 2,556
0 - SUCCESS: Load processed without errors

-- 确认数据已导入
SQL> SELECT COUNT(*) FROM date_dim_bak;
   COUNT(*)
-----------
       2556
```

alias 命令可以为常用的 SQL、PL/SQL 和 SQL Plus 脚本定义别名，从而可以节省输入时间，在测试或监控场景中可以方便快捷地调用：

```
SQL> ALIAS imseg = SELECT segment_name, inmemory_size, bytes, bytes_not_populated
 FROM v$im_segments;

SQL> imseg
   SEGMENT_NAME      INMEMORY_SIZE        BYTES      BYTES_NOT_POPULATED
---------------   ---------------    -------------  ----------------------
LINEORDER              593035264       1409155072                       0

SQL> ALIAS LIST imseg
imseg
-----
select segment_name, inmemory_size, bytes, bytes_not_populated from v$im_segments

SQL> ALIAS DROP imseg
Alias imseg dropped
```

history 命令可以查看命令历史和执行时间，或重新执行历史命令：

```
SQL> HISTORY
History:
  1  info customer
 ...
 19  select * from v$im_segments;
 20  select employee_id, first_name, salary from hr.employees;
 ...
SQL> HISTORY 20
  1* select employee_id, first_name, salary from hr.employees;
SQL> /
   EMPLOYEE_ID     FIRST_NAME     SALARY
---------------   ------------   --------
```

```
            100 Steven                    24000
            101 Neena                     17000
...

SQL> HISTORY TIME
...
 92  (00.005) select employee_id, first_name, salary from hr.employees;

SQL> history clear
History Cleared
```

5.3.3 Data Pump

Data Pump 也称为数据泵，可以在不同版本、不同字节序的 Oracle 数据库之间实现高速、并行的批量数据和元数据移动，并可以实施压缩和加密，是 DBA 常用的数据备份和迁移工具。Data Pump 可以导入导出部分或整个数据库、表空间、指定用户下的对象或指定的表。Data Pump 包括 expdp 和 impdp 两个客户端实用程序，分别实现数据库数据的导出和导入。这两个程序已经包含在数据库默认安装中，通过以下命令可获得语法帮助：

```
$ expdp help=y
$ impdp help=y
```

Data Pump 基于服务器运行，expdp 和 impdp 只是与其交互的客户端程序。因此需要使用服务器上的目录作为中介来存取数据文件和日志文件，以便实施和保证相应的文件安全性。在数据库中可以定义目录对象来映射服务器上的目录名。例如，以下命令在数据库 orclpdb1 中创建目录对象 dumpdir，然后赋予读写权限给用户 ssb。目录/opt/oracle/oradata 必须实际存在，如果不存在则需要预先创建：

```
CONNECT / AS SYSDBA
ALTER SESSION SET CONTAINER=orclpdb1;
CREATE OR REPLACE DIRECTORY dumpdir AS '/opt/oracle/oradata';
GRANT READ, WRITE ON DIRECTORY dumpdir TO ssb;
```

以下为将测试表 d1 从数据库中导出并导入的过程。我们在导入前将测试表 d1 删除，以此验证导出文件中不仅包含了数据，还包含了表结构等元数据。

```
# 创建测试表 d1
$ sqlplus ssb@orclpdb1 <<< "create table d1 as select 1 as c1 from dual connect by level <= 1000;"

# 将临时表从数据库中导出
$ expdp ssb@orclpdb1 tables=d1 directory=dumpdir dumpfile=d1.dmp

# 删除测试表 d1
$ sqlplus ssb@orclpdb1 <<< "drop table d1;"

# 导入测试表 d1
$ impdp ssb@orclpdb1 tables=d1 directory=dumpdir dumpfile=d1.dmp
```

对于定义了 INMEMORY 属性的对象，expdp 在导出时会保留这些属性。使用 impdp 导入时，可以选择保留、忽略已有 INMEMORY 属性或重新定义 INMEMORY 属性，这可以通过 impdb 实用程序的 TRANSFORM 参数实现，如表 5-1 所示。

表 5-1　impdp TRANSFORM 参数对 INMEMORY 对象的支持

TRANSFORM 参数	说　明
INMEMORY:Y	默认选项，保留原有 INMEMORY 属性
INMEMORY:N	忽略 INMEMORY 原有属性
INMEMORY_CLAUSE: "INMEMORY 属性"	使用新定义的 INMEMORY 属性

以上参数对于表和表空间均有效。需要注意，当采用 INMEMORY:N 参数时，导入的对象会继承表空间 INMEMORY 属性。这特别适合在数据迁移时，希望新的数据库采用 Database In-Memory 的情形。此时可以预先建立设置了 INMEMORY 属性的表空间，然后使用 INMEMORY:N 选项导入。另外，无论之前导出的对象是否具有 INMEMORY 属性，INMEMORY_CLAUSE 参数都将为对象附加其定义的属性。例如，尽管测试表 d1 最初并未定义 INMEMORY 属性，仍可以在导入时添加内存压缩级别、发布优先级等属性：

```
$ impdp ssb@orclpdb1 tables=d1 directory=dumpdir dumpfile=d1.dmp
 transform=inmemory_clause:\"inmemory priority medium\" table_exists_action=replace

$ sqlplus ssb@orclpdb1
SQL> select inmemory_compression, inmemory_priority from user_tables where table_name = 'D1';
INMEMORY_COMPRESS INMEMORY_PRIORITY
----------------- -----------------
FOR QUERY LOW     MEDIUM
```

此时测试表 d1 已经具有 INMEMORY 属性。导出文件将保留这些属性，但可以在导入时忽略这些属性：

```
$ expdp ssb@orclpdb1 tables=d1 directory=dumpdir dumpfile=d1.dmp

$ impdp ssb@orclpdb1 tables=d1 directory=dumpdir dumpfile=d1.dmp transform=inmemory:n
table_exists_action=replace

$ sqlplus ssb@orclpdb1
SQL> select inmemory_compression, inmemory_priority from user_tables where table_name = 'D1';
INMEMORY_COMPRESS INMEMORY_PRIORITY
----------------- -----------------
```

5.4　统计信息与执行计划

5.4.1　优化器统计信息

优化器统计信息是描述数据库和其中对象的一组数据，这些统计信息被优化器用来为 SQL 语句选择最佳的执行计划。

最基础的统计信息包含了表和列的描述。表的统计信息包括表中的行数、使用的数据块数量和行的平均长度。列的统计信息包括列中唯一值的数量、列中的最小和最大值。优化器使用表和列的统计信息来估算 SQL 语句返回的行数。其他的统计信息包括直方图、列组和索引统计信息等。直方图描述了列中不同值的分布情况，列组描述了同一表中不同列之间的关联情况，索引统计信息包括索引中唯一值的数量、索引深度等。这些信息有助于优化器生成更为准确的执行计划。

Oracle 会在预定义的维护窗口期内自动执行统计信息生成任务，在特定情形下，也可以使用 Oracle 建议的 DBMS_STATS PL/SQL 包来手工生成统计信息：

```
-- 生成当前数据库中所有对象的统计信息
EXEC dbms_stats.gather_database_stats();

-- 生成指定用户 SSB 下所有对象的统计信息
EXEC dbms_stats.gather_schema_stats('SSB');

-- 全量读取生成表的统计信息
EXEC dbms_stats.gather_table_stats('SSB', 'LINEORDER');

-- 采样生成表的统计信息
EXEC dbms_stats.gather_table_stats('SSB', 'LINEORDER',
    estimate_percent => dbms_stats.auto_sample_size);
```

如图 5-10 所示，在 Enterprise Manager 中，可以查看到自动和手动执行的统计信息生成任务。

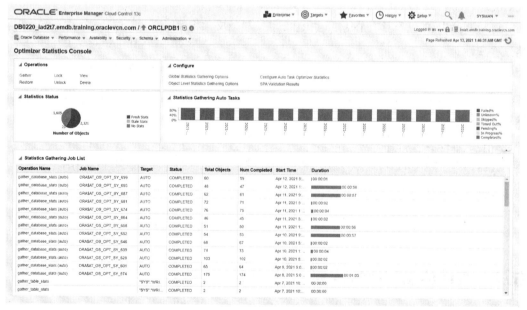

图 5-10　在 Enterprise Manager 中查看统计信息生成任务

优化器统计信息存放在数据字典中，例如以下 SQL 可以查询表和列的统计信息：

```
-- 表统计信息
```

```
SQL> SELECT table_name, num_rows, blocks, avg_row_len, sample_size FROM
 user_tab_statistics;

TABLE_NAME     NUM_ROWS    BLOCKS AVG_ROW_LEN SAMPLE_SIZE
-----------   ---------- --------- ----------- -----------
SUPPLIER           16000       244         102       16000
CUSTOMER          240000      3772         107      240000
LINEORDER       11997996    167199          96    11997996
PART             1000000     13157          87     1000000
DATE_DIM            2556        43         100        2556

-- 列统计信息
SQL> SELECT column_name, num_distinct, low_value, high_value, avg_col_len FROM
user_tab_col_statistics WHERE table_name = 'COUNTRIES';

COLUMN_NAME  NUM_DISTINCT LOW_VALUE            HIGH_VALUE           AVG_COL_LEN
-----------  ------------ -------------------- -------------------- -----------
COUNTRY_ID             25 4152                 5A57                           3
COUNTRY_NAME           25 417267656E74696E61   5A696D6261627765               9
REGION_ID               4 C102                 C105                           3
```

5.4.2 SQL 执行统计信息

SQL 执行统计信息实际是数据库统计信息的一个子类，其他类型的统计信息包括用户、操作系统和缓存等。SQL 执行统计信息指 DML 类型的 SQL 语句在执行时采集的性能统计数据，对于确定和解决性能问题非常有用。获取 SQL 执行统计信息最简单的方式是通过 SQL Plus 中的 SET AUTOTRACE 命令，但其输出的统计信息非常有限：

```
SQL> SET AUTOTRACE ON STATISTICS
SQL> SELECT COUNT(*) FROM lineorder;

  COUNT(*)
----------
  11997996

Statistics
----------------------------------------------------------
        230  recursive calls
          0  db block gets
        209  consistent gets
          8  physical reads
          0  redo size
        553  bytes sent via SQL*Net to client
        392  bytes received via SQL*Net from client
          2  SQL*Net roundtrips to/from client
         19  sorts(memory)
          0  sorts(disk)
          1  rows processed
```

所有数据库统计信息都存放在动态性能视图中，其中 V$MYSTAT 存放了当前会话期间的统计信息，V$SYSSTAT 存放了数据库系统中所有的统计信息。通过这两个表的联合查询可以获得和 In-Memory 相关的 SQL 执行统计信息，也可以修改以下语句中的 WHERE 条件直接查询感兴趣的统计项目：

```
$ cat ~/dbimbook/imstats.sql
SET PAGES 9999
SET LINES 120

SELECT
    t1.name,
    t2.value
FROM
    v$sysstat   t1,
    v$mystat    t2
WHERE
    (t1.name IN('CPU used by this session',
                'physical reads',
                'physical reads direct',
                'physical reads cache',
                'session logical reads',
                'session logical reads - IM',
                'session pga memory',
                'table scans(long tables)',
                'table scans(IM)',
                'table scan disk IMC fallback')
      OR(t1.name LIKE 'IM scan%'))
    AND t1.statistic# = t2.statistic#
    AND t2.value != 0
ORDER BY
    t1.name;
```

所有的统计项存放在动态性能视图 V$STATNAME 中，与 Database In-Memory 相关的统计项超过 500 个：

```
SQL> SELECT name FROM v$statname WHERE name LIKE 'IM %'
NAME
------------------------------------------------------------
IM populate blocks invalid
IM populate transactions check
IM populate undo segheader rollback
IM populate undo records applied
IM populate transactions active
...
IM CUs hwm mismatch drop
IM CUs hwm dropped

523 rows selected.
```

表 5-2 列出了比较常用的与 Database In-Memory 相关的统计信息，完整的统计项及说明可参见 Oracle 数据库参考手册附录中的统计信息描述[①]。

[①] https://docs.oracle.com/en/database/oracle/oracle-database/21/refrn/statistics-descriptions.html

表 5-2　常用的 In-Memory 相关 SQL 执行统计信息

统计项名称	描述
CPU used by this session	当前会话使用的 CPU 时间，单位为 10ms
physical reads	从磁盘读取的数据块数量
physical reads cache	从磁盘读取到 Buffer Cache 的数据块数量
session pga memory	当前会话的 PGA 大小
session logical reads	从 Buffer Cache 和进程私有内存读取的数据块数量
session logical reads - IM	从内存列式存储中读取的数据库块数量
table scans (IM)	使用 In-Memory 扫描的 Segment 数量
table scan disk IMC fallback	由于内存列式存储中不存在而去 Buffer Cache 中获取的行数
table scan disk non-IMC rows gotten	从非内存扫描中获取的行数
IM scan segments minmax eligible	有资格通过存储索引进行裁剪的 CU 数量
IM scan rows	在 IMCU 中扫描的行数
IM scan rows optimized	在 IMCU 扫描时因优化而略高的行数
IM scan rows projected	IMCU 扫描后返回的行数
IM scan CUs pruned	被存储索引裁剪掉的 CU 数量
IM scan CUs columns accessed	In-Memory 扫描的列的数量
IM scan (dynamic) rows	In-Memory 动态扫描所处理的行数
IM scan CUs memcompress for <压缩级>	在内存列式存储中扫描的相应压缩级 CU 的数量
IM scan rows pcode aggregated	应用了过滤谓词后余下的行数
IM scan CUs pcode aggregation pushdown	应用了聚合下推的 IMCU 的数量
IM scan bytes in-memory	从内存列式存储中读取数据的字节数
IM scan bytes uncompressed	从磁盘中读取数据的字节数
IM scan CUs predicates received	接收到的谓词数量
IM scan CUs predicates applied	应用到存储索引中 WHERE 语句中谓词的数量
IM scan rows optimized	因为采用存储索引或下推等优化手段而在内存列式存储中没有被扫描的行数

5.4.3　解读执行计划

当查询被发往数据库时，SQL 优化器会根据优化器统计信息判断如何执行查询，执行计划就是数据库运行 SQL 语句时所执行的一系列操作。

可用通过 SQL Plus、SQL Developer 或 Enterprise Manager 等工具获取文本或图形格式的执行计划。图 5-11 为典型的执行计划输出，其中最上方为描述信息，包括 SQL ID，执行的 SQL 语句和执行计划的哈希值。中间部分称为行源树（row source tree），是执行计划中最核心的部分。行源树反映了 SQL 执行的具体步骤和顺序，包括访问方法、JOIN 方法、过滤和排序操作等。正如其名称所示，行源树为树状结构。例如第 0 行为根节点，第 1 和第 2 行为父子关系，第 3 和第 4 行均为叶子节点，并且是兄弟关系。

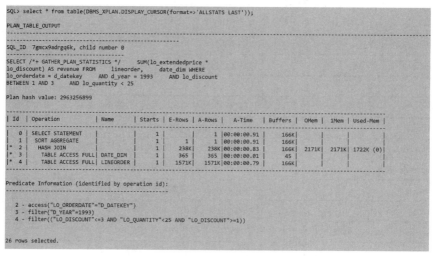

图 5-11　SQL 执行计划的组成部分

行源树采用深度优先的顺序执行，以图 5-11 为例，其执行顺序如下。

（1）从根节点（第 0 行）遍历到第一个叶子节点（第 3 行），对 DATE_DIM 表进行全表扫描。

（2）结果返回上一级（第 2 行）。

（3）移动到下一个叶子节点（第 4 行），对 LINEORDER 表进行全表扫描。

（4）结果返回到上一级（第 2 行）。

（5）无其他叶子节点，针对所有叶子节点返回数据执行哈希联结操作。

（6）结果返回上一级（第 1 行），执行聚合操作。

（7）聚合操作结果返回上一级，即输出查询结果。

行源树中还包括其他一些统计信息，例如 E-Rows 为优化器估算此操作返回的行数，A-Rows 是操作实际返回的行数。如果这两个值相差较大，通常都不会是好的执行计划。

最下方是谓词（predicate）信息，是对行源树中执行计划的补充说明。其中的标号与行源树中带星号的操作相对应。谓词这个术语来自数理逻辑，本质是返回一个是或者否的判定，简单来说就是 WHERE 条件。谓词信息部分中的 access 表示存在一种方法，让优化器可以直接定位到需要访问的数据，例如通过索引。而 filter 则表示访问的数据比需要的多，获取数据后需要进行过滤，最终返回满足条件的数据。

5.4.4　生成和显示执行计划

生成执行计划的第一种方法是通过 SQL Plus 中的 SET AUTOTRACE 命令。这种方法的好处是比较简单，并且可以同时生成 SQL 执行统计信息。例如：

```
Execution Plan
----------------------------------------------------------
Plan hash value: 1388734953
```

```
| Id | Operation        | Name | Rows | Cost (%CPU)| Time     |
-------------------------------------------------------------------
|  0 | SELECT STATEMENT |      |   1  |   2   (0) | 00:00:01 |
|  1 |  FAST DUAL       |      |   1  |   2   (0) | 00:00:01 |
-------------------------------------------------------------------

Statistics
-------------------------------------------------------------
          0  recursive calls
          0  db block gets
          0  consistent gets
          0  physical reads
          0  redo size
        543  bytes sent via SQL*Net to client
        601  bytes received via SQL*Net from client
          2  SQL*Net roundtrips to/from client
          0  sorts (memory)
          0  sorts (disk)
          1  rows processed
```

SET AUTOTRACE 命令包含多个选项来控制输出的内容，常用的选项组合如下：

```
SET AUTOTRACE ON              -- 显示查询结果、执行计划和 SQL 执行统计信息
SET AUTOTRACE TRACEONLY       -- 不显示查询结果，只显示执行计划和 SQL 执行统计信息
SET AUTOTRACE ON STATISTICS   -- 不显示执行计划，只显示查询结果和 SQL 执行统计信息
SET AUTOTRACE OFF             -- 关闭 AUTOTRACE 选项
```

SQL Developer 也支持 AUTOTRACE，如图 5-12 所示，单击上方的 AUTOTRACE 图标或按 F6 快捷键即可启用。SQL Developer 中的 AUTOTRACE 支持展开收起、热点标注和保存为 HTML 文件等操作，因此比较直观和易于理解。

图 5-12　SQL Developer 中的 AUTOTRACE

AUTOTRACE 的不足之处是执行计划的输出过于简单，一些重要的统计项并没有包含在内。另外，AUTOTRACE 不能实时监控 SQL 执行过程，必须等查询结束才能显示执行计划。

5.6 节介绍的实时 SQL 监控也可以用来生成执行计划。和 AUTOTRACE 相比，实时 SQL 监控可以实时查看 SQL 的执行过程，以便于更准确地定位问题；同时可以自动捕获 SQL 命令，不用像 AUTOTRACE 那样手工执行。

如果需要指定执行计划的来源（执行计划表、AWR 或 SQL 调优集）或控制执行计划显示的项目，可以使用 DBMS_XPLAN PL/SQL 包，实际上这也是最常用和最灵活的方式。最简单的调用方法是在执行完 SQL 后，执行以下的语句：

```
SELECT * FROM TABLE(dbms_xplan.display_cursor);
```

可以通过加减号控制在执行计划输出中需要额外显示和隐藏的项目，例如：

```
SQL> SELECT * FROM TABLE(dbms_xplan.display_cursor(FORMAT=>'-COST +NOTE'));

SQL_ID  7c1rnh08dp922, child number 0
-------------------------------------
select count(*) from employees

Plan hash value: 3964236718

---------------------------------------------------------------
| Id | Operation              | Name         | Rows | Time     |
---------------------------------------------------------------
|  0 | SELECT STATEMENT       |              |      |          |
|  1 |  SORT AGGREGATE        |              |    1 |          |
|  2 |   INDEX FAST FULL SCAN | EMP_EMAIL_UK |  107 | 00:00:01 |
---------------------------------------------------------------

Note
-----
   - dynamic statistics used: dynamic sampling (level=2)

18 rows selected.
```

如果需要显示额外的运行时统计信息，如 Starts（此步骤执行的次数）、A-Rows（实际操作的行数），可以使用以下方式执行：

```
ALTER SESSION SET statistics_level = 'ALL';
<执行你的 SQL 语句>
SELECT * FROM TABLE(dbms_xplan.display_cursor(FORMAT=>'+COST ALLSTATS LAST'));
```

也可以在 SQL 语句中加入 /*+ gather_plan_statistics */ 提示，等同于将 STATISTICS_LEVEL 初始化参数设为 ALL。例如：

```
SQL> SELECT /*+ gather_plan_statistics */ COUNT(*) FROM lineorder;

  COUNT(*)
----------
  11997996
```

```
SQL> SELECT * FROM TABLE(dbms_xplan.display_cursor(FORMAT=>'+COST ALLSTATS LAST'));

-----------------------------------------------------------------------------
| Id  | Operation           | Name     | Starts | E-Rows | A-Rows |   A-Time    |
-----------------------------------------------------------------------------
|   0 | SELECT STATEMENT    |          |      1 |        |      1 | 00:00:01.13 |
|   1 |  SORT AGGREGATE     |          |      1 |      1 |      1 | 00:00:01.13 |
|   2 |   TABLE ACCESS FULL | LINEORDER|      1 |    11M |    11M | 00:00:02.40 |
-----------------------------------------------------------------------------
```

不过，获取完整统计信息最简单和最直观的方式还是使用实时 SQL 监控，5.6 节将对其进行详细介绍。

5.5 Oracle 数据库建议器

Oracle 数据库包含内置的警报机制，用于通知数据库即将出现的问题和处理建议。此外，Oracle 数据库也包含一系列建议器，用于对数据库进行规划、管理和性能调优。一般而言，生成数据库警报的成本较低，性能影响较小。建议器需消耗更多的资源，但进行的分析更详细，生成的建议更全面。和 Database In-Memory 直接相关的建议器包括 In-Memory 建议器和压缩建议器，分别用于确定适合使用 Database In-Memory 的数据库对象和估算内存列式存储的容量，这些都是在规划 Database In-Memory 时必须回答的重要问题。

5.5.1 In-Memory 建议器

In-Memory 建议器通过对数据库中的分析型工作负载进行综合分析，来确定哪些数据库对象能从 Database In-Memory 中受益，分析基于 ASH（活动会话历史）和 AWR（自动工作负载仓储）中存放的运行统计数据。最终分析结果将以 HTML 格式呈现，其中包含了分析型工作负载的时长和比例，不同内存大小时的性能提升估算，推荐置入内存列式存储的数据库对象，以及将其置入内存的 SQL 语句。输出报告中的建议可通过本章介绍的 SQL 性能分析器或实时 SQL 监控等工具进行验证。

In-Memory 建议器是数据库调优包的一部分，支持 11.2.0.3 及以上版本的 Oracle 数据库。建议器程序需要从 Oracle 售后支持网站[①]下载，大小约为 300KB。In-Memory 建议器可以安装在 CDB 或 PDB 一级，建议安装在 CDB 一级，这样所有的 PDB 均可以使用。以下为安装的简要过程，解压程序包 imadvisor.zip，然后运行 instimadv.sql 脚本开始安装：

```
$ mkdir imadvisor
$ unzip /vagrant/imadvisor.zip -d imadvisor
$ cd imadvisor
$ sqlplus / as sysdba @instimadv
```

[①] https://support.oracle.com/epmos/faces/DocContentDisplay?id=1965343.1

安装过程中需要指定永久表空间和临时表空间，分别按 Enter 键可以使用默认值。根据主机的性能，安装时间在几分钟到十几分钟不等：

```
TABLESPACE_NAME
------------------------------
SYSAUX
SYSTEM (default permanent tablespace)
TEMP (default temporary tablespace)
UNDOTBS1
USERS


Enter value for permanent_tablespace:<输入回车以使用默认值 SYSTEM>

Permanent tablespace to be used with C##IMADVISOR: SYSTEM


Enter value for temporary_tablespace: <输入回车以使用默认值 TEMP>

Temporary tablespace to be used with C##IMADVISOR: TEMP
…
DBMS_INMEMORY_ADVISOR installation successful.

Users who will use the DBMS_INMEMORY_ADVISOR package must be granted
the ADVISOR privilege.

DBMS_INMEMORY_ADVISOR installation and setup complete.

To uninstall:
SQL> @catnoimadv.sql
```

In-Memory 建议器可以直接在生产数据库中运行，为了避免影响生产数据库，也可以将 AWR 等数据导出到另一个测试数据库中进行分析。其他支持的场景包括基于 PDB 和基于长时间运行的批处理工作负载等。

通过一个示例了解一下 In-Memory 建议器生成建议报告的过程。首先需要在数据库中模拟分析型工作负载，运行的分析型 SQL 语句如下：

```sql
$ cat imadvisor_workload.sql
SELECT
    d.d_year,
    p.p_brand1,
    SUM(lo_revenue)        tot_rev,
    COUNT(p.p_partkey)     tot_parts
FROM
    lineorder   l,
    date_dim    d,
    part        p,
    supplier    s
WHERE
    l.lo_orderdate = d.d_datekey
```

```
    AND l.lo_partkey = p.p_partkey
    AND l.lo_suppkey = s.s_suppkey
    AND p.p_category = 'MFGR#12'
    AND s.s_region = 'AMERICA'
    AND d.d_year = 1997
GROUP BY
    d.d_year,
    p.p_brand1;
```

然后执行以下脚本，重复运行以上 SQL 工作负载 100 次。需要注意，工作负载运行时，涉及的表不能位于内存列式存储中，否则后续运行 In-Memory 建议器时会报错：

```
$ cat imadvisor_workload.sh
for i in {1..100};
do
        sqlplus /nolog <<-EOF
                @../userlogin.sql
                @imadvisor_workload.sql
                exec dbms_session.sleep(1);
        EOF
done
```

打开另一个终端，运行 In-Memory 建议器。需要指定的参数为任务名、分析的数据库名、开始时间、延续时间和假定的内存列式存储的大小。其中任务名、开始时间和延续时间均有默认值。

```
$ sqlplus / as sysdba @imadvisor_recommendations
...
Default task_name (new task): im_advisor_task_20210324215805
Enter value for task_name:<指定任务名，输入回车以接受默认任务名>

Advisor task name specified: im_advisor_task_20210324215805 (default)

New Advisor task will be named: im_advisor_task_20210324215805...

Analyzing and reporting on a live workload on this database (DBID=2837271925)...
Enter value for pdb_name: <指定需分析的数据库名，此处输入 orclpdb1>

orclpdb1

The In-Memory Advisor optimizes the In-Memory configuration for a specific
In-Memory size that you choose.

After analysis, the In-Memory Advisor can provide you a list of performance
benefit estimates for a range of In-Memory sizes.  You may then choose the
In-Memory size for which you wish to optimize.

If you already know the specific In-Memory size you wish, please enter
the value now.  Format: nnnnnnn[KB|MB|GB|TB]

Or press <ENTER> to get performance estimates first.
```

```
Enter value for inmemory_size: <指定假定的内存列式存储大小，此处输入 1500MB>

The In-Memory Advisor will optimize for this In-Memory size: 1500MB


Enter begin time for report:

--    Valid input formats:
--       To specify absolute begin time:
--          [MM/DD[/YY]] HH24:MI[:SS]
--          Examples: 02/23/03 14:30:15
--                    02/23 14:30:15
--                    14:30:15
--                    14:30
--       To specify relative begin time: (start with '-' sign)
--          -[HH24:]MI
--          Examples: -1:15   (SYSDATE - 1 Hr 15 Mins)
--                    -25     (SYSDATE - 25 Mins)

Default begin time: -60
Enter value for begin_time:<开始时间，按 Enter 键接受默认值，即当前时间往前 60 分钟>

Report begin time specified:

Enter duration in minutes starting from begin time:
(defaults to SYSDATE - begin_time)

Enter value for duration:<延续时间，按 Enter 键接受默认值，即一个小时>

Report duration specified:

Using 2021-MAR-24 20:58:21.000000000 as report begin time
Using 2021-MAR-24 21:58:22.000000000 as report end time


You may optionally specify a comma separated list of object owner
and name patterns to be considered for In Memory Placement.
Example:

GEEK_SUMMARY.%,%.GEEK_%

Press ENTER to consider all objects.


Enter value for consider_objects_like:<指定对象，按 Enter 键接受默认值，即所有对象>

Considering all objects for In Memory placement.

In-Memory Advisor: Adding statistics…

In-Memory Advisor: Finished adding statistics.
```

```
In-Memory Advisor: Analyzing statistics…

In-Memory Advisor: Finished analyzing statistics.

The Advisor is optimizing for an In-Memory size of 1500MB…
Fetching recommendation files for task: im_advisor_task_20210324215805
Placing recommendation files in: the current working directory

Fetched file: imadvisor_im_advisor_task_20210324215805.html <-生成的建议报告
Purpose:       recommendation report primary html page

Fetched file: imadvisor_im_advisor_task_20210324215805.sql <-生成的 SQL 脚本
Purpose:       recommendation DDL sqlplus script

You can re-run this task with this script and specify a different an In-Memory
size. Re-running a task to optimize for a different In-Memory size is faster
than creatng and running a new task from scratch.
```

　　In-Memory 建议器执行完成后，会生成两个文件。一个是 HTML 格式的详细建议报告，另一个是实现此建议的 SQL 脚本。

　　In-Memory 建议报告如图 5-13 所示。其中显示分析型工作负载执行时间为 144 秒。如果使用 Database In-Memory，执行时间可减少 55 秒，性能将提升 1.6 倍。在报告中，还给出了每张表可以获得的性能提升以及所需要的内存空间。实际上，性能提升和内存空间都是估算值，要获取准确的内存空间需求，可以使用 5.5.2 节介绍的压缩建议器。In-Memory 建议器最重要的功能还是确定分析型负载是否能从 Database In-Memory 中受益，以及为实现性能提升，需要发布哪些数据库对象。

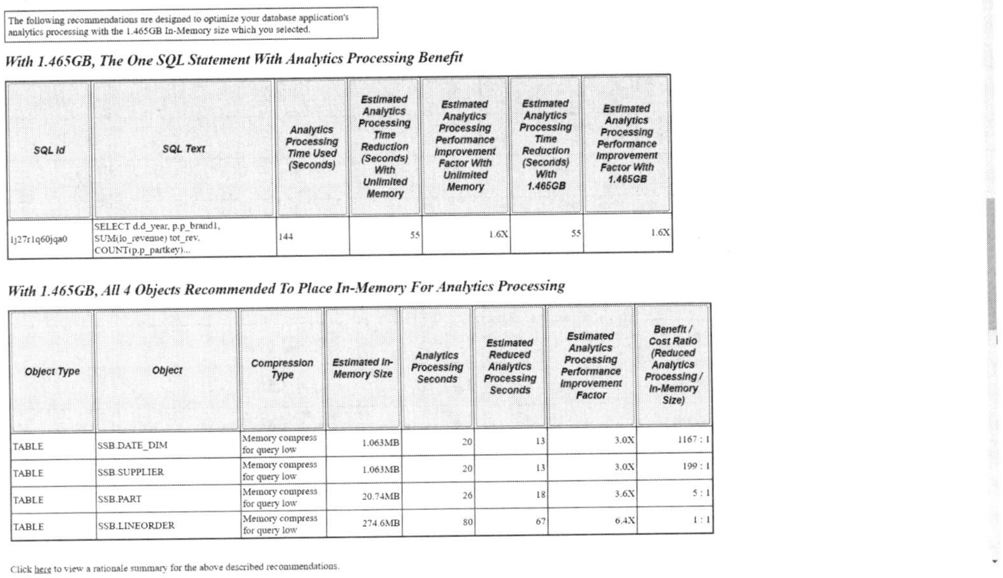

The following recommendations are designed to optimize your database application's analytics processing with the 1.465GB In-Memory size which you selected.

With 1.465GB, The One SQL Statement With Analytics Processing Benefit

SQL Id	SQL Text	Analytics Processing Time Used (Seconds)	Estimated Analytics Processing Time Reduction (Seconds) With Unlimited Memory	Estimated Analytics Processing Performance Improvement Factor With Unlimited Memory	Estimated Analytics Processing Time Reduction (Seconds) With 1.465GB	Estimated Analytics Processing Performance Improvement Factor With 1.465GB
1j27r1q60jqa0	SELECT d.d_year, p.p_brand1, SUM(lo_revenue) tot_rev, COUNT(p.p_partkey)...	144	55	1.6X	55	1.6X

With 1.465GB, All 4 Objects Recommended To Place In-Memory For Analytics Processing

Object Type	Object	Compression Type	Estimated In-Memory Size	Analytics Processing Seconds	Estimated Reduced Analytics Processing Seconds	Estimated Analytics Processing Performance Improvement Factor	Benefit / Cost Ratio (Reduced Analytics Processing / In-Memory Size)
TABLE	SSB.DATE_DIM	Memory compress for query low	1.063MB	20	13	3.0X	1167 : 1
TABLE	SSB.SUPPLIER	Memory compress for query low	1.063MB	20	13	3.0X	199 : 1
TABLE	SSB.PART	Memory compress for query low	20.74MB	26	18	3.6X	5 : 1
TABLE	SSB.LINEORDER	Memory compress for query low	274.6MB	80	67	6.4X	1 : 1

Click here to view a rationale summary for the above described recommendations.

图 5-13　In-Memory 建议器生成的建议报告局部

在 SQL 脚本中，包含了为分析型工作负载所涉及的表设置 INMEMORY 属性的命令，其中内存压缩采用了默认的 FOR QUERY LOW 级别。

```
$ cat imadvisor_im_advisor_task_20210324215805.sql
Rem Copyright (c) 2014, 2016, Oracle and/or its affiliates. All rights reserved.
ALTER TABLE "SSB"."LINEORDER" INMEMORY MEMCOMPRESS FOR QUERY LOW;
ALTER TABLE "SSB"."PART" INMEMORY MEMCOMPRESS FOR QUERY LOW;
ALTER TABLE "SSB"."DATE_DIM" INMEMORY MEMCOMPRESS FOR QUERY LOW;
ALTER TABLE "SSB"."SUPPLIER" INMEMORY MEMCOMPRESS FOR QUERY LOW;
```

5.5.2 压缩建议器

压缩建议器基于对实际数据样本的分析，提供了使用行压缩、混合列压缩和内存列压缩等多种方式所获得压缩比的估算值。此估算值可帮助用户对使用各种压缩选项所需的存储或内存容量进行规划。压缩建议器是 Oracle 提供的免费工具，以 PL/SQL 包形式提供。如果 Oracle 数据库版本为 9iR2 到 11gR1，压缩建议器需要额外下载[①]。在 Oracle 数据库 11gR2 及以上，压缩建议器所使用的 DBMS_COMPRESSION PL/SQL 包已经包含在数据库中。

压缩建议器要求进行评估的表至少为 100 万行。基于近 1200 万行的 LINEORDER 表，表 5-3 列出了在各种压缩类型下评估的压缩比，以及与实际压缩比的差异。其中混合列压缩级别要求底层的存储支持混合列压缩功能，例如 Exadata 和 Oracle 数据库云服务等。可以看到，评估压缩比与实际值非常接近，偏差大多在 5%之内。这是由于在评估过程中，在指定的表空间基于源评估表采样后生成了临时表，然后在临时表上进行了实际的压缩，评估完毕后立即将其删除。

表 5-3 压缩建议器压缩评估与实际值比较

压 缩 类 型	评估时间/秒	评估压缩比	实际压缩比	偏差/%
高级行压缩	12.43	1.6	1.90	-16.00
混合列压缩（查询，低）	194.63	2.8	2.95	-5.00
混合列压缩（查询，高）	253.68	4.3	4.42	-3.00
混合列压缩（归档，低）	285.55	4.5	4.67	-4.00
混合列压缩（归档，高）	366.61	5.6	5.79	-3.00
内存列压缩（无压缩）	160.90	1.2	1.26	-5.00
内存列压缩（DML）	153.65	1.2	1.26	-5.00
内存列压缩（查询，低）	184.57	2.3	2.37	-3.00
内存列压缩（查询，高）	196.16	3.4	3.55	-4.00
内存列压缩（容量，低）	202.42	3.9	4.05	-4.00
内存列压缩（容量，高）	195.54	5.0	5.12	-2.00

---

① https://www.oracle.com/technetwork/database/options/compression/downloads/index.html

以下为使用压缩建议器生成表 5-3 结果的 SQL 脚本：

```sql
SET SERVEROUTPUT ON

DECLARE
    p_blks_cmp      PLS_INTEGER;
    p_blks_uncmp    PLS_INTEGER;
    p_rows_cmp      PLS_INTEGER;
    p_rows_uncmp    PLS_INTEGER;
    p_cmp_ratio     NUMBER;
    p_comptype_str  VARCHAR2(256);
    p_sample_rows NUMBER := dbms_compression.comp_ratio_allrows;
--  p_sample_rows NUMBER := dbms_compression.comp_ratio_minrows;
    t1              PLS_INTEGER;
    TYPE array_t IS VARRAY(11) OF NUMBER;
    comptypes array_t := array_t(
        dbms_compression.COMP_ADVANCED,
        dbms_compression.COMP_QUERY_HIGH,
        dbms_compression.COMP_QUERY_LOW,
        dbms_compression.COMP_ARCHIVE_HIGH,
        dbms_compression.COMP_ARCHIVE_LOW,
        dbms_compression.COMP_INMEMORY_NOCOMPRESS,
        dbms_compression.COMP_INMEMORY_DML,
        dbms_compression.COMP_INMEMORY_QUERY_LOW,
        dbms_compression.COMP_INMEMORY_QUERY_HIGH,
        dbms_compression.COMP_INMEMORY_CAPACITY_LOW,
        dbms_compression.COMP_INMEMORY_CAPACITY_HIGH
    );
BEGIN
    FOR i IN 1..comptypes.count LOOP
        t1 := dbms_utility.get_time;
        dbms_compression.get_compression_ratio(
            scratchtbsname => 'USERS',
            ownname => 'SSB',
            objname => 'LINEORDER',
            subobjname => NULL,
            comptype => comptypes(i),
            blkcnt_cmp => p_blks_cmp,
            blkcnt_uncmp => p_blks_uncmp,
            row_cmp => p_rows_cmp,
            row_uncmp => p_rows_uncmp,
            cmp_ratio => p_cmp_ratio,
            comptype_str => p_comptype_str,
            subset_numrows => p_sample_rows,
            objtype => dbms_compression.objtype_table
        );

        dbms_output.put_line('-------------------------------------------------');
        dbms_output.put_line('Compression type: ' || p_comptype_str);
        dbms_output.put_line('Evaluation time: ' || (dbms_utility.get_time - t1) / 100
|| ' seconds');
```

```
        dbms_output.put_line('Compression ratio: ' || p_cmp_ratio);
        dbms_output.put_line('');
    END LOOP;
END;
/
```

压缩建议器在进行评估时可以选择源表采样的行数，图 5-14 为采样了所有行的评估结果。如果使用最小采样值（100 万行）运行压缩建议器，可以发现其评估压缩比与完全采样差别不大，都非常接近于真实的压缩比。评估时间对于内存列压缩节省有限，但对于混合列压缩，评估时间下降明显，为之前的 1/4 到 1/3。因此，通常情况下建议使用默认的最小采样值进行压缩评估，因为这样既可以得到相对准确的评估结果，又可以节省评估时间。

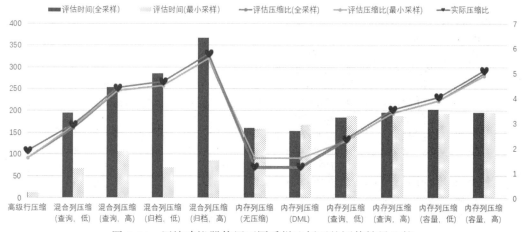

图 5-14　压缩建议器使用不同采样比例下的评估结果比较

5.6　实时 SQL 监控

实时 SQL 监控属于 Oracle 数据库调优包，是 Oracle 11.1 引入的新特性，主要用于对长时间运行、资源消耗严重的 SQL 语句的监控和性能问题诊断。实时 SQL 监控提供运行时交互式查看方式，执行计划的每一步骤都提供了细粒度的关键性能指标，包括运行时间、CPU 时间、读写次数和 I/O 等待时间等。这使得 DBA 可以更快速和准确地定位性能问题，并及时采取应对措施，如终止 SQL 执行或对 SQL 进一步调优。

启用实时 SQL 监控需要首先设置两个初始化参数，其中，STATISTICS_LEVEL 要求设为 TYPICAL 或 ALL，CONTROL_MANAGEMENT_PACK_ACCESS 要求设为 DIAGNOSTIC+TUNING，因为此功能属于 Oracle 调优包。以下命令显示了这两个参数的默认值：

```
SQL> SHOW PARAMETER statistics_level

NAME                                 TYPE        VALUE
------------------------------------ ----------- ------------------------------
statistics_level                     string      TYPICAL
```

```
SQL> SHOW PARAMETER control_management_pack_access

NAME                                 TYPE        VALUE
------------------------------------ ----------- ------------------------------
control_management_pack_access       string      DIAGNOSTIC+TUNING
```

然后，需要指定被监控的 SQL 语句。默认情况下，实时 SQL 监控会对并行运行或单次执行的 CPU 或 I/O 时间超过 5 秒的 SQL 语句启用监控。也可以使用以下方式手工启用监控：

（1）在 SQL 语句中加入 /*+ MONITOR */ 提示。

（2）在 sql_monitor 事件中指定需要监控的 SQL ID，例如：

```
ALTER SYSTEM SET EVENTS 'sql_monitor [sql: g0t052az3rx44|sql: cq8s0cpgt7xn9] force=true'
```

此外，还可以调整以下 3 个隐含参数控制 SQL Monitor 的行为。

（1）_sqlmon_threshold：监控 SQL 语句 CPU 或 I/O 时间的最小时长，单位为秒。默认值为 5，设为 0 表示禁用 SQL 监控。

（2）_sqlmon_max_plan：最大 SQL 监控条目数，默认为 20 × CPU_COUNT。

（3）_sqlmon_max_planlines：最多监控的执行计划行数，默认值为 300。

可以通过以下 3 种方式获取和查看实时 SQL 监控报告。

（1）DBMS_SQLTUNE 或 DBMS_SQL_MONITOR PL/SQL（19c 及以后）包中的 REPORT_SQL_MONITOR 函数。

（2）Oracle Enterprise Manager Cloud Control 或 Enterprise Manager Database Express。

（3）Oracle SQL Developer。

无须借助图形化工具，DBMS_SQL_MONITOR 是最简单快捷的方式。以下脚本为最近一个运行的 SQL 生成实时 SQL 监控报告，并存于文件 sqlmon_active.html 中。

```
-- 在 SQL 语句中加入 MONITOR 提示
<运行你的 SQL 语句>

SET TRIMSPOOL ON
SET TRIM ON
SET PAGESIZE 0
SET LINESIZE 32767
SET LONG 1000000
SET LONGCHUNKSIZE 1000000
SPOOL sqlmon_active.html

SELECT DBMS_SQL_MONITOR.REPORT_SQL_MONITOR(report_level => 'ALL', type => 'ACTIVE')
FROM dual;

SPOOL OFF
```

将输出文件 sqlmon_active.html 中的第一行和最后一行删除（因为 SELECT 和 SPOOL 命令也输出到了此文件），然后即可在浏览器中显示，如图 5-15 所示。将鼠标悬浮在某些项目（如 Activity、Timeline）或单击某些项目（如 Information 下的图标），可显示更详细的信息。

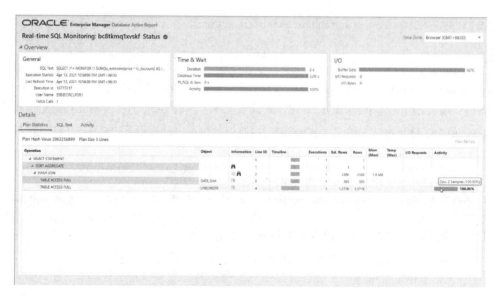

图 5-15　交互式的实时 SQL 监控报告

在上例中，指定 ACTIVE 参数可以生成交互式的 SQL 监控报告，此外报告也可以支持文本、XML 和简单 HTML 格式。以下为生成文本格式 SQL 监控报告的示例：

```
SQL> SELECT DBMS_SQL_MONITOR.REPORT_SQL_MONITOR(report_level => 'ALL',
    type => 'TEXT') FROM dual;

SQL Monitoring Report

SQL Text
------------------------------
SELECT /* 1.1 SSB_SAMPLE_SQL */ /*+ MONITOR */ SUM(lo_extendedprice * lo_discount) AS
revenue FROM lineorder, date_dim WHERE lo_orderdate = d_datekey AND d_year = 1993 AND
lo_discount BETWEEN 1 AND 3 AND lo_quantity < 25

Global Information
------------------------------
 Status              :  DONE(ALL ROWS)
 Instance ID         :  1
 Session             :  SSB(253:54027)
 SQL ID              :  62gb0a7n7xvdm
 SQL Execution ID    :  16777216
 Execution Started   :  06/23/2021 15:27:47
 First Refresh Time  :  06/23/2021 15:27:47
 Last Refresh Time   :  06/23/2021 15:27:47
 Duration            :  .068931s
 Module/Action       :  SQL*Plus/-
 Service             :  orclpdb1
 Program             :  sqlplus@oracle-19c-vagrant(TNS V1-V3)
 Fetch Calls         :  1

Global Stats
```

```
=================================================================
| Elapsed | Cpu     | IO       | Other    | Fetch | Buffer | Read | Read  |
| Time(s) | Time(s) | Waits(s) | Waits(s) | Calls | Gets   | Reqs | Bytes |
=================================================================
|    0.07 |    0.06 |     0.00 |     0.00 |     1 |     95 |   10 | 81920 |
=================================================================

SQL Plan Monitoring Details (Plan Hash Value=2403472142)
=================================================================
| Id |          Operation           |   Name   |   Rows   | Cost |  Time      |...
|    |                              |          | (Estim)  |      | Active(s)  |...
=================================================================
|  0 | SELECT STATEMENT             |          |          |      |     1      |...
|  1 |  SORT AGGREGATE              |          |       1  |      |     1      |...
|  2 |   HASH JOIN                  |          |    238K  | 1942 |     1      |...
|  3 |    JOIN FILTER CREATE        | :BF0000  |     365  |   13 |     1      |...
|  4 |     TABLE ACCESS INMEMORY FULL | DATE_DIM |     365  |   13 |     1      |...
|  5 |    JOIN FILTER USE           | :BF0000  |      2M  | 1924 |     1      |...
|  6 |     TABLE ACCESS INMEMORY FULL | LINEORDER |    2M  | 1924 |     1      |...
=================================================================
```

SQL Developer 中的实时 SQL 监控报告位于菜单 Tools>Real Time SQL Monitor。如图 5-16 所示，被监控的 SQL 显示在上方，单击感兴趣的 SQL，下方会显示执行计划和性能指标等信息。与 Enterprise Manager 相比，SQL Developer 中的实时 SQL 监控报告显示的信息要简单一些，格式也有所不同。

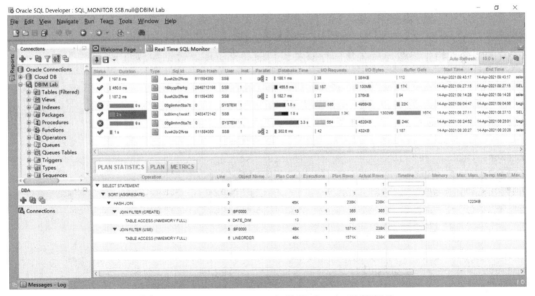

图 5-16　SQL Developer 中的实时 SQL 监控报告

Enterprise Manager 中的实时 SQL 监控报告位于 Performance 标签页下的菜单 Performance Hub>SQL Monitoring，其功能最强大，显示的信息也最丰富。如图 5-17 所示，其中显示了被监控 SQL 执行期间 CPU、内存和 I/O 的使用情况。

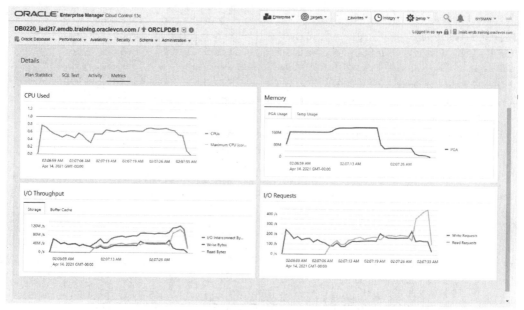

图 5-17　Enterprise Manager 中的实时 SQL 监控报告

5.7　Database In-Memory 初始化参数

与 Database In-Memory 相关的初始化参数均以 inmemory 开头，可配置 Database In-Memory 和控制其使用行为。其中大多为动态参数，无须重启即可生效。根据参数不同，可以指定在实例或会话一级设置，或在 CDB 或 PDB 一级设置，详情可查阅 Oracle 数据库参考手册[①]。包含 xmem 的 3 个初始化参数与持久化内存有关，目前尚未实现。常见参数如下：

```
SQL> SHOW PARAMETER inmemory
NAME                                   TYPE         VALUE
------------------------------------   -----------  -------
inmemory_adg_enabled                   boolean      TRUE
inmemory_automatic_level               string       OFF
inmemory_clause_default                string
inmemory_deep_vectorization            boolean      TRUE
inmemory_expressions_usage             string       ENABLE
inmemory_force                         string       DEFAULT
inmemory_max_populate_servers          integer      0
inmemory_optimized_arithmetic          string       DISABLE
inmemory_prefer_xmem_memcompress       string
inmemory_prefer_xmem_priority          string
inmemory_query                         string       ENABLE
inmemory_size                          big integer  0
inmemory_trickle_repopulate_servers_percent integer 1
inmemory_virtual_columns               string       MANUAL
```

---

① https://docs.oracle.com/en/database/oracle/oracle-database/21/refrn/initialization-parameter-descriptions.html

```
inmemory_xmem_size                              big integer    0
optimizer_inmemory_aware                        boolean        TRUE
```

（1）inmemory_adg_enabled：在 Oracle ADG 的备用数据库端是否启用 Database In-Memory，默认为启用，从 Oracle 12.2.0.1 版本开始支持。此参数只能在备用数据库端设置，如果备用数据库端为 RAC，则需为每一个实例设置相同的值，例如：

```
SQL> ALTER SYSTEM SET inmemory_adg_enabled = true SID = '*';
```

由于 ADG 会接受来自主数据库端的数据，如果此参数设为 TRUE，ADG 会执行额外的操作获取内存对象的数据或标记对象无效。

（2）inmemory_automatic_level：Oracle 从 18.1 版本开始支持自动 In-Memory 管理。此参数决定了是否启用自动 In-Memory 管理，以及启用后的自动化层级设置。自动化层级可设为 LOW、MEDIUM 和 HIGH。LOW 表示在内存紧张时会将不活跃的对象从内存中移除。MEDIUM 包含了 LOW 的操作，此外还会优先发布之前因内存紧张未能发布的对象。HIGH 是 21c 新增选项，对象无须再设置 INMEMORY 属性，系统自动决定加载哪些对象到内存列式存储，或从内存列式存储中移除。

（3）inmemory_clause_default：当为表启用 INMEMORY 时，此参数指定了默认的 INMEMORY 属性，包括内存压缩级、发布优先级等。例如：

```
SQL> ALTER SESSION SET inmemory_clause_default="memcompress for capacity low";
Session altered.

SQL> CREATE TABLE t1(a INT) INMEMORY;
Table created.

SQL> SELECT INMEMORY, inmemory_compression FROM user_tables WHERE table_name = 'T1';
INMEMORY  INMEMORY_COMPRESS
--------  -----------------
ENABLED   FOR CAPACITY LOW
```

设置为空字符串即恢复其默认值：

```
SQL> ALTER SESSION SET inmemory_clause_default='';
Session altered.
```

此参数对外部表和混合分区表无效。

（4）inmemory_deep_vectorization：In-Memory 深度向量化是 21c 版本开始支持的功能，可以将高阶、复杂的 SQL 操作（例如联结）拆解为适合于 SIMD 向量处理的单元。此参数控制是否启用 In-Memory 深度向量化，默认为启用。

（5）inmemory_expressions_usage：此参数控制是否发布以及哪些 In-Memory 表达式会被发布到内存列式存储中。DISABLE 表示不会发布任何 In-Memory 表达式。STATIC_ONLY 表示仅发布静态表达式，即 OSON（JSON 的二进制形式）列。DYNAMIC_ONLY 表示仅发布动态表达式，系统将自动捕获这些表达式并发布到内存中。ENABLE 为默认值，表示将发布静态和动态表达式。

（6）inmemory_force：默认值为 DEFAULT，表示开启了 INMEMORY 属性的对象将会被发

布。OFF 表示开启了 INMEMORY 属性的对象也不会被发布。BASE_LEVEL 表示开启 Base Level 特性，可以免费试用 Database In-Memory 的部分功能。CELLMEMORY_LEVEL 启用本地部署 Exadata 上的 CellMemory 特性。

（7）inmemory_max_populate_servers：用于控制后台用于发布的进程数量，以避免发布时对系统造成影响。默认值为 CPU_COUNT/2 和 PGA_AGGREGATE_TARGET/512 中较小的值。

（8）inmemory_optimized_arithmetic：18c 版本以后，Oracle 支持在内存中以优化的格式存储 NUMBER 列。此参数可控制是否使用这种优化格式，默认为禁用。如果设置为 ENABLE，则内存压缩级为 QUERY LOW 的表中的 NUMBER 列将被编码为等宽的整数，以便后续利用 SIMD 实现更快的计算。

（9）inmemory_query：此参数控制是否使用内存列式存储进行 SQL 查询，默认为使用。此参数可以在数据库或会话一级设置，常用于在测试中对比使用和禁用内存列式存储时的性能。

（10）inmemory_size：此参数设置了数据库 In-Memory Area 的大小，默认值为 0，即禁用内存列式存储。开启内存列式存储需将此值设为 100MB 以上。首次启用内存列式存储时需要重启实例，后续可以动态调整大小。

（11）inmemory_trickle_repopulate_servers_percent：为 inmemory_max_populate_servers 的百分比，可限制涓流式重新发布进程的数量，如果设为 0，则禁止涓流式重新发布。此参数最大可设为 50，表示至少留出一半的处理能力用于其他工作负载。

（12）inmemory_virtual_columns：控制表的虚拟列是否会被发布到内存列式存储中。默认为 MANUAL，表示需要为此列单独开启 INMEMORY 属性后才能发布。ENABLE 表示虚拟列将和其他的普通列一样自动发布。DISABLE 则表示不会发布任何虚拟列。

（13）optimizer_inmemory_aware：此参数控制优化器在对 SQL 进行优化时是否会考虑对象的 INMEMORY 属性，默认值为 TRUE。

5.8 Database In-Memory 视图

本节介绍和 Database In-Memory 相关的数据字典视图和动态性能视图。数据字典包括表和视图，它们是静态的，因为它们只在数据字典改变时才改变。动态性能视图是由数据库内部维护的视图，属于 SYS 用户。之所以称为动态，是因为在数据库运行期间会被不断更新。它们也被称为 V$视图，因为它们的命名均以 V$开头。

5.8.1 数据字典视图

（1）DBA_EXPRESSION_STATISTICS 提供表中的表达式跟踪统计信息。

（2）DBA_EXTERNAL_TABLES 提供外部表信息。INMEMORY 列指示此外部表是否启用了内存列式存储，INMEMORY_COMPRESSION 列表示其内存压缩级别。

（3）DBA_FEATURE_USAGE_STATISTICS 提供数据库特性使用信息统计，其中也包括

Database In-Memory 相关特性，如内存列式存储、In-Memory 联结组、In-Memory 聚合等。

（4）DBA_HEAT_MAP_SEGMENT 提供存储段对象最近一次读、写、全表扫描和索引扫描的时间。

（5）DBA_INMEMORY_AIMTASKS 提供自动 In-Memory 任务创建时间和状态。

（6）DBA_INMEMORY_AIMTASKDETAILS 提供自动 In-Memory 对象和其上执行的操作。

（7）DBA_ILMDATAMOVEMENTPOLICIES 提供自动优化策略中与数据移动相关的属性，自动数据优化是 Oracle 12.2 开始支持的功能。

（8）DBA_IM_EXPRESSIONS 提供 In-Memory 表达式（即设置了 INMEMORY 属性，命名以 SYS_IME 开头的虚拟列）信息，基于此信息可决定是否后续将其发布到内存列式存储中。

（9）DBA_JOINGROUPS 提供存储 In-Memory 联结组的信息。In-Memory 联结组是一组共享全局字典的列，这些列通过全局字典实现更高效的 SQL 联结。

（10）DBA_SEGMENTS，INMEMORY 列指示此段对象是否启用了内存列式存储，但并不表示其已经发布到内存列式存储。INMEMORY_PRIORITY、INMEMORY_DISTRIBUTE、INMEMORY_COMPRESSION 和 INMEMORY_DUPLICATE 列分别表示发布优先级、RAC 数据分布方式、内存压缩级和内存复制方式。

（11）DBA_TABLES 可参考 DBA_SEGMENTS。此数据字典可认为是 DBA_SEGMENTS 的子集，因为段对象包括表、分区和物化视图等。

5.8.2 动态性能视图

通过以下 SQL 可以查询绝大部分和 In-Memory 相关的动态性能视图：

```
SQL> SELECT object_name FROM dba_objects WHERE
    REGEXP_LIKE(object_name, '^V\$IM|^V\$INMEM')ORDER BY 1;

OBJECT_NAME
--------------------------------------------------------------------------------
V$IMEU_HEADER
V$IMHMSEG
V$IM_ADOELEMENTS
V$IM_ADOTASKDETAILS
V$IM_ADOTASKS
V$IM_COLUMN_LEVEL
V$IM_COL_CU
V$IM_DELTA_HEADER
V$IM_GLOBALDICT
V$IM_GLOBALDICT_PIECEMAP
V$IM_GLOBALDICT_SORTORDER
V$IM_GLOBALDICT_VERSION
V$IM_HEADER
V$IM_IMECOL_CU
V$IM_SEGMENTS
V$IM_SEGMENTS_DETAIL
V$IM_SEG_EXT_MAP
```

```
V$IM_SMU_CHUNK
V$IM_SMU_DELTA
V$IM_SMU_HEAD
V$IM_TBS_EXT_MAP
V$IM_USER_SEGMENTS
V$INMEMORY_AREA
V$INMEMORY_FASTSTART_AREA
V$INMEMORY_XMEM_AREA

25 rows selected.
```

以下为与 In-Memory 相关的部分动态性能视图的简介。

（1）V$IMEU_HEADER：IMEU 元数据信息。

（2）V$IMHMSEG：Automatic In-Memory 对象的热图统计信息，包括增删改查和扫描的统计。

（3）V$IM_ADOELEMENTS：Automatic In-Memory 各对象的状态和统计信息。

（4）V$IM_ADOTASKDETAILS：Automatic In-Memory 任务详情，输出类似 V$IM_ADOTASKDETAILS。

（5）V$IM_ADOTASKS：Automatic In-Memory 任务概览，输出类似 DBA_INMEMORY_AIMTASKS。

（6）V$IM_COLUMN_LEVEL：默认情况下，列的内存压缩级别继承自其父对象，如表和分区。如果为列单独指定了内存压缩级，此视图显示这些列的信息。

（7）V$IM_COL_CU：IMCU 中每个 CU 的元数据信息。包括本地字典的条目数、列的最小和最大值等。

（8）V$IM_HEADER：IMCU 元数据信息。每一个 IMCU 一条记录，包括每一个 IMCU 的地址、分配和已用空间、发布和重新发布标志、包含的列数、行数和数据块数量。

（9）V$IM_IMECOL_CU：和 V$IM_COL_CU 类似，只不过这里的列为 In-Memory 表达式。

（10）V$IM_SEGMENTS：所有用户发布到内存列式存储中的段对象（表、分区和子分区）信息，每一个对象一条记录。包括段对象占用的磁盘空间、分配的内存空间、发布状态和各种 INMEMORY 属性。

（11）V$IM_SEGMENTS_DETAIL：发布到内存列式存储中的段对象的详细信息，每一个对象一条记录。是对 V$IM_SEGMENTS 的补充，主要补充了存储分配信息。

（12）V$IM_SEG_EXT_MAP：In-Memory Area 中的列式数据池和元数据池与磁盘上 Extent 的映射信息。

（13）V$IM_SMU_HEAD：发布到内存列式存储中对象的 SMU 元数据信息。包括每一个 IMCU 包含的行数，在磁盘上分配的 Extent 数量和数据块数量。

（14）V$IM_TBS_EXT_MAP：数据库对象分配的 Extent 与 IMCU 的映射信息，每一个 Extent 一条记录。

（15）V$IM_USER_SEGMENTS：当前用户发布到内存列式存储中的段对象信息，参见

V$IM_SEGMENTS。

（16）V$INMEMORY_AREA：In-Memory Area 空间分配信息。包括列式数据池和元数据池的分配空间、已用空间和发布状态。

（17）V$INMEMORY_FASTSTART_AREA：In-Memory FastStart 区域的信息。包括其对应表空间的信息、状态、分配和已用空间和最近一次检查点时间等。

（18）V$ACTIVE_SESSION_HISTORY：在会话期间的活动采样信息，其中 INMEMORY_QUERY 和 INMEMORY_POPULATE 表示在会话期间是否有与内存列式存储相关的查询和发布活动。

（19）V$HEAT_MAP_SEGMENT：包括实时的段的访问信息。

（20）V$SGA：包含了 In-Memory Area 的大小。

5.8.3　In-Memory 视图使用示例

通过一个示例演示如何在 In-Memory 视图中查找重要的配置和发布信息。为简化和易于查找，整个实例只在一个 PDB 中配置了内存列式存储，并且只发布了 LINEORDER 一张表。

通过 V$SGA 视图，可以查看到实例的 In-Memory Area 配置为 1.5GB：

```
SQL> SELECT * FROM v$sga;

NAME                      VALUE      CON_ID
-------------------- ----------  ----------
Fixed Size              9150296           0
Variable Size        3305111552           0
Database Buffers     1929379840           0
Redo Buffers            7626752           0
In-Memory Area       1577058304           0
```

从 USER_SEGMENTS 中，可以查询到 LINEORDER 表所有 INMEMORY 相关属性，如发布优先级、内存压缩级等。

```
SQL> SELECT inmemory, inmemory_priority AS im_prio, inmemory_compression AS im_comp,
inmemory_distribute AS im_dist, inmemory_duplicate AS im_dup, cellmemory FROM
user_segments WHERE segment_name = 'LINEORDER';

INMEMORY IM_PRIO  IM_COMP           IM_DIST         IM_DUP          CELLMEMORY
-------- -------- ----------------- --------------- --------------- ----------
ENABLED  NONE     FOR QUERY LOW     AUTO            NO DUPLICATE
```

从此视图还可以查询到原始表的存储分配信息。例如，LINEORDER 表占用磁盘空间为 1344MB，总共分配了 172 032 个数据库块、204 个 Extent。

```
SQL> SELECT segment_name, segment_type, tablespace_name, bytes/1024/1024 AS disk_mb,
blocks, extents FROM user_segments WHERE segment_name = 'LINEORDER';

SEGMENT_NAME  SEGMENT_TYPE  TABLESPACE_NAME   DISK_MB    BLOCKS   EXTENTS
------------  ------------  ---------------  --------  --------  --------
LINEORDER     TABLE         USERS                1344    172032       204
```

从 V$IM_COLUMN_LEVEL 中可以查询到 LINEORDER 表每一列设置的内存压缩级：

```
SQL> SELECT table_name, column_name, inmemory_compression FROM v$im_column_level WHERE
table_name = 'LINEORDER';

TABLE_NAME    COLUMN_NAME          INMEMORY_COMPRESSION
-----------   ------------------   -------------------------
LINEORDER     LO_ORDERKEY          DEFAULT
LINEORDER     LO_LINENUMBER        DEFAULT
LINEORDER     LO_CUSTKEY           DEFAULT
LINEORDER     LO_PARTKEY           DEFAULT
LINEORDER     LO_SUPPKEY           DEFAULT
LINEORDER     LO_ORDERDATE         DEFAULT
LINEORDER     LO_ORDERPRIORITY     DEFAULT
LINEORDER     LO_SHIPPRIORITY      DEFAULT
LINEORDER     LO_QUANTITY          DEFAULT
LINEORDER     LO_EXTENDEDPRICE     DEFAULT
LINEORDER     LO_ORDTOTALPRICE     DEFAULT
LINEORDER     LO_DISCOUNT          DEFAULT
LINEORDER     LO_REVENUE           DEFAULT
LINEORDER     LO_SUPPLYCOST        DEFAULT
LINEORDER     LO_TAX               DEFAULT
LINEORDER     LO_COMMITDATE        DEFAULT
LINEORDER     LO_SHIPMODE          DEFAULT

17 rows selected.
```

从 V$INMEMORY_AREA 视图可以查看到 In-Memory Area 的分配与使用情况。本例中，1.5GB 的 In-Memory Area 被划分为 1039MB 的列式数据池和 448MB 的元数据池。LINEORDER 表在这两个数据池中分别占用了 560MB 和 4.5625MB。由于数据池的划分比例是系统自动确定的，用户无法指定，因此在规划内存时需考虑为元数据池预留空间。

```
SQL> SELECT pool, alloc_bytes/1024/1024 AS alloc_mb, used_bytes/1024/1024 AS used_mb,
populate_status FROM v$inmemory_area;

POOL          ALLOC_MB    USED_MB  POPULATE_STATUS
-----------   ---------   -------- -------------------------
1MB POOL          1039        560  DONE
64KB POOL          448     4.5625  DONE
```

从 V$IM_SEGMENTS 中可以查看到对象实际发布的状态，是否部分发布等。注意到 DISK_MB 比之前 USER_SEGMENTS 中的值（1344）小，这是由于剔除了一些无法在内存中表示的磁盘元数据块。

```
SQL> SELECT segment_name, inmemory_size/1024/1024 AS im_mb, bytes/1024/1024 AS disk_mb,
bytes_not_populated, populate_status, inmemory_compression AS im_comp FROM
v$im_segments;

SEGMENT_NAME      IM_MB       DISK_MB  BYTES_NOT_POPULATED POPULATE_STAT IM_COMP
---------------   ---------   -------- ------------------- ------------- -------------
LINEORDER         564.5625    1301.14844                 0 COMPLETED     FOR QUERY LOW
```

从 V$IM_SEGMENTS_DETAIL 可获取发布对象的概要信息,包括发布时间、分配的 IMCU 数量、内存和磁盘存储分配信息。

```sql
SQL> SELECT drammembytes/1024/1024 AS im_mb, extents, blocks, blocksinmem, imcusinmem,
BYTES/1024/1024 AS disk_mb, to_char(createtime, 'DD-MON-YYYY HH24:MI') AS createtime
FROM v$im_segments_detail;
```

IM_MB	EXTENTS	BLOCKS	BLOCKSINMEM	IMCUSINMEM	DISK_MB	CREATETIME
564.5	204	172032	166547	24	1344	24-APR-2021 15:47

V$IM_HEADER 中包含了 IMCU 的存储分配信息。从以下 SQL 可知,LINEORDER 表共分配了 24 个 IMCU。除最后一个 IMCU 外,每个 IMCU 大小为 24MB,存储的行数为 50 万左右。

```sql
SQL> SELECT head_piece_address, SUM(allocated_len)/1024/1024 AS im_alloc_mb,
round(SUM(used_len)/1024/1024,2) AS im_used_mb, SUM(num_rows), SUM(num_disk_extents)
FROM v$im_header GROUP BY head_piece_address;
```

HEAD_PIECE_ADDRESS	IM_ALLOC_MB	IM_USED_MB	SUM(NUM_ROWS)	SUM(NUM_DISK_EXTENTS)
0000000079FFFF28	24	23.43	516875	9
000000008ADFFF88	24	23.41	514065	8
00000000935FFFB8	24	23.41	514065	8
00000000825FFF58	24	23.4	514072	8
0000000088DFFF70	24	23.42	514063	8
0000000085FFFF70	24	23.41	514059	8
0000000095FFFFD0	24	23.47	515495	2
000000007B5FFF28	26	25.56	565884	77
000000007DFFFF40	24	23.42	514066	8
000000008FDFFFA0	24	23.42	514075	8
00000000875FFF70	24	23.41	514060	8
0000000091FFFFB8	24	23.42	514069	8
000000009A5FFFE8	24	23.48	515504	2
000000007EDFFF40	24	23.41	514064	8
000000008E5FFFA0	24	23.41	514063	8
0000000083DFFF58	24	23.41	514054	8
000000008C5FFF88	24	23.41	514061	8
0000000089FFFF88	24	23.41	514062	8
00000000985FFFD0	24	23.46	515437	3
0000000099FFFFE8	6	5.76	112757	1
0000000094DFFFB8	24	23.41	514065	8
00000000805FFF40	24	23.42	514070	8
0000000096DFFFD0	24	23.47	515503	2
000000009BDFFFE8	24	23.47	515508	2

24 rows selected.

从此视图中还可以得到 LINEORDER 表的发布时间为 31 秒,在列式数据池中分配的空间为 560MB 等信息。

```sql
SQL> SELECT SUM(allocated_len)/1024/1024 AS total_alloc_mb, SUM(used_len)/1024/1024 AS
total_used_mb, SUM(num_rows), SUM(num_blocks), SUM(time_to_populate)/1000 AS
"pop_time(s)" FROM v$im_header;
```

```
TOTAL_ALLOC_MB TOTAL_USED_MB SUM(NUM_ROWS) SUM(NUM_BLOCKS) POP_TIME(S)
-------------- -------------- ------------- --------------- -----------
           560      546.68497      11997996          166547      31.231
```

视图 V$IM_COL_CU 包含了 CU 信息。以下示例显示 LINEORDER 表第 1 列 LO_ORDERKEY 在每一个 IMCU 中的最小值、最大值及字典条目数。

```
SQL> SELECT utl_raw.cast_to_number(minimum_value) AS min,
utl_raw.cast_to_number(maximum_value) AS max, dictionary_entries FROM v$im_col_cu
WHERE column_number = 1;

       MIN        MAX DICTIONARY_ENTRIES
---------- ---------- ------------------
         1     565510             141382
    565510    1082785             129316
   1082785    1597216             128608
   1597216    2110464             128313
   2110464    2625473             128754
   2625473    3139555             128523
   3139555    3652230             128172
   3652230    4166658             128605
   4166659    4680260             128402
   4680261    5194694             128610
   5194695    5709063             128593
   5709063    6222785             128427
   6222785    6737474             128674
   6737475    7253184             128926
   7253184    7766307             128284
   7766307    8279808             128374
   8279808    8793697             128474
   8793697    9307489             128449
   9307489    9823139             128915
   9823140   10339328             129045
  10339328   10855174             128967
  10855175   11371236             129014
  11371237   11887460             129056
  11887460   12000000              28133

24 rows selected.
```

5.9　优化器提示

　　SQL 优化器制定的执行计划适用于大多数情形，但有时优化器了解的信息并不全面，或者有时需要针对特定类型的语句或工作负载调整优化器，此时可以通过一些技术来影响优化器对执行路径的选择，以达到最佳效果。提示即是其中一种手段，其他的技术包括初始化参数、SQL Profile 等。对于性能问题，提示通常只是临时解决方法，更好的方式是采用 SQL 调优建议器和 SQL 计划管理等工具来定位问题根源，然后对 SQL 甚至数据库设计进行重构。以下介绍的提示可以在一定程度上影响基于内存列式存储运行的 SQL 行为，主要用于 Database In-Memory 学习过程中的测试和概念理解。

可以在 DML 语句中使用提示，提示以特殊的注释形式嵌入到 SQL 语句中，以 "/*+" 开始，以 "*/" 结束。如果有多个提示，这些提示需要加到同一个注释中。

以下仅介绍和 Database In-Memory 相关的主要提示，更多提示和详情请参考 Oracle SQL 语言参考手册[①]。

（1）FULL 提示通知优化器对指定的表实施全表扫描，对于设置了 INMEMORY 属性的表，将使其发布到内存列式存储中。例如：

```
SELECT /*+ FULL(t) */ FROM lineorder t;
```

（2）INMEMORY 提示启用基于内存列式存储的查询，加上前缀 NO_ 则为禁用。在下例中，即使内存列式存储已配置，以下的 SQL 查询也不会使用：

```
SELECT /*+ NO_INMEMORY */ COUNT(*) FROM lineorder;
```

（3）INMEMORY_PRUNING 提示通知优化器使用内存存储索引，加上前缀 NO_ 则为禁用。

（4）MONITOR 提示强制对 SQL 启用实时 SQL 监控，即使 SQL 并非并行运行或长时间运行。

（5）PX_JOIN_FILTER 强制优化器使用布隆过滤器将 SQL 联结操作转换为过滤操作。加上前缀 NO_ 则禁用。此特性可用于测试 In-Memory 联结。

（6）VECTOR_TRANSFORM 提示启用向量转换，此特性可用于 In-Memory 聚合，加上前缀 NO_ 则为禁用。

---

① https://docs.oracle.com/en/database/oracle/oracle-database/21/sqlrf/Comments.html#GUID-D316D545-89E2-4D54-977F-FC97815CD62E

第6章

Database In-Memory 基础性能优化

6.1 列格式组织

对于分析型负载，Database In-Memory 最基础的性能优化来自于数据的列格式组织形式。行格式组织将记录的相关属性集结在一起，对于交易性负载，查询一条记录的多个属性非常方便。而分析型负载的特点是访问大多数行中的少数几列，列格式组织可以避免访问一行中不需要的字段。

在图 6-1 的示例中，需要查询 MANAGER_ID 为 108 的员工的工资（SALARY）汇总。采用行式访问的大致步骤如下。

（1）读取一行数据。
（2）从第一个字段开始一直向前移动到 MANAGER_ID 字段，读取数据。
（3）如果 MANAGE_ID 为 108，读取 SALARY 字段。
（4）重复步骤（1），直到表中的所有行处理完毕。

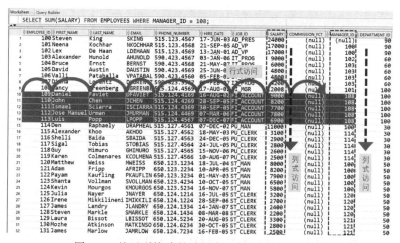

图 6-1 基于列格式的访问与基于行格式的访问

而列式访问的大致步骤如下。

（1）读取 MANAGER_ID 字段。

（2）如果 MANAGE_ID 为 108，直接读取对应行的 SALARY 字段。

对比两种方式，可以发现列式访问避免了处理不必要的数据，而行式访问则花费了大量开销在无关数据上，这种开销在全表扫描时尤其明显。虽然可通过建立索引快速定位 MANAGER_ID 为 108 的记录，但开销只是减少，并不能完全消除。以上只是一个简单的查询，现实生产环境中的查询可能更复杂，种类可能更多。这意味着需要建立更多的索引来保证性能。另外，虽然固定查询可以事先建立索引，但对于查询模式无法预知的即席查询，则无法通过提前建立索引实施优化。

行式访问中的额外开销与 I/O 无关，也就是说，即使把表全部置入内存中，其针对分析型负载的查询效率也会低于列式访问。下面我们通过一个实验来证明这一点。

查询涉及 LINEORDER 和 DATE_DIM 两张表，它们的数据量约为 1.5GB，将 KEEP 缓存池的内存大小设置为 1.8GB：

```
SQL> SELECT bytes FROM user_segments WHERE segment_name = 'LINEORDER';
    BYTES
----------
1414529024

SQL> SELECT bytes FROM user_segments WHERE segment_name = 'DATE_DIM';
    BYTES
----------
393216

SQL> ALTER SYSTEM SET db_keep_cache_size = 1800M SCOPE = SPFILE;
```

然后将这两张表存放在 KEEP 缓存池中：

```
SQL> ALTER TABLE lineorder STORAGE(BUFFER_POOL KEEP);
SQL> ALTER TABLE lineorder CACHE;
SQL> ALTER TABLE date_dim STORAGE(BUFFER_POOL KEEP);
SQL> ALTER TABLE date_dim CACHE;
SQL> SELECT table_name, cache, buffer_pool FROM user_tables WHERE table_name
IN('LINEORDER', 'DATE_DIM');

TABLE_NAME            CACHE                BUFFER_POOL
-------------------   -------------------  -------------------
DATE_DIM              Y                    KEEP
LINEORDER             Y                    KEEP
```

测试使用了 SSB 示例 SQL 1_1：

```
$ cat ~/dbimbook/chap02/SSB/sql/1_1.sql
SELECT /* 1.1 SSB_SAMPLE_SQL */
    SUM(lo_extendedprice * lo_discount) AS revenue
FROM
    lineorder,
    date_dim
WHERE
    lo_orderdate = d_datekey
    AND d_year = 1993
    AND lo_discount BETWEEN 1 AND 3
```

```
AND lo_quantity < 25;
```

首先在传统行式存储中查询,第一次查询会将磁盘中的数据载入 KEEP 缓存池,因此需要多执行几次,确保这两张表的数据已经全部进入内存,以确保后续比较的公平性。以下查询显示了两张表加载到内存缓存中的数据块数量:

```
SQL> select o.object_name, count(*) number_of_blocks from dba_objects o, v$bh bh
where o.data_object_id = bh.objd
and o.owner = 'SSB'
group by o.object_name
order by count(*);

OBJECT_NAME                              NUMBER_OF_BLOCKS
---------------------------------------- ----------------
DATE_DIM                                               44
LINEORDER                                         167,200
```

在此前提下,得到了基于行式查询的时间和性能统计信息。结果显示查询时间为 0.64 秒,并且没有物理读,这表明数据全部从内存中读取:

```
SQL> @1_1

    REVENUE
----------
8.9428E+11

Elapsed: 00:00:00.64

Statistics
----------------------------------------------------------
         4  recursive calls
         0  db block gets
    166615  consistent gets
         0  physical reads
         0  redo size
       554  bytes sent via SQL*Net to client
       574  bytes received via SQL*Net from client
         2  SQL*Net roundtrips to/from client
         0  sorts(memory)
         0  sorts(disk)
         1  rows processed
```

然后将 LINEORDER 和 DATE_DIM 全部发布到内存列式存储中,再次执行查询。结果显示查询时间为 0.03 秒,比基于行式的查询快了 21 倍:

```
SQL> @1_1

    REVENUE
----------
8.9428E+11

Elapsed: 00:00:00.03

Statistics
----------------------------------------------------------
         4  recursive calls
         0  db block gets
```

```
          5  consistent gets
          0  physical reads
          0  redo size
        554  bytes sent via SQL*Net to client
        574  bytes received via SQL*Net from client
          2  SQL*Net roundtrips to/from client
          0  sorts(memory)
          0  sorts(disk)
          1  rows processed
```

当然，这 21 倍的性能提升并不能完全归功于列格式组织。下节开始介绍其他重要的基础性能优化技术，包括数据压缩、SIMD 向量处理和内存存储索引等。正是由于这些技术的相互配合，才最终实现了分析查询性能的全面优化。

实验结束后，恢复这两张表的默认缓存设置，并解除其与 KEEP 缓存池的绑定关系。

```
ALTER TABLE lineorder NOCACHE;
ALTER TABLE date_dim NOCACHE;
ALTER TABLE lineorder STORAGE(BUFFER_POOL DEFAULT);
ALTER TABLE date_dim STORAGE(BUFFER_POOL DEFAULT);
```

6.2 内存存储索引

第 3 章已经介绍过，在 IMCU 的每一个 CU 的头部信息中，保留了此列数据的最小值和最大值，我们称之为内存存储索引。对于某一列而言，有多少个 IMCU，就有多少个存储索引。存储索引可以实现 IMCU 裁剪，非常适合于等值查询和范围查询，即通过 SQL 查询条件过滤掉不满足条件的 IMCU，从而避免不必要的扫描。

如图 6-2 所示，销售额列在内存列式存储中被存为 3 个 IMCU，每一个 IMCU 中都建立了此列的存储索引，包含了此列在此 IMCU 中的最小和最大值。当查询销售额在 5000 和 6000 之间的店铺时，从存储索引中可知 IMCU 1 和 IMCU 3 没有满足条件的记录，因此可以快速略过，从而避免了不必要的扫描，达到查询提速的目的。

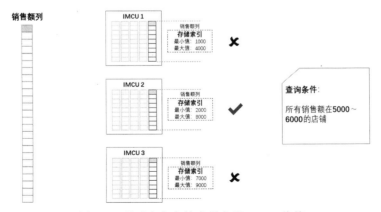

图 6-2　通过内存存储索引实现 IMCU 裁剪

内存存储索引并不是一项新的技术，早在 2008 年推出的 Exadata 数据库云平台中使用的

Exadata 存储索引，以及 Oracle 12.1.0.2 推出的 Zone Map，在实现上都采用了类似的原理。这类技术本质上都是一组过滤器，通过将数据分割为多个区域，然后维护每个区域的最小值和最大值，就可以避免扫描不必要的区域来实现查询提速的目的。和索引相比，存储索引的目的并不一样，或者也可以称其为反索引技术。因为索引的目的是为了快速找到数据，而存储索引的目的则是为了快速过滤数据。数据库分区选件也可以通过分区裁剪减少 I/O，但需要在查询中指定分区键才能生效，而存储索引类技术的粒度更细，也更灵活。虽然存储索引需要初始建立和后期维护，但这些开销相当于其带来的好处是微不足道的。更重要的是，存储索引是后台自动维护的，对于应用完全透明。

通过一个例子来了解一下内存存储索引的工作原理。首先创建一个 200 万行的表并发布到内存列式存储。第 1 列为顺序递增的 ID，第 2 列为 $1 \sim 1 \times 10^6$ 的随机数。

```
-- 创建一个 200 万行的表
CREATE TABLE t1 AS SELECT ROWNUM-1 k, MOD(ROWNUM, 1000000) v
FROM dual CONNECT BY LEVEL <= 2000000;

-- 启用 INMEMORY，并通过全表扫描发布
ALTER TABLE t1 INMEMORY;
SELECT /*+ FULL(p) NO_PARALLEL(p) */ COUNT(*) FROM t1 p;
```

从视图 V$IM_COL_CU 中，可以查看到第 2 列在每一个 IMCU 中的存储索引：

```
SQL>
SELECT head_piece_address AS imcu, column_number,
utl_raw.cast_to_number(minimum_value) min, utl_raw.cast_to_number(maximum_value) max
FROM v$im_col_cu WHERE objd IN
(SELECT objd FROM v$im_header WHERE table_objn IN (SELECT object_id FROM user_objects
WHERE object_name = 'T1')) AND column_number = 2
ORDER BY 2,1;

IMCU                COLUMN_NUMBER          MIN         MAX
----------------    -------------      ----------  ----------
00000000763FFF10                2          787084      847820
00000000765FFF10                2          301187      787083
00000000770FFF10                2               0      999999
00000000780FFF10                2               0      999999
```

从存储索引中的信息可知，如果在第 2 列中查找大于 900 000 的行，前两个 IMCU 都可以略过；如果查找等于 800 000 的行，则第 2 个 IMCU 可以略过。通过查看 SQL 执行统计信息可以验证假设，其中，IM scan CUs pruned 表示被略过的 IMCU：

```
SQL> SET AUTOTRACE ON
SQL> SELECT COUNT(*) FROM t1 WHERE v >= 900000;

  COUNT(*)
----------
    200000
...
Statistics
----------------------------------------------------------
```

```
           4  IM scan CUs predicates applied
           2  IM scan CUs predicates optimized
           4  IM scan CUs predicates received
           2  IM scan CUs pruned
SQL> SELECT COUNT(*) FROM t1 WHERE v = 800000;

  COUNT(*)
----------
         2
…
Statistics
----------------------------------------------------------
           4  IM scan CUs predicates applied
           1  IM scan CUs predicates optimized
           4  IM scan CUs predicates received
           1  IM scan CUs pruned
```

从以上查询结果可知，等于 800 000 的记录只有两行。这意味着在 4 个 IMCU 中至少有两个不包含等于 800 000 的记录。而统计信息显示只有 1 个 IMCU 被略过。那么是否有可能进一步优化以略过更多的 IMCU 呢？实际上，可以将近似的数据组织在一起以实现更精准的控制，这就是聚集（clustering）技术。Oracle 很早就支持聚集表、索引聚集等功能，这些功能使得原先分散的存取操作变得相对集中，从而减少了昂贵的 I/O 操作。在 12.1.0.2 版本又新增了属性聚集（attribute clustering）功能，此功能可以使 Exadata 存储索引，内存存储索引和 Zone Map 技术的使用更为高效。

通过一个示例来演示属性聚集所带来的好处。首先为表启用属性聚集功能。由于需要重新组织数据，表会从内存列式存储中清除，因此需要重新发布此表：

```
-- 启用属性聚集
ALTER TABLE t1 ADD CLUSTERING BY LINEAR ORDER (v) WITHOUT MATERIALIZED ZONEMAP;

-- 重新组织数据，将导致表 t1 从内存列式存储中移除
ALTER TABLE t1 MOVE;

-- 通过全表扫描重新发布表
SELECT /*+ FULL(p) NO_PARALLEL(p) */ COUNT(*) FROM t1 p;
```

查看存储索引中的 MIN 和 MAX，发现数据的组织变得更为集中和有序：

```
IMCU                  COLUMN_NUMBER          MIN         MAX
----------------      -------------      ----------  ----------
000000008A3FFF88                  2          852317      999999
000000008A9FFF88                  2          616949      852317
000000008B2FFF88                  2          263897      616949
000000008C0FFF88                  2               0      263897
```

再次执行查询，统计信息显示被略过的 IMCU 由之前的 1 个增加到了 3 个，效率明显提升：

```
SQL> SELECT COUNT(*) FROM t1 WHERE v = 800000;

  COUNT(*)
----------
```

```
----------
         2
...
Statistics
----------------------------------------------------------
         4  IM scan CUs predicates applied
         3  IM scan CUs predicates optimized
         4  IM scan CUs predicates received
         3  IM scan CUs pruned
```

6.3 SIMD 向量处理

大规模的数据处理例如数据分析对性能的要求越来越高，在硬件层面，CPU 起着非常关键的作用。在过去，性能的提升主要依赖于 CPU 的主频速度。现在，软件更多地利用了现代 CPU 所具备的并行能力，这其中最重要的就是多线程和 SIMD。

SIMD 也称为单指令多数据，即可以用一条指令操作多个数据值的处理方式，也称为向量处理。向量是相对于标量而言的，例如在数据类型的定义中，标量数据类型只存储单个值，并且没有内部成员，例如数字、字符串、日期和布尔值。而向量数据类型具有内部成员，且内部成员可以单独操作，例如数组。SIMD 并不是 Oracle 的发明，早在 20 世纪 60 年代就已经被 ILLIAC IV、CDC Star-100 等超级计算机使用，被广泛应用于图形图像处理和科学计算等领域。如今，从桌面系统到后台服务器，均已广泛支持 SIMD。目前普遍使用的 SIMD 指令集为 SSE 和 AVX。Intel 最初于 1999 年支持 SSE，当时的 SIMD 寄存器为 128 bit。AVX 是 SSE 的延伸架构，SIMD 寄存器扩展到 256 bit，计算效率也随之提高一倍。最新的 AVX-512 支持 32 个 512 bit 的 SIMD 寄存器。

图 6-3 是使用 Intel CPU 鉴定程序对笔者所用笔记本电脑 CPU 的查询结果，可以看到 SSE 和 AVX（Advanced Vector Extensions）均支持。

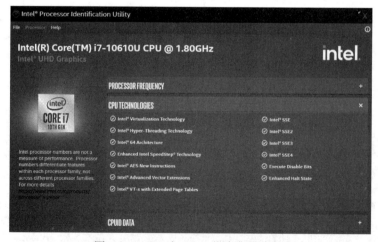

图 6-3　CPU 对 SIMD 指令集的支持

以下是针对 Oracle 公有云中基于 AMD CPU 的计算实例的查询结果，输出显示从 SSE 到最新的 AVX512 均支持。

```
$ cat /proc/cpuinfo|tr " " "\n"|egrep -i "sse|avx"|sort|uniq
avx
avx2
avx512bw
avx512cd
avx512dq
avx512f
avx512vl
sse
sse2
sse4_1
sse4_2
ssse3
```

图 6-4 演示了对于一组数值乘以 3 的操作，分别采用向量处理和标量处理时的处理过程。在标量处理中，每次只从内存中加载一个值，然后循环处理下一个值，总共需要加载 4 次、运算 4 次和保存 4 次。假设每一个数值的宽度是 32 bit，SIMD 寄存器的宽度是 128 bit。SIMD 向量处理一次可以同时加载 4 个值，使用一个指令就可以完成乘法运算。加载、运算和保存均只需要一次，效率提高了 4 倍。

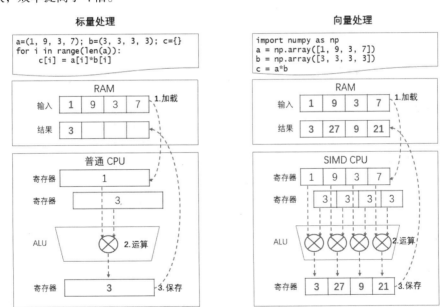

图 6-4 SIMD 标量处理与向量处理

SQL 语句中的比较操作、条件分支操作、聚合操作和 GROUP BY 操作，都可以很方便地转换为 SIMD 操作，并下推到数据扫描时一起执行。图 6-5 演示了将一个 SQL 语句 WHERE 条件转换为 SIMD 操作的过程。此 SQL 语句 WHERE 条件中的两个比较操作和一个逻辑操作，均被转换为 SIMD 向量处理，每一个 CPU 指令都可以处理多个值。Oracle 数据库中单精度浮点数的宽度是 32 bit，双精度浮点数的宽度是 64 bit。假设 SIMD 寄存器的宽度为 512 bit，那么一个 SIMD 指令就可以处理 16 个单精度或 8 个双精度浮点数。结合多核处理器和多线程处理，实现在 Database In-Memory 白皮书中所宣称的每秒 10 亿行的扫描速度并不是一件难事。

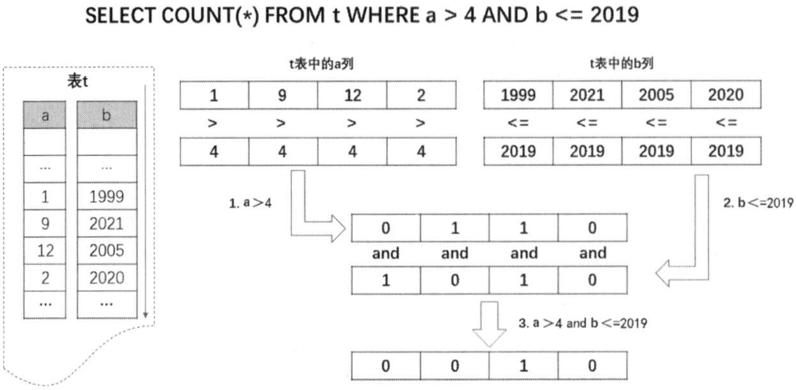

图 6-5　SQL 操作到 SIMD 向量操作的转换

SIMD 非常适合针对大规模数据集进行批量计算。SIMD 指令集支持丰富的操作，包括转换、数学运算、逻辑比较和模式匹配等。同时，应用软件需要经过特殊的编译才能调用 SIMD 指令。实际上，Oracle 数据库是支持 SIMD 的。在下例中，我们可以看到 Oracle 数据库进程使用了支持 AVX 的动态链接库：

```
$ ps -ef|grep ora_smon|grep -v grep
oracle      1063     1  0 08:55 ?        00:00:00 ora_smon_ORCLCDB

$ pmap -x 1063|grep avx
00007f744fb6d000     3044     328       0 r-x-- libshpkavx219.so
00007f744fe66000     2044       0       0 ----- libshpkavx219.so
00007f7450065000      132     124     124 r---- libshpkavx219.so
00007f7450086000        4       4       4 rw--- libshpkavx219.so

$ cd $ORACLE_HOME/lib
$ ls *.so|egrep "avx|sse"
libmkl_avx2.so
libmkl_avx512_mic.so
libmkl_avx512.so
libmkl_avx.so
libmkl_vml_avx2.so
libmkl_vml_avx512_mic.so
libmkl_vml_avx512.so
libmkl_vml_avx.so
libshpkavx19.so
libshpkavx219.so
libshpkavx51219.so
libshpksse4219.so
```

由于数据分析涉及大量重复的扫描和过滤操作，内存纯列式格式连续的数据存放非常适合于顺序处理和批量处理，一些在行存储中变长的数据类型经内存列压缩后可转换为等宽的编码。

这些都使得 Database In-Memory 可以很好地与 SIMD 向量处理技术相结合。事实上，除了结合 SQL 谓词的扫描和过滤重度使用 SIMD 操作外，压缩与解压，以及后续介绍到的在聚合中使用的 Vector Group By、In-Memory Optimized Arithmetic、In-Memory JSON 和 In-Memory Vector Join，无一例外都使用了 SIMD 技术。正是由于软件和硬件的紧密结合，使得硬件能力可以被充分利用，这也为 Database In-Memory 在海量数据下的实时分析提供了性能保障。

在 Oracle 19c 版本，和 SIMD 相关的系统统计参数共有 21 个：

```
SQL> SELECT name FROM v$statname WHERE NAME LIKE 'IM simd%';

NAME
----------------------------------------------------------------
IM simd compare calls
IM simd decode calls
IM simd rle burst calls
IM simd set membership calls
IM simd bloom filter calls
IM simd xlate filter calls
IM simd decode unpack calls
IM simd decode symbol calls
IM simd compare selective calls
IM simd decode selective calls
IM simd rle burst selective calls
IM simd set membership selective calls
IM simd bloom filter selective calls
IM simd xlate filter selective calls
IM simd decode unpack selective calls
IM simd decode symbol selective calls
IM simd compare HW offload calls
IM simd decode HW offload calls
IM simd rle burst HW offload calls
IM simd set membership HW offload calls
IM simd decode unpack HW offload calls

21 rows selected.
```

Oracle 21c 版本，新增了 9 个 SIMD 相关系统统计参数，其中 7 个包含"IM simd KV"的参数与深度向量化功能有关，深度向量化将在第 7 章详细介绍。

```
IM simd KV add calls
IM simd KV add rows
IM simd KV probe calls
IM simd KV probe keys
IM simd KV probe rows
IM simd KV probe serial_buckets
IM simd KV probe chain_buckets
IM simd hash calls
IM simd hash rows
```

6.4 数据压缩

Oracle 数据库有着悠久的数据压缩技术支持历史。最早对压缩技术的支持可以追溯到 9iR2 版本推出的基本表压缩，随后在 11gR1 版本推出了高级行压缩功能，并成为高级压缩选件的一部分。2008 年，随着 Exadata 数据库云平台的推出，Oracle 开始支持混合列压缩技术。2014 年，Oracle 在 Database In-Memory 选件中支持内存纯列式压缩。Oracle 数据库支持丰富的数据压缩算法和压缩级别，可以从行、列、表、表分区和表空间多个层面进行压缩，可以支持磁盘、内存和网络传输数据的压缩。表 6-1 列出了 Oracle 支持的主要压缩方法及其特征。

表 6-1　Oracle 数据库支持的压缩方法及比较

压缩类型	压缩比	CPU 开销	应用	CREATE/ALTER TABLE 语法
基本表压缩	一般	低	DSS	ROW STORE COMPRESS [BASIC]
高级行压缩	一般	低	OLTP，DSS	ROW STORE COMPRESS ADVANCED
混合列压缩	非常高	高	DSS，归档	COLUMN STORE COMPRESS FOR …
内存列压缩	很高	最高	OLTP，DSS	MEMCOMPRESS FOR …

众所周知，压缩可以节省磁盘空间。对于 Oracle 数据库而言，压缩节省的空间并不限于数据库本身，还包括数据库备份、数据库复制和网络传输数据等。除了空间节省，Oracle 数据库使用压缩技术的另一个好处是提升性能，包括查询分析的性能、备份恢复的性能、网络传输的性能。尽管压缩会在数据加载和执行 DML 操作时消耗额外的 CPU 进行压缩和解压，但空间和数据传输的节省，以及内存中可以缓存更多的数据带来的磁盘 I/O 减少，远远抵消了压缩带来的开销。另外非常重要的一点，所有的压缩技术对于应用而言都是完全透明的，也就是说，应用无须做任何修改。

6.4.1　行级压缩

基本表压缩或标准压缩，是 Oracle 最早支持的行级压缩技术，通过消除数据库块中的重复值来压缩数据。基本表压缩仅适用于直接路径操作，例如 CTAS（CREATE TABLE AS SELECT）或 IAS（INSERT AS SELECT）。如果使用 UPDATE 对数据进行更新，则该数据库块中的数据将被解压缩以进行修改，并以未压缩的方式写回到磁盘上。基本压缩只适合于数据相对静止的情形，如数据仓库环境或 OLTP 环境下的"非活动"表分区。

以下为使用 CTAS 创建基本压缩表和对已有表进行基本表压缩的命令示例：

```
SQL> CREATE TABLE t1 COMPRESS BASIC AS SELECT * FROM user_tables;
Table created.

SQL> ALTER TABLE t1 MOVE COMPRESS BASIC;
Table altered.
```

基本表压缩只能在数据加载等直接路径操作时进行压缩，可用的场景非常有限。Oracle 在 11gR1 版本推出了高级行压缩，允许数据在所有类型的数据操作（如 INSERT 和 UPDATE 之类的常规 DML）时进行压缩，因此也被称为 OLTP 压缩。高级行压缩和高级索引压缩、RMAN 备份压缩、高级网络压缩、Data Pump 导出数据压缩和自动数据优化等功能均为 Oracle 高级压缩选件的一部分。

高级行压缩在节省空间的同时，还可以提升数据读取的速度。这是由于其采用了一种独特的专门用于 OLTP 应用程序的压缩算法，该算法通过消除数据库块中跨多个列的重复值来发挥作用。如图 6-6 所示，在每一个数据块的表头部分，包含一种称为符号表的结构，用于维护压缩元数据。压缩数据块时，会先将重复值的单个副本添加到符号表，然后建立与一个更简短的符号之间的对应关系。之后，每个重复值将替换为符号表中相应的条目，从而实现空间的节省。由于压缩数据和元数据均包含在数据块中，和其他维护全局符号表的压缩算法相比，这种自包含的结构减少了在数据访问时引发的额外 I/O。另外，数据块在内存缓存中可以保持其压缩格式，Oracle 可以直接读取这些未经解压的数据，数据访问时读取的数据量更少，内存缓存中可容纳的数据块更多，相应地缓存命中率也更高。所有这些 I/O 优化使得在读取数据时不仅没有负面影响，反而可以得到性能的提升。

ID	Shape	Color
1	Triangle	Red
2	Circle	Yellow
3	Triangle	Blue
4	Circle	Yellow

（a）表的逻辑结构

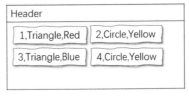

（b）表的物理结构（未压缩）

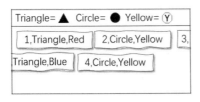

（c）建立符号表

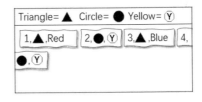

（d）表的物理结构（未压缩）

图 6-6 使用符号表实现高级行压缩

高级行压缩使用另一套机制来减少写入时的性能开销，我们称之为批处理压缩。也就是说，并非在每次写操作时都压缩数据，而是在块中的数据超过内部设定的阈值时进行批处理压缩。因此，大部分的写操作的性能和未采用高级行压缩时是一样的，只有少数几个引发批处理压缩的操作才会引发额外的性能开销。如图 6-7 所示，最初插入的数据会保持未压缩状态，就和在未采用压缩的表上操作一样。直到数据块中的数据达到内部设定阈值时，才会触发批处理操作压缩数据块中的所有内容。由于压缩后空闲空间再次低于阈值，随后插入的数据仍保持未压缩状态，直到再次达到阈值时触发下一轮批处理压缩操作。该过程会反复进行，直到此数据块数据占满为止，除非删除操作释放出额外的剩余空间。更新操作与插入操作类似，也不会立即进行压缩。在更新时，未涉及的列无须解压，这些列仍保持其压缩格式并可使用符号表进行读取。

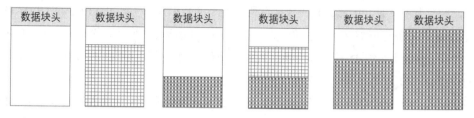

图 6-7　采用批处理压缩减少写操作性能影响

以上措施使得高级行压缩在节省空间外，还可以同时满足 OLTP 应用对性能的要求。引用 Oracle 技术白皮书[①]中的数据，和未采用压缩相比，高级行压缩在表扫描时的性能可提升 2.5 倍，DML 操作的性能开销低于 3%。

对于已有的表，可以采用 3 种方法启用高级行压缩。

（1）ALTER TABLE … ROW STORE COMPRESS ADVANCED，表中现有数据保持未压缩状态，未来的 DML 操作将启用高级行压缩。

（2）ALTER TABLE … MOVE ROW STORE COMPRESS ADVANCED，立即压缩现有表中数据，并为未来的 DML 操作启用高级行压缩。在启用压缩过程中，表可以被读取，但所有的 DML 操作将被阻塞。

（3）在线重定义（DBMS_REDEFINITION），和（2）相同，但在启用压缩过程中允许数据读写，这是对业务影响最小的压缩方式。

以下示例演示了通过在线重定义启用高级行压缩的过程。首先创建一个 100 万行的测试表：

```
DROP TABLE t1;

CREATE TABLE t1(
    id      INT,
    singer  VARCHAR(16)
);

CREATE OR REPLACE PROCEDURE filltab(
    p_rows NUMBER DEFAULT 10000
)AS
    TYPE list_of_singers_t IS
        TABLE OF VARCHAR2(16);
    singers list_of_singers_t := list_of_singers_t();
BEGIN
    singers.extend(15);
    singers(1) := 'Michael Jackson';
```

---

① https://www.oracle.com/cn/a/tech/docs/technical-resources/advanced-compression-wp-12c.pdf

```
        singers(2) := 'Taylor Swift';
        singers(3) := 'Mariah Carey';
        singers(4) := 'Josh Turner';
        singers(5) := 'Andrea Bocelli';
        singers(6) := 'Celine Dion';
        singers(7) := 'Whitney Houston';
        singers(8) := 'Alan Walker';
        singers(9) := 'Avril Lavigne';
        singers(10) := 'Backstreet Boys';
        singers(11) := 'Air Supply';
        singers(12) := 'Roxette';
        singers(13) := 'Carpenters';
        singers(14) := 'Bob Dylan';
        singers(15) := 'Declan Galbraith';
        FOR i IN 1..p_rows LOOP
            INSERT INTO t1 VALUES(
                i,
                singers(floor(dbms_random.value(1, 16)))
            );

            IF MOD(i, 1000) = 0 THEN
                COMMIT;
            END IF;
        END LOOP;

        COMMIT;
END;
/

EXEC filltab(1000000);
```

然后为此表创建主键,以便后续验证索引是否也能被迁移。另外,在线重定义默认的迁移方法需要使用主键。如果没有主键,也可以使用 ROWID。

```
SQL> ALTER TABLE t1 ADD PRIMARY KEY(id);
Table altered.

-- 验证目前表处于未压缩状态
SQL> SELECT table_name, compression, compress_for FROM user_tables WHERE table_name IN
('T1');

TABLE_NAME           COMPRESS COMPRESS_FOR
-------------------- -------- ------------------------------
T1                   DISABLED
```

创建空的临时表 t1_tmp,并启用高级行压缩。在线重定义要求使用临时表作为过渡:

```
CREATE TABLE t1_tmp ROW STORE COMPRESS ADVANCED AS SELECT * FROM t1 WHERE 1=2;
```

以下为在线重新定义的过程:

```
-- 确认源表是否满足在线重定义的要求
SQL> EXEC dbms_redefinition.can_redef_table('ssb', 't1');
PL/SQL procedure successfully completed.
```

```
-- 启动在线重定义，在迁移过程中，源表处于在线状态，可以读写
SQL> EXEC dbms_redefinition.start_redef_table('ssb', 't1', 't1_tmp');
PL/SQL procedure successfully completed.

-- 迁移表中的索引
SQL>
DECLARE
  num_errs PLS_INTEGER;
BEGIN
    dbms_redefinition.copy_table_dependents('ssb', 't1', 't1_tmp',
    dbms_redefinition.cons_orig_params,TRUE, TRUE, TRUE, TRUE, num_errs);
END;
/
PL/SQL procedure successfully completed.

-- 同步迁移过程中可能发生的数据更改
SQL> EXEC dbms_redefinition.sync_interim_table('ssb', 't1', 't1_tmp');
PL/SQL procedure successfully completed.

-- 结束在线重定义过程
SQL> EXEC dbms_redefinition.finish_redef_table('ssb', 't1', 't1_tmp');
PL/SQL procedure successfully completed.
```

最后，通过查询系统表，可以验证高级行压缩成功启用，索引也随之迁移，实现的压缩比为 2。

```
-- 验证压缩效率
SQL> SELECT segment_name, bytes FROM user_segments WHERE segment_name IN ('T1', 'T1_TMP');

SEGMENT_NAME           BYTES
-------------------- ----------
T1                   13631488
T1_TMP               27262976

-- 确认高级行压缩已启动
SQL> SELECT table_name, compression, compress_for FROM user_tables WHERE table_name IN ('T1', 'T1_TMP');

TABLE_NAME           COMPRESS  COMPRESS_FOR
-------------------- --------  --------------------
T1                   ENABLED   ADVANCED
T1_TMP               DISABLED

-- 验证索引已迁移
SQL> SELECT index_name, table_name, uniqueness FROM user_indexes WHERE table_name IN ('T1', 'T1_TMP');

INDEX_NAME           TABLE_NAME           UNIQUENES
-------------------- -------------------- ---------
SYS_C008061          T1                   UNIQUE
```

```
TMP$$_SYS_C0080610    T1_TMP                    UNIQUE
-- 删除用于过渡的临时表
SQL> DROP TABLE t1_tmp;
Table dropped.
```

6.4.2 混合列压缩

混合列压缩（Hybrid Columnar Compression，HCC）是 Exadata 等工程化系统独有的功能。顾名思义，混合列压缩采用行列结合的方式来组织数据。这种格式具有比行格式更高的压缩比，相较纯列格式又可以减少在单行数据访问时的性能下降。混合列压缩主要针对数据仓库类分析型负载或数据归档应用，适合于静态或不太活跃的数据。

混合列压缩使用称为压缩单元（compression unit）的逻辑结构来存储数据。如图 6-8 所示，压缩单元由一组行集合组成，包含这些行中的所有列。和纯列压缩格式只包含列不同，混合列压缩的压缩单元同时包含行和列，这也是其被称为"混合"的原因。压缩单元对应于一组数据块，多行中相同的列被组织到一起压缩存放，并有可能跨数据块。由于具有相同数据类型和相似特征的列组织在一起，因此可以实现较高的压缩比。

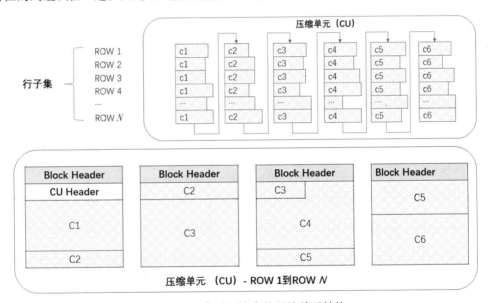

图 6-8　混合列压缩中的压缩单元结构

混合列压缩共支持 4 种压缩级别，适用于数据仓库查询分析和历史数据归档两类场景。混合列压缩是与存储相关的数据库特性，需要底层存储系统的支持。最早支持此特性的是 2008 年推出的 Exadata 数据库云平台。由于 Exadata 特殊的工程整合设计，可以利用其存储服务器上的处理能力、内存和高速网络来加速数据仓库查询。之后混合列压缩在 ZFS、Axiom 和 FS1 存储系统上也提供支持，这些系统主要用于数据归档。在 Oracle 公有云中，数据库云服务高性能版和数据库云服务超强性能版也可以支持混合列压缩。

以查询 LINEORDER 表为例，表 6-2 列出了各种混合列压缩类型在空间和查询性能方面的比较结果。可以看到，随着混合列压缩级别的提高，压缩比和压缩时间也不断增加。一个有趣的现象是，高压缩级别的查询速度比无压缩时要快，而消耗的 CPU 除最高的 ARCHIVE HIGH 类型外，均比无压缩时要少。另外，查询引发的逻辑读更少，也就是说 I/O 更少。以 QUERY LOW 压缩类型为例，和未压缩时相比，其压缩比约为 3，CPU 消耗下降一半，而查询速度提高近 3 倍。

表 6-2 混合列压缩不同压缩类型比较—查询负载

压缩类型	压缩时间/秒	压缩比	查询 CPU	查询时间/秒	查询逻辑读
无压缩	N/A	1	814	11.06	333824
QUERY LOW	67	2.947	422	03.60	202078
QUERY HIGH	115	4.421	440	03.16	144224
ARCHIVE LOW	141	4.667	410	02.93	89127
ARCHIVE HIGH	240	5.793	1050	05.92	68865

针对更新负载，混合列压缩的表现又如何呢？以 LINEORDER 表为例，批量更新其中某一列（近 40 万行，占比 3.3%），测试结果如表 6-3 所示。很明显，压缩级别越高，更新消耗的 CPU 越大。但同时也看到，有压缩时的更新时间除 ARCHIVE HIGH 外，均比未压缩时要低。另外，在更新后，所有采用混合列压缩的表压缩比均有不同程度的下降，也就是其空间占用更大。稍后会解释产生此现象的原因。

表 6-3 混合列压缩不同压缩类型比较—更新负载

压缩类型	更新后压缩比	原压缩比	空间增长/%	更新 CPU	更新时间/秒	逻辑读
无压缩	0.997	1	0.30	1645	52.45	1244135
QUERY LOW	2.847	2.947	3.51	2268	31.30	1072155
QUERY HIGH	4.097	4.421	7.91	3085	37.24	1005119
ARCHIVE LOW	4.308	4.667	8.33	2390	27.49	988646
ARCHIVE HIGH	5.25	5.793	10.34	7574	81.26	965235

混合列压缩只在数据以直接路径加载时生效，如 IMPDP 直接路径导入，直接路径 INSERT 等。如果对启用混合列压缩的表中某行数据进行更改，这行数据可能会被迁移出压缩单元，并使用更低的压缩级别，如高级行压缩或无压缩。此种混合列压缩和其他压缩级别共存的情况会在数据访问时引发额外的开销，并且被迁出的行不会自动回迁到压缩单元。为减少性能的影响，需要手动定期将这些行数据回迁，使其重新使用混合列压缩。

再回到前面的更新负载测试，由于满足条件的行采用常规路径更新，因此无法使用混合列压缩。这些行从压缩单元中迁出，并采用更低的压缩级别甚至无压缩，最终导致了 0.3%～10% 的空间增长。

通过一个示例了解修改行数据导致迁移和后续回迁的过程：

```
-- 创建测试表，启用混合列压缩，使用直接路径插入数据
SQL> CREATE TABLE t1 COMPRESS FOR QUERY LOW AS SELECT * FROM all_objects;
Table created.
```

```
-- 任选一行用于后续测试，记录其 ROWID
SQL> SELECT ROWID FROM t1 WHERE ROWNUM = 1;

ROWID
------------------
AAASORAAMAABYK7AAA

-- 获取压缩类型，8 表示 COMP_QUERY_LOW，即混合列压缩中用于查询的低压缩级别
SQL> SELECT dbms_compression.get_compression_type ('SSB', 'T1', 'AAASORAAMAABYK7AAA')
complevel FROM dual;

 COMPLEVEL
----------
         8

-- 记录其数据块 ID
SQL> SELECT dbms_rowid.rowid_block_number('AAASORAAMAABYK7AAA') AS blkid FROM dual;

     BLKID
----------
    361147

-- 修改此行，并提交
SQL> UPDATE t1 SET object_name = 'ABCDEDFH' WHERE ROWID = 'AAASORAAMAABYK7AAA';
1 row updated.

SQL> COMMIT;
Commit complete.

-- 被修改的行 ROWID 发生变化，表明发生了行迁移，记录其新的数据块 ID
SQL> SELECT ROWID FROM t1 WHERE object_name = 'ABCDEDFH';

ROWID
------------------
AAASORAAMAABYR5AAA

-- 此行的当前压缩类型变为了 1，即 COMP_NOCOMPRESS，表示没有压缩
SQL> SELECT dbms_compression.get_compression_type ('SSB', 'T1', 'AAASORAAMAABYR5AAA')
complevel FROM dual;

 COMPLEVEL
----------
         1

-- 重新将所有的数据置为混合列压缩
SQL> ALTER TABLE t1 MOVE COMPRESS FOR QUERY LOW;
Table altered.

-- 被修改行的 ROWID 再次发生变化
SQL> select rowid from t1 where object_name = 'ABCDEDFH';
```

```
ROWID
------------------
AAASOSAAMAABYYJAFN

-- 确认此行又重新回到混合列压缩级别
SQL> SELECT dbms_compression.get_compression_type ('SSB', 'T1', 'AAASOSAAMAABYYJAFN')
complevel FROM dual;

 COMPLEVEL
----------
         8
```

6.4.3　内存列压缩

Oracle 数据库在 12.1.0.2 版本推出 Database In-Memory 选件，首次使用了内存列压缩技术。和传统的压缩技术不同，内存列压缩除了可以节省昂贵的内存空间外，更侧重于性能的提升，也就是在内存列上的扫描速度。

这一点可以从内存列压缩与混合列压缩的扫描效率比较中得到验证。Database In-Memory 和混合列压缩中都有压缩单元的概念。在组织形式上，Database In-Memory 采用的是纯列式，而混合列压缩是行列混合的格式。另一方面，由于有内存性能做保证，Database In-Memory 采用了比混合列压缩更大的压缩单元尺寸，也就是在一个压缩单元内可以存放更多行的数据。从这两点来看，内存列压缩的压缩效率理应高于混合列压缩。但从第 5 章给出的压缩建议器的测试结果看，内存列压缩的压缩比要略低于混合列压缩。这是由于内存列压缩没有像混合列压缩那样选择更为激进的算法，这也是其为保证扫描性能所做的折衷。

从低压缩级到高压缩级，混合列压缩依次采用了 LZO、ZLIB 和 BZIP 等多种算法。LZO 是 Lempel-Ziv-Oberhumer 的缩写，此算法由两位以色列计算机科学家 Abraham Lempel 和 Jacob Ziv 发明，由奥地利的开发者 Markus F. X. J. Oberhumer 实现。LZO 算法的压缩和解压速度快，但压缩效率不高。ZLIB 的作者是法国人 Jean-loup Gailly 和美国人 Mark Adler，他们分别实现了压缩和解压算法，并凭借此算法获得了 USENIX 2009 STUG 大奖。ZLIB 已成为事实上的业界标准，对各种数据均能提供很好的压缩效果，同时消耗很少的系统资源。BZIP 算法由英国人 Julian Seward 发明，此算法压缩速度较慢，但压缩效率非常高。

通过以下 Linux 命令，可以了解这 3 种算法的压缩效率。从结果可以看出，按照执行顺序，其压缩时间越来越长，但压缩效率越来越高。

```
# LZO 压缩算法
$ time lzop part.tbl
real    0m0.221s
user    0m0.188s
sys     0m0.032s

# ZLIB 压缩算法
$ time gzip part.tbl
real    0m2.405s
```

```
user    0m2.376s
sys     0m0.027s

# BZIP 压缩算法
$ time bzip2 part.tbl
real    0m5.695s
user    0m5.646s
sys     0m0.035s

$ ls -lhS
total 138M
-rw-r--r--. 1 oracle oinstall 83M  Mar 28 21:14 part.tbl
-rw-r--r--. 1 oracle oinstall 31M  Mar 28 19:15 part.tbl.lzo
-rw-r--r--. 1 oracle oinstall 16M  Mar 28 21:16 part.tbl.gz
-rw-r--r--. 1 oracle oinstall 9.2M Mar 28 19:15 part.tbl.bz2
```

内存列压缩采用了一系列为查询性能优化的压缩算法，包括字典编码、RLE（游程编码）和 Oracle 拥有专利的 OZIP 算法。混合列压缩的压缩效率虽然很高，但是数据必须先解压才能进行查询和修改。内存列压缩中在部分压缩级别下（如 MEMCOMPRESS FOR QUERY）可以直接在压缩格式下进行扫描，这其中最重要的原因是因为采用了字典编码。字典编码是列式数据库中最为常用的压缩算法，其满足了数据库压缩所期望的 3 个目标：

（1）无损压缩，和图形图像压缩不同，数据库压缩要求能完整还原，不允许有信息丢失。

（2）将变长的数据转换为定长的数据，定长等宽的数据便于循环批处理，能高效地与 SIMD 向量处理结合。

（3）延迟物化，尽可能晚地解压数据，直到客户端需要展示这些数据。延迟物化是列式数据库常用的性能优化技术，指以数字编码替代实际的列值，这些编码在查询中可以直接使用，并尽可能晚地翻译为原始的列值。

字典编码的原理如图 6-9 所示。压缩前的源表包括变长的 Song 列和定长的 Date 列。字典编码经过 3 个步骤为每一列生成本地字典。

（1）提取每列的唯一值集合。

（2）对这些唯一值进行排序。

（3）对排序后的值进行编码。实际使用的编码值与图 6-9 示例中采用的可能有所不同，但目标都是用等宽的、占用空间更少的编码来替代原始数据。

字典编码使得数据无须解压即可进行查询操作。例如寻找歌名为 Close to You 的等值查询可以替换为寻找压缩值等于 1 的记录，又例如寻找日期在 1971-05-14 和 1973-09-17 之间的范围查询可以替换为寻找压缩值在 1 和 3 之间的记录。另外，如果想知道一共有多少首不同的歌曲，只需要直接查询本地字典即可。还有一点需要强调的是，每一列都拥有自己的本地字典，即使对于同一列，不同 IMCU 中的本地字典也是不一样的。因此数据的解压是自包含的，无须涉及其他的列和其他的 IMCU。

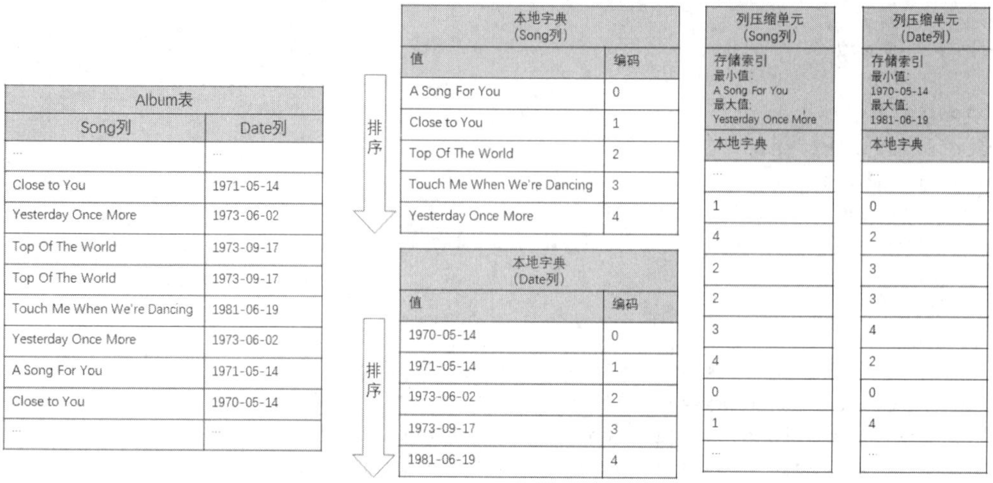
Album表
Song列
Date列
...
...
Close to You
1971-05-14
Yesterday Once More
1973-06-02
Top Of The World
1973-09-17
Top Of The World
1973-09-17
Touch Me When We're Dancing
1981-06-19
Yesterday Once More
1973-06-02
A Song For You
1971-05-14
Close to You
1970-05-14
...
...
排序
本地字典
(Song列)
值
编码
A Song For You
0
Close to You
1
Top Of The World
2
Touch Me When We're Dancing
3
Yesterday Once More
4
排序
本地字典
(Date列)
值
编码
1970-05-14
0
1971-05-14
1
1973-06-02
2
1973-09-17
3
1981-06-19
4
列压缩单元
(Song列)
存储索引
最小值:
A Song For You
最大值:
Yesterday Once More
本地字典
...
1
4
2
2
3
4
0
1
...
列压缩单元
(Date列)
存储索引
最小值:
1970-05-14
最大值:
1981-06-19
本地字典
...
0
2
3
3
4
2
0
4
...

（a）压缩前　　（b）本地字典　　（c）压缩后

图 6-9　字典编码压缩示意图

字典编码的压缩效率与基数关系密切，基数是指一组数据中不同值的数量。例如阿拉伯数字的基数是 10（0 ~ 9），英文字母的基数为 26（A ~ Z）。基数越小，表示数据的重复度越高，压缩的效率也越高。另外，基数越小，编码表的长度和编码的宽度越小。例如阿拉伯数字只需要 4bit（2^4>10）进行编码，英文字母则需要 5bit 进行编码（2^5>26）。

通过实验来对比一下不同基数下的压缩效率。首先通过脚本建立两张测试表 T1 和 T2，记录数为 500 万行。它们均只有一个宽度为 16 字节的列，并且每一行都用 16 字节的字符串填满。因此在未压缩前，这两个表占用的空间完全相同。

```
CREATE TABLE t1(a VARCHAR(16));
CREATE TABLE t2(a VARCHAR(16));

BEGIN
   FOR i IN 1..&&rowcnt LOOP

      INSERT INTO t1 VALUES(
         decode(floor(dbms_random.value(1, 11)),
         1, '1111111111111111',
         2, '2222222222222222',
         3, '3333333333333333',
         4, '4444444444444444',
         5, '5555555555555555',
         6, '6666666666666666',
         7, '7777777777777777',
         8, '8888888888888888',
         9, '9999999999999999',
         10, '0000000000000000')
      );

   END LOOP;
```

```
    COMMIT;
END;
/
BEGIN
    FOR i IN 1..&&rowcnt LOOP
        INSERT INTO t2 VALUES(dbms_random.string('p', 16));
    END LOOP;

    COMMIT;
END;
/
```

通过以下脚本可以查询到表 T1 的基数为 10，而表 T2 的基数较高，为 500 万：

```
SQL> SELECT COUNT(DISTINCT(A)) AS cardinality FROM t1;

CARDINALITY
-----------
         10

SQL> SELECT COUNT(DISTINCT(a)) AS cardinality FROM t2;

CARDINALITY
-----------
    5000000
```

实际数据的特征对压缩比和发布时间都有明显的影响。从表 6-4 的测试结果来看，具有较低基数的列压缩比较高，为较高基数列的 7~9 倍，发布时间为较高基数列的一半。

表 6-4 不同基数下内存列压缩效率对比

压 缩 级 别	T1 压缩比	T2 压缩比	T1 发布时间/秒	T2 发布时间/秒
QUERY LOW	7.17	0.88	3.01	7.02
QUERY HIGH	7.17	1	3.01	5.01
CAPACITY LOW	11.24	1.18	4	13.01
CAPACITY HIGH	11.69	1.59	3.01	7.01

通过以下的 SQL 脚本可以查看每一列在不同 IMCU 中本地字典的条目数、列的最小和最大值：

```
SET PAGES 9999
COL MIN FOR A18
COL MAX FOR A18
BREAK ON objd SKIP 1 DUPLICATES

SELECT
    a.objd,
    a.num_rows,
    b.dictionary_entries,
    utl_raw.cast_to_varchar2(b.minimum_value)          min,
    utl_raw.cast_to_varchar2(b.maximum_value)          max
```

```
FROM
    v$im_header  a,
    v$im_col_cu  b
WHERE
        a.objd = b.objd
    AND a.imcu_addr = b.head_piece_address
ORDER BY 1;
```

以下为基数较高的表 T2 在默认压缩级 QUERY LOW 时的输出。其中 MIN 和 MAX 为存储索引的信息，DICTIONARY_ENTRIES 表示此列在其 IMCU 中的基数：

```
      OBJD   NUM_ROWS DICTIONARY_ENTRIES MIN                  MAX
---------- ---------- ------------------ -------------------- --------------------
     76267     496320             496320 0001676DSAKPW3JX     ZZZVG0S3GFKSS7XP
     76267     623700             623700 000175M27R5MBJMB     ZZZU000U2MSMGDGP
     76267     665280             665280 0002FLSXJX8Q0G39     ZZZX1FD5RU46ZPAZ
     76267     532790             532790 0004SRIXJEUR067J     ZZZJWF4MLUYND965
     76267     670230             670230 0000XL2AN486S9AZ     ZZZZEXOPQHL8VA41
     76267     673200             673200 00017PUAQY9JFW71     ZZZZ4SY4BGSAAN2J
     76267     673200             673200 00005D62XYAAA93T     ZZZXVY91SRSIL977
     76267     665280             665280 0004J6WPTD29Q0BT     ZZZVXYNDYAFTBM6K

8 rows selected.
```

以下为两个表在最高压缩级 CAPACITY HIGH 时的输出。可以看到，较低基数的表仍然使用了字典编码，而较高基数的表则没有使用（DICTIONARY_ENTRIES 为 0）。这表明，虽然字典编码是压缩级为 FOR QUERY LOW 时默认的压缩算法，但当需要更高的压缩比时，系统会决定是在字典编码的基础上再叠加其他压缩算法还是使用其他的压缩算法替代：

```
      OBJD   NUM_ROWS DICTIONARY_ENTRIES MIN                  MAX
---------- ---------- ------------------ -------------------- --------------------
     76266     558030                 10 0000000000000000     9999999999999999
     76266     280670                 10 0000000000000000     9999999999999999
     76266     673200                 10 0000000000000000     9999999999999999
     76266     673200                 10 0000000000000000     9999999999999999
     76266     630960                 10 0000000000000000     9999999999999999
     76266     665610                 10 0000000000000000     9999999999999999
     76266     665280                 10 0000000000000000     9999999999999999
     76266     623700                 10 0000000000000000     9999999999999999
     76266     229350                 10 0000000000000000     9999999999999999
     76267     673200                  0 00005D62XYAAA93T     ZZZXVY91SRSIL977
     76267     673200                  0 00017PUAQY9JFW71     ZZZZ4SY4BGSAAN2J
     76267     670230                  0 0000XL2AN486S9AZ     ZZZZEXOPQHL8VA41
     76267     665280                  0 0004J6WPTD29Q0BT     ZZZVXYNDYAFTBM6K
     76267     665280                  0 0002FLSXJX8Q0G39     ZZZX1FD5RU46ZPAZ
     76267     623700                  0 000175M27R5MBJMB     ZZZU000U2MSMGDGP
     76267     532790                  0 0004SRIXJEUR067J     ZZZJWF4MLUYND965
     76267     496320                  0 0001676DSAKPW3JX     ZZZVG0S3GFKSS7XP

17 rows selected.
```

除了默认压缩级使用的字典编码压缩，Oracle 还为内存列式存储提供了 RLE 和 OZIP 等压缩算法。图 6-10 为这些算法原理的简单示例。从最左侧开始，首先根据列中的原始值生成字典和编码，字典为列中唯一值的排序集合。然后根据字典在数据区中存放实际的数据，例如最开始两个连续的 Blue 被置换为 1,1。在下一个压缩级 QUERY HIGH，会在字典编码的基础上叠加RLE 压缩算法。RLE 常用于图形图像压缩，它将连续出现的值替换为值和其重复次数。例如两个连续出现的 4 被替换为 4 和 2。显然，连续出现的次数越多，压缩比越高。在下一个压缩级CAPACITY LOW，会在之前两种压缩算法基础上再施加 OZIP 压缩。OZIP 对数据进行模式替换，例如经 RLE 压缩过的数据有多个 1,2,3 模式和 1,2 模式，OZIP 将其替换为 0 和 2。

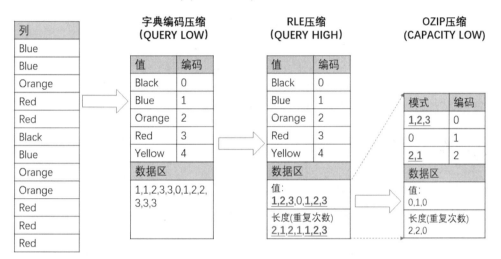

图 6-10　内存列式存储中的压缩算法

在生产环境中，实际的压缩比与采用的压缩算法，数据类型和数据的实际值相关。因此建议先针对不同的压缩级别进行测试，建立比较基线，然后根据性能和内存空间的要求选择最合适的压缩级。可使用第 5 章介绍的 Enterprise Manager 和压缩建议器等工具来完成这些任务。

当表进行发布时，如果由于内存不足导致只有部分数据发布到内存列式存储中，称为不完整发布。本节最后要来回答几个问题：查询是否能利用不完整发布的内存中数据？不完整发布是建议的吗？如果不建议，有什么解决方法？

采用以下的 SQL 进行测试，此查询只涉及到 LINEORDER 这一张表，共 11 997 996 行：
SELECT MAX(lo_ordtotalprice), SUM(lo_quantity)FROM lineorder;

首先在表未发布时进行测试，建立性能基线。此时查询时间为 2.24 秒。由于没有利用到内存列式存储，执行计划显示为全表扫描，并存在物理读。

```
Elapsed: 00:00:02.24
---------------------------------------------------------------------------
| Id  | Operation          | Name      | Rows  | Bytes | Cost (%CPU)| Time     |
---------------------------------------------------------------------------
|   0 | SELECT STATEMENT   |           |       |       | 46772 (100)|          |
|   1 |  SORT AGGREGATE    |           |     1 |     9 |            |          |
|   2 |   TABLE ACCESS FULL| LINEORDER |   11M |  102M | 46772   (1)| 00:00:02 |
---------------------------------------------------------------------------
```

```
NAME                                                          VALUE
------------------------------------------------------- -----------
CPU used by this session                                         73
physical reads                                               166680
session logical reads                                        166774
```

在内存足够时,将表 LINEORDER 进行完整发布。使用以下 SQL 查询发布状态,从输出可知,LINEORDER 表在磁盘中空间占用为 1343.88MB,在内存列式存储中占用 565.56MB。发布状态为 COMPLETED,未发布数据(UNPOPED_MB)为 0 表示完整发布:

```
SQL>
SELECT
    POPULATE_STATUS as POP_STATUS,
    inmemory_compression,
    round(bytes/1024/1024,2) as "DISK_SIZE(MB)",
    round(BYTES_NOT_POPULATED/1024/1024,2) as UNPOPED_MB,
    round(inmemory_size/1024/1024,2) as "IM_SIZE(MB)",
    round(bytes / inmemory_size,2) AS compress_ratio
FROM
    v$im_segments
WHERE
segment_name = 'LINEORDER';


POP_STATUS    INMEMORY_COMPRESS DISK_SIZE(MB) UNPOPED_MB IM_SIZE(MB) COMPRESS_RATIO
------------- ----------------- ------------- ---------- ----------- --------------
COMPLETED     FOR QUERY LOW          1343.88           0      565.56           2.38
```

在完整发布后性能测试结果如下。性能统计信息中的 IM scan rows 表示从内存中扫描的行数,其值为 11 997 996,与表的总记录数一致,侧面说明了整个表都已发布到内存中。查询速度为 0.02 秒,比未发布时提升了 112 倍。

```
Elapsed: 00:00:00.02
---------------------------------------------------------------------------
| Id | Operation                   | Name      | Rows | Bytes | Cost (%CPU)|
---------------------------------------------------------------------------
|  0 | SELECT STATEMENT            |           |      |       | 1848 (100) |
|  1 |  SORT AGGREGATE             |           |    1 |     9 |            |
|  2 |   TABLE ACCESS INMEMORY FULL| LINEORDER |  11M |  102M | 1848   (7) |
---------------------------------------------------------------------------


NAME                                                          VALUE
------------------------------------------------------- -----------
CPU used by this session                                          4
IM scan rows                                               11997996
session logical reads                                        172106
session logical reads - IM                                   172016
```

将 INMEMORY_SIZE 下调到 500MB,此时的内存不足以容纳需求为 565.56MB 的 LINEORDER 表。重新发布 LINEORDER 表,虽然发布状态为 COMPLETED,但输出显示还有 172.51MB 数据(约 13%)未发布到内存中,表明这是一个不完整发布。

```
POP_STATUS    INMEMORY_COMPRESS DISK_SIZE(MB) UNPOPED_MB IM_SIZE(MB) COMPRESS_RATIO
```

```
------------- ----------------- ------------- ---------- ------------ -------------
COMPLETED     FOR QUERY LOW         1343.88      172.51       491.19         2.74
```

不完整发布时的性能测试结果如下。此时执行计划中仍然出现 TABLE ACCESS INMEMORY FULL。这也说明此标识只能说明使用了内存列式存储，并不表示数据完全在内存中。执行统计信息中的 table scan disk IMC fallback 表示由于数据在内存列式存储中不存在，从而去往 Buffer Cache 中读取的行数，约为总行数的 15%。虽然完整发布时的查询速度比不完整发布时快了 8 倍，但不完整发布仍比未发布时快了 14 倍。

```
Elapsed: 00:00:00.16

--------------------------------------------------------------------------------
| Id  | Operation                    | Name      | Rows  | Bytes | Cost (%CPU)|
--------------------------------------------------------------------------------
|  0  | SELECT STATEMENT             |           |       |       | 7522 (100) |
|  1  |  SORT AGGREGATE              |           |    1  |    9  |            |
|  2  |   TABLE ACCESS INMEMORY FULL | LINEORDER |  11M  |  102M | 7522   (3) |
--------------------------------------------------------------------------------


NAME                                                                VALUE
------------------------------------------------------------    ------------
CPU used by this session                                                  17
IM scan rows                                                        10206080
session logical reads                                                 168170
session logical reads - IM                                            142895
table scan disk IMC fallback                                         1810348
```

尽管不完整发布时的查询性能要优于完全不发布，但仍应尽量避免。因为其可能会带来复杂性和不确定性。避免的方法包括添加物理内存，或者在查询性能可以接受的前提下调高压缩级。例如，通过将压缩级从默认的 QUERY LOW 调整为最高压缩级 CAPACITY HIGH，可以实现对象的完整发布。

```
POP_STATUS    INMEMORY_COMPRESS DISK_SIZE(MB) UNPOPED_MB IM_SIZE(MB) COMPRESS_RATIO
------------- ----------------- ------------- ---------- ------------ -------------
COMPLETED     FOR CAPACITY HIGH      1343.88           0      262.56           5.12
```

从性能测试结果可知，调高压缩比后，查询性能比默认压缩级完整发布时仅慢 0.1 秒，但比不完整发布仍快 0.04 秒。

```
Elapsed: 00:00:00.12

--------------------------------------------------------------------------------
| Id  | Operation                    | Name      | Rows  | Bytes | Cost (%CPU)|
--------------------------------------------------------------------------------
|  0  | SELECT STATEMENT             |           |       |       | 1864 (100) |
|  1  |  SORT AGGREGATE              |           |    1  |    9  |            |
|  2  |   TABLE ACCESS INMEMORY FULL | LINEORDER |  11M  |  102M | 1864   (8) |
--------------------------------------------------------------------------------


NAME                                                                VALUE
------------------------------------------------------------    ------------
```

```
CPU used by this session                              15
IM scan rows                                    11997996
session logical reads                             172106
session logical reads - IM                        172016
```

6.5 操作下推

传统的 SQL 查询将处理层和数据层分开。查询数据时，通常的执行逻辑是将数据推向处理层，即将数据读取回来后进行处理。对于分析型系统，这可能会导致在两层之间大量的数据传输，从而造成性能瓶颈。通过一些数据库优化技术可以减少数据的传输，如索引可以直达所需数据，分区可以排除不需要的数据。但对于行列过滤、联结和聚合等操作，仍需要将数据取回后进行处理。

操作下推（Pushdown），是指将过滤、聚合等操作下放到数据层执行，以尽可能减少向上一层返回的数据。如图 6-11 所示，由于返回的数据减少，网络资源占用更少，网络传输速度加快，上一层处理时间减少，因此整个处理的性能得以提升。加之 Database In-Memory 具有内存的性能，以及适合于分析的纯列格式，因此操作下推可以和 SIMD 向量处理及内存存储索引等技术配合，实现进一步的性能提升。

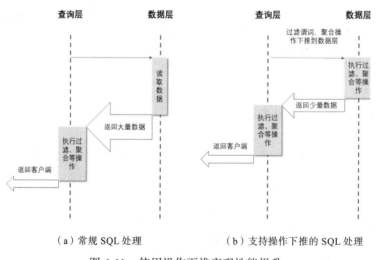

（a）常规 SQL 处理　　（b）支持操作下推的 SQL 处理

图 6-11　使用操作下推实现性能提升

下推的概念比较简单，是一项常用的 SQL 查询优化技术。对于 Oracle 而言，下推技术是继 Exadata 存储层软件、Big Data SQL 后，在内存列式存储中的又一次应用。

内存列式存储支持过滤谓词和聚合操作的下推，以下通过两个示例来加以说明。为简化，测试用例仅使用 LINEORDER 一张表，此表近 1200 万行，发布到内存列式存储后共 24 个 IMCU。

```
SQL> ALTER TABLE lineorder INMEMORY;

SQL> SELECT COUNT(*) FROM lineorder;
```

```
  COUNT(*)
----------
  11997996
SQL>
SELECT COUNT(*) AS number_of_imcus FROM v$im_header WHERE is_head_piece = 1 AND
table_objn IN (SELECT object_id FROM user_objects WHERE object_name = 'LINEORDER');

NUMBER_OF_IMCUS
---------------
             24
```

6.5.1 过滤谓词下推

在 SQL 语句中，对列的选择称为投影（Projection），对行的选择则称为过滤。过滤是通过谓词实现的，SQL 中常见的谓词包括比较操作符（如=、<=和>）、IN、BETWEEN 和 LIKE。

Database In-Memory 支持将过滤谓词下推到内存列式存储，并在其中进行过滤操作。下面我们通过一个示例来了解其过程，测试 SQL 返回合作伙伴 ID 为 53 439 的所有行，本例返回 55 行：

```
SQL> SELECT * FROM lineorder WHERE lo_partkey = 53439;
...
55 rows selected.
```

在未启用内存列式存储时的执行计划如下。在 Predicate Information 部分，只能看到 filter 关键字，表明此表因没有适合的手段（如索引）直接访问数据，只能采用过滤谓词来排除不需要的数据：

```
--------------------------------------------------------------------------
| Id  | Operation         | Name      | Rows  | Bytes | Cost (%CPU)| Time     |
--------------------------------------------------------------------------
|   0 | SELECT STATEMENT  |           |       |       | 45431 (100)|          |
|*  1 |  TABLE ACCESS FULL| LINEORDER |    59 |  5664 | 45431   (1)| 00:00:02 |
--------------------------------------------------------------------------

Predicate Information (identified by operation id):
---------------------------------------------------

   1 - filter("LO_PARTKEY"=53439)
```

当 LINEORDER 表发布到内存列式存储后，再次执行以上 SQL，执行计划和执行统计信息如下：

```
-----------------------------------------------------------------------------
| Id  | Operation                  | Name      | Rows  | Bytes | Cost (%CPU)|
-----------------------------------------------------------------------------
|   0 | SELECT STATEMENT           |           |       |       |  2432 (100)|
|*  1 |  TABLE ACCESS INMEMORY FULL| LINEORDER |    59 |  5664 |  2432  (31)|
-----------------------------------------------------------------------------
```

```
Predicate Information (identified by operation id):
---------------------------------------------------

   1 - inmemory("LO_PARTKEY"=53439)
       filter("LO_PARTKEY"=53439)


NAME                                                         VALUE
---------------------------------------------------------- ----------
IM scan CUs columns accessed                                  360
IM scan CUs columns theoretical max                           408
IM scan CUs current                                            27
IM scan CUs memcompress for query low                          24
IM scan CUs no cleanout                                        24
IM scan CUs pcode pred evaled                                  24
IM scan CUs predicates applied                                 24
IM scan CUs predicates received                                24
IM scan CUs readlist creation accumulated time                  6
IM scan CUs readlist creation number                           24
IM scan CUs split pieces                                       32
IM scan bytes in-memory                                   573227195
IM scan bytes uncompressed                               1140734066
IM scan delta - only base scan                                 24
IM scan rows                                             11997996
IM scan rows projected                                         55
IM scan rows valid                                       11997996
IM scan segments minmax eligible                               24
```

在执行计划的 Predicate Information 部分，会发现多了一个 inmemory 关键字，此即表明过滤谓词被下推到了内存列式存储。在执行统计信息部分，IM scan CUs predicates received 和 IM scan CUs predicates applied 分别表示接收和应用了过滤谓词的 IMCU 数量。IM scan rows projected 指从内存列式存储向上一层返回的行数，其输出值表明近 1200 万条数据通过过滤后返回 55 条数据，近 99.9995% 的数据被过滤掉，从而极大地节省了网络带宽，提升了查询性能。

6.5.2 聚合下推

聚合（Aggregation）操作指针对一组行进行计算，然后得到一行结果。常见的聚合操作包括求和（SUM）、求平均值（AVG）和计数（COUNT）等。聚合操作通常和 SELECT 语句中的 GOURP BY 一起使用。

Database In-Memory 支持将聚合操作下推到内存列式存储中执行，下面通过一个示例了解其过程。

测试 SQL 返回除水运外各运输方式（共 5 种）的运输量。其中，SUM 和 GROUP BY 为聚合操作，NOT IN 为过滤谓词：

```
SQL>
SELECT /*+ MONITOR */
    lo_shipmode,
    SUM(lo_quantity)
```

```
FROM
    lineorder
WHERE
    lo_shipmode NOT IN ( 'SHIP', 'FOB' )
GROUP BY
lo_shipmode;

LO_SHIPMOD SUM(LO_QUANTITY)
---------- ----------------
REG AIR            43760999
TRUCK              43694498
AIR                43784780
RAIL               43675817
MAIL               43692312
```

当 LINEORDER 表发布到内存列式存储后，执行以上 SQL，执行计划和执行统计信息如下：

```
--------------------------------------------------------------------------
| Id | Operation                  | Name     | Rows  | Bytes | Cost (%CPU)|
--------------------------------------------------------------------------
|  0 | SELECT STATEMENT           |          |       |       | 1997 (100)|
|  1 |  HASH GROUP BY             |          |     7 |    98 | 1997  (16)|
|* 2 |   TABLE ACCESS INMEMORY FULL| LINEORDER| 8814K |  117M | 1996  (16)|
--------------------------------------------------------------------------

Predicate Information (identified by operation id):
---------------------------------------------------

   2 - inmemory(("LO_SHIPMODE"<>'SHIP' AND "LO_SHIPMODE"<>'FOB'))
       filter(("LO_SHIPMODE"<>'SHIP' AND "LO_SHIPMODE"<>'FOB'))


NAME                                                        VALUE
----------------------------------------------------        ----------
IM scan CUs columns accessed                                48
IM scan CUs columns theoretical max                         408
IM scan CUs current                                         24
IM scan CUs memcompress for query low                       24
IM scan CUs no cleanout                                     24
IM scan CUs pcode aggregation pushdown                      24
IM scan CUs pcode pred evaled                               72
IM scan CUs predicates applied                              48
IM scan CUs predicates received                             48
IM scan CUs readlist creation accumulated time              8
IM scan CUs readlist creation number                        24
IM scan CUs split pieces                                    32
IM scan bytes in-memory                                     573227179
IM scan bytes uncompressed                                  1140734066
IM scan delta - only base scan                              24
IM scan dict engine results reused                          24
IM scan rows                                                11997996
IM scan rows pcode aggregated                               8571673
```

```
IM scan rows projected                                              120
IM scan rows valid                                              11997996
IM scan segments minmax eligible
                                                                    24
```

在执行计划部分，inmemory 关键字表明 SQL 中的 NOT IN 操作被转换为过滤谓词并下放到内存列式存储中，此外并没有任何与聚合操作相关的信息。但是在执行统计信息部分，IM scan CUs pcode aggregation pushdown 表明聚合操作被下推到内存列式存储中执行，其值 24 正是 LINEORDER 表的 IMCU 数量。另一个统计项 IM scan rows pcode aggregated 表示应用了聚合操作的行数，其值为 8 571 673，应该是过滤了水运方式后的所有行，可以通过以下的 SQL 进行验证：

```
SQL> SELECT COUNT(*) FROM lineorder WHERE lo_shipmode NOT IN ('FOB', 'SHIP');

  COUNT(*)
----------
   8571673
```

如图 6-12 所示，在实时 SQL 监控报告中，单击 Information 列下方的望远镜图标，其中的 Rowsets Leveraging Storage Aggregation 也表示其使用了聚合下推。

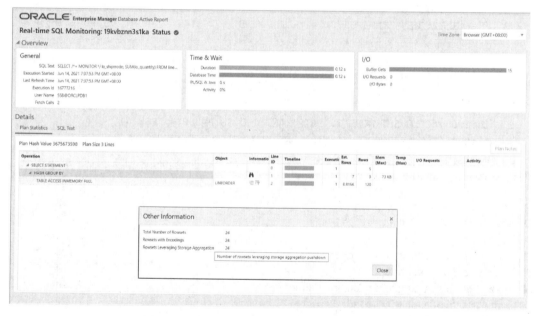

图 6-12　实时 SQL 监控报告中的聚合下推信息

聚合操作被下推到 IMCU 后，每一个 IMCU 都会分别执行求和（SUM）与分组（GROUP BY）操作，然后将结果返回到查询层，在查询层只需将来自各 IMCU 的结果合并。由于除水运方式外，还有 5 种运输方式，因此从内存列式存储中返回的行数应该为 IMCU 的数量乘以 5，也就是 120。这可以从 IM scan rows projected 的值为 120 得到印证。

在本例中，近 1200 万条数据通过聚合下推，仅返回上层 120 条数据，极大地减少了网络传输，SQL 执行的效率也大为提升。

6.5.3 下推与内存存储索引

操作下推可以与其他优化技术一起配合。结合 SIMD 向量处理，可以实现更快的扫描和过滤；与内存存储索引配合，可过滤掉更多的数据，减少向上一层返回的数据。

下面给出一个下推与内存存储索引结合的例子。示例 SQL 返回订单号小于 100 万的订单数量，结果为 1 000 048。

```
SQL> SELECT COUNT(*) FROM lineorder WHERE lo_orderkey < 1000000;

  COUNT(*)
----------
   1000048
```

其在内存列式存储中的执行统计信息如下：

```
IM scan CUs columns accessed                          2
IM scan CUs columns theoretical max                 408
IM scan CUs current                                  24
IM scan CUs memcompress for query low                24
IM scan CUs no cleanout                              24
IM scan CUs optimized read                            1
IM scan CUs pcode aggregation pushdown                2
IM scan CUs pcode pred evaled                         1
IM scan CUs predicates applied                       24
IM scan CUs predicates optimized                     22
IM scan CUs predicates received                      24
IM scan CUs pruned                                   22
IM scan CUs readlist creation accumulated time        1
IM scan CUs readlist creation number                 24
IM scan CUs split pieces                             33
IM scan bytes in-memory                       573227210
IM scan bytes uncompressed                   1140734066
IM scan delta - only base scan                       24
IM scan dict engine results reused                    2
IM scan rows                                   11997996
IM scan rows optimized                         10915237
IM scan rows pcode aggregated                   1000048
IM scan rows projected                                2
IM scan rows valid                              1082759
IM scan segments minmax eligible                     24
```

在以上的统计信息中，IM scan CUs pruned 的值表示根据存储索引的信息，22 个不满足条件的 IMCU 被排除。IM scan CUs pcode aggregation pushdown 的值表示 COUNT(*) 这个聚合操作被下推到余下的 2 个 IMCU 执行。IM scan rows projected 的值表示内存列式存储只向上一层返回了 2 条数据（每一个 IMCU 中的计算合计），上一层只需将这两条数据相加即可完成整个 SQL 查询。

6.6 In-Memory 联结

6.6.1 联结方法

联结（Join）是 SQL 中非常核心的操作，是数据仓库等分析型工作负载不可或缺的一部分，联结将多个行集合(如表和视图)的数据进行关联并返回一个结果集。在 SQL 语句中，在 FROM 子句后指定联结的表与视图，在 WHERE 子句后指定联结条件。

联结方法是指两个行集合实现联结的具体实现方式。最常见的联结方法包括嵌套循环、排序合并和哈希联结。数据库基于统计信息、优化器选择最低成本的方式实现联结。

图 6-13 为两张表分别通过三种联结方法进行等值联结的简明原理示意。首先来看最左侧的嵌套循环方法。对于外层表中的每一行，嵌套循环需要遍历内层表中的所有行。例如对于外部表的第一行数据 7，从内部表的第一行开始向下扫描，在扫描到第 3 行时发现了匹配的记录。但此时并不能停止扫描，因为不能确保后续没有匹配的记录。再来看一下排序合并方法，此方法通过将外层表和内层表排序减少了行遍历的次数。对于外层表中的每一行，在内层表中开始扫描的位置为上一行数据停止扫描的位置，而其结束位置为内层表中首个不匹配行。例如，对于外层表中的 2，扫描开始位置为内层表的第 1 行，而结束位置为第 2 行，即不匹配的数据 3。对于外层表中接下来的数据 4 和 6，由于在内层表中没有匹配，因此仍停留在内层表中数据 3 的位置。再来看一下最后一种方式，哈希联结首先基于外层表（通常是维度表）中的数据建立哈希表，然后对于内层表（通常是事实表）中的每一行数据计算哈希值，如果哈希值匹配则表示满足联结条件。

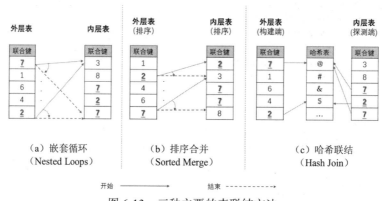

图 6-13　三种主要的表联结方法

排序合并需要两次排序，哈希联结需要构建和探测哈希表。为补偿这些额外的开销，排序后的结果集及哈希表均存放在 PGA 中，从而可以避免资源争用和减少逻辑 I/O。这三种联结方式各自有自己适合与受限的场景。例如嵌套循环非常适合于两个小表的联结；当一个较小的表和一个非常大的表进行联结时，哈希联结就是理想的选择，特别是当小表可以完全置入内存时。但哈希联结只支持等值联结。当联结条件并非等值联结，并且其他的操作正好也需要排序时，则可以考虑采用排序合并。

不过对于内存列式存储而言，最常用的方式还是哈希联结。这是因为内存列式存储适合于分析型负载，而数据分析中的星型模型，通常是在一个大的事实表与多个较小的维度表间进行联结。另外，在内存列式存储中，哈希联结还可以与另一项优化技术相结合，这就是下一节要介绍的布隆过滤器。

6.6.2　In-Memory 联结与布隆过滤器

布隆过滤器（Bloom filter）是 1970 年由美国计算机科学家 Burton H. Bloom 发明的一种兼具时间和空间效率的数据结构，用于检测一个元素是否为集合中的一员。布隆过滤器的主要特点如下。

（1）空间效率，与其所表示的数据集相比，布隆过滤器占用的空间非常小。

（2）时间效率，检测一个元素是否为集合中成员所需的时间为常数 $O(n)$，与集合中元素的数量无关。

（3）概率数据结构，存在误判，但不会漏判。也就是可以确定元素肯定不在集合中，但不能确定元素肯定在集合中。也就是说，当布隆过滤器的结论是元素位于集合中时，有可能数据在集合中并不存在。

布隆过滤器的基本原理如图 6-14 所示。m 表示布隆过滤器所使用的比特数组长度，最初比特数组中所有的比特均置为 0。k 表示哈希函数的个数，集合中的每一个值均会通过这些哈希函数进行计算，并依据结果将比特数组中对应位置的比特置为 1。假设集合中有 3 个元素 x、y 和 z，首先通过 3 个哈希函数计算出位于中部的比特数组。接下来需要判断 m 和 n 是否在集合中。对于 m，3 个哈希函数结果对应位置的值均为 1，这表示 m 可能在集合中。如果哈希函数结果对应位置的值中至少有 1 个 0，就可以确定其肯定不在集合中，而 n 就是这种情况。

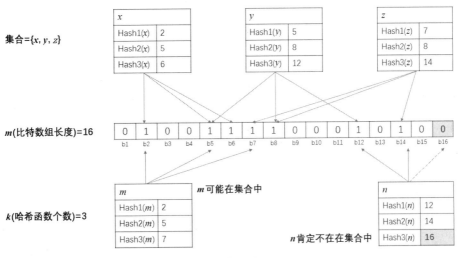

图 6-14　布隆过滤器原理

布隆过滤器出现时，Oracle 数据库还未诞生。Oracle 数据库在 10.2 版本首次引入布隆过滤器，用来减少并行哈希联结各进程间的数据传输。Oracle 11.1 版本利用布隆过滤器实现了分区裁剪和结果集缓存。Exadata 数据库云平台中的存储索引可以与布隆过滤器结合以在表联结时跳过不需要的数据。

在执行计划中可以确认布隆过滤器是否被使用，通过一个分区裁剪的示例来说明。首先创建两个分区表 T1 和 T2：

```
CREATE TABLE t1(x, y)
```

```
PARTITION BY RANGE(x)
(
    PARTITION p1 VALUES LESS THAN(2500),
    PARTITION p2 VALUES LESS THAN(5000),
    PARTITION p3 VALUES LESS THAN(7500),
    PARTITION p4 VALUES LESS THAN(10000)
)
AS SELECT ROWNUM, ROWNUM FROM dual CONNECT BY LEVEL < 10000;

CREATE TABLE t2(x, y)
PARTITION BY RANGE(x)
(
    PARTITION p1 VALUES LESS THAN(5000),
    PARTITION p2 VALUES LESS THAN(10000),
    PARTITION p3 VALUES LESS THAN(15000),
    PARTITION p4 VALUES LESS THAN(20000)
)
AS SELECT ROWNUM, ROWNUM FROM dual CONNECT BY LEVEL < 20000;
```

执行以下查询并查看执行计划。由于这两个表的分区不能完全对齐，不能使用智能分区联结（partition-wise join），因此优化器选择通过布隆过滤器进行分区裁剪。可以看到，布隆过滤器在执行计划的第 3 行构建，在第 6 和第 7 行被使用。

```
SQL> SELECT COUNT(*) FROM t1, T2 WHERE t1.x = t2.x;
...
Plan hash value: 666786458

---------------------------------------------------------------------------------
| Id  | Operation                    | Name   | Rows  | Bytes | Pstart| Pstop |
---------------------------------------------------------------------------------
|   0 | SELECT STATEMENT             |        |       |       |       |       |
|   1 |  SORT AGGREGATE              |        |     1 |       |       |       |
|*  2 |   HASH JOIN                  |        |  9999 | 89991 |       |       |
|   3 |    PART JOIN FILTER CREATE   | :BF0000|  9999 | 89991 |       |       |
|   4 |     PARTITION RANGE ALL      |        |  9999 | 39996 |     1 |     4 |
|   5 |      TABLE ACCESS FULL       | T1     |  9999 | 39996 |     1 |     4 |
|   6 |    PARTITION RANGE JOIN-FILTER|       | 19999 | 99995 |:BF0000|:BF0000|
|   7 |     TABLE ACCESS FULL        | T2     | 19999 | 99995 |:BF0000|:BF0000|
---------------------------------------------------------------------------------

Predicate Information(identified by operation id):
---------------------------------------------------

   2 - access("T1"."X"="T2"."X")

Note
-----
   - this is an adaptive plan
```

Oracle 12.1.0.2 版本推出的 Database In-Memory 也支持在 In-Memory 联结时使用布隆过滤器。在 In-Memory 联结时，联结中较小维度表上的谓词被转换为布隆过滤器，作为额外的过滤条件应用到较大的事实表上。由于维度表上联结键的基数通常较低，而事实表的基数较高，也就是事实表的联结键有大量重复值，因此过滤的效果非常好。加之布隆过滤器可以与 SIMD 向量处理结合，因此过滤的速度很快。这些优化手段的结合极大提升了 In-Memory 联结的性能。

通过一个示例了解 In-Memory 联结的过程，测试语句为 SSB 示例中的 SQL 1_2。首先在未启用内存列式存储时执行，以建立比较基准。SQL 执行时间为 1.14 秒，执行计划第 2~4 行显示这是一个 LINEORDER 表与 DATE_DIM 表之间的哈希联结。

```
SQL> @1_2

       REVENUE
---------------
 192,370,252,105

Elapsed: 00:00:01.14

------------------------------------------------------------------------------------
| Id  | Operation           | Name     | Rows  | Bytes | Cost (%CPU)| Time     |
------------------------------------------------------------------------------------
|   0 | SELECT STATEMENT    |          |       |       | 45531 (100)|          |
|   1 |  SORT AGGREGATE     |          |     1 |    29 |            |          |
|*  2 |   HASH JOIN         |          |  8434 |  238K | 45531   (1)| 00:00:02 |
|*  3 |    TABLE ACCESS FULL| DATE_DIM |    31 |   341 |    13   (0)| 00:00:01 |
|*  4 |    TABLE ACCESS FULL| LINEORDER|  654K |   11M | 45516   (1)| 00:00:02 |
------------------------------------------------------------------------------------

Predicate Information(identified by operation id):
---------------------------------------------------

   2 - access("LO_ORDERDATE"="D_DATEKEY")
   3 - filter("D_YEARMONTHNUM"=199401)
   4 - filter(("LO_QUANTITY">=26 AND "LO_DISCOUNT">=4 AND
              "LO_DISCOUNT"<=6 AND "LO_QUANTITY"<=35))
```

将 LINEORDER 和 DATE_DIM 发布到内存列式存储后再次执行 SQL。此时执行时间下降到 0.01 秒，性能提升超过 110 倍。从两处可以确认布隆过滤器被使用，首先是在执行计划的第 3 行和第 5 行，JOIN FILTER CREATE 和 JOIN FILTER USE 分别表示创建和使用了布隆过滤器。然后在 Predicate Information 部分，SYS_OP_BLOOM_FILTER 表示布隆过滤器被作为额外的过滤条件施加到 LINEORDER 表上。

```
SQL> @1_2

       REVENUE
---------------
 192,370,252,105

Elapsed: 00:00:00.01

PLAN_TABLE_OUTPUT
------------------------------------------------------------------------------------
------------------------------------
```

```
SQL_ID  3vs7vp0f87st0, child number 0
-------------------------------------
SELECT /* 1.2 SSB_SAMPLE_SQL */ /*+ MONITOR */    SUM(lo_extendedprice
* lo_discount) AS revenue FROM     lineorder,    date_dim WHERE
 lo_orderdate = d_datekey      AND d_yearmonthnum = 199401     AND
lo_discount BETWEEN 4 AND 6     AND lo_quantity BETWEEN 26 AND 35

Plan hash value: 2403472142

----------------------------------------------------------------------
| Id  | Operation                    | Name     | Starts | E-Rows | A-Rows |
----------------------------------------------------------------------
|   0 | SELECT STATEMENT             |          |      1 |        |      1 |
|   1 |  SORT AGGREGATE              |          |      1 |      1 |      1 |
|*  2 |   HASH JOIN                  |          |      1 |   8434 |   8406 |
|   3 |    JOIN FILTER CREATE        | :BF0000  |      1 |     31 |     31 |
|*  4 |     TABLE ACCESS INMEMORY FULL| DATE_DIM |     1 |     31 |     31 |
|   5 |    JOIN FILTER USE           | :BF0000  |      1 |   654K |   8406 |
|*  6 |     TABLE ACCESS INMEMORY FULL| LINEORDER|     1 |   654K |   8406 |
----------------------------------------------------------------------

Predicate Information (identified by operation id):
---------------------------------------------------

   2 - access("LO_ORDERDATE"="D_DATEKEY")
   4 - inmemory("D_YEARMONTHNUM"=199401)
       filter("D_YEARMONTHNUM"=199401)
   6 - inmemory(("LO_QUANTITY">=26 AND "LO_DISCOUNT">=4 AND "LO_DISCOUNT"<=6 AND
              "LO_QUANTITY"<=35 AND SYS_OP_BLOOM_FILTER(:BF0000,"LO_ORDERDATE")))
       filter(("LO_QUANTITY">=26 AND "LO_DISCOUNT">=4 AND "LO_DISCOUNT"<=6 AND
              "LO_QUANTITY"<=35 AND SYS_OP_BLOOM_FILTER(:BF0000,"LO_ORDERDATE")))
```

按照深度优先的规则,整个行源树的遍历顺序为(4,3,6,5,2,1,0)。为更好地理解 In-Memory 联结的过程,将上述执行计划中的步骤加以解读。

(1)第 4 行,对 DATE_DIM 进行内存中全表扫描和过滤。在 Predicate Information 部分的注释 4,可以看到对应的过滤条件。从 A-Rows 列可以看出过滤后返回了 31 行,也可以通过以下 SQL 验证:

```
SQL> SELECT COUNT(*) FROM date_dim WHERE d_yearmonthnum = 199401;

  COUNT(*)
----------
        31
```

(2)第 3 行,根据上一步过滤后的结果建立布隆过滤器。集合中的元素为联结键 d_datekey 的值,元素的数量为 31。在如图 6-15 所示的实时 SQL 监控报告中,单击第 3 行的望远镜图标,通过弹出窗口的 Bits Set 可以确认元素的数量。另一个属性 Bitmap Size 表示布隆过滤器的宽度,其值为 8192。

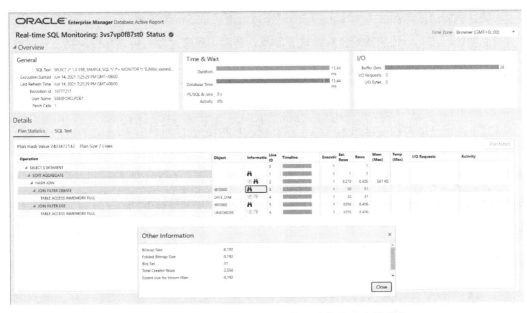

图 6-15　在实时 SQL 监控报告中查看布隆过滤器信息

（3）第 6 行，对 LINEORDEER 进行内存中全表扫描和过滤。从 Predicate Information 中对应的注释 6 可知，布隆过滤器被作为附加的过滤条件，用来判断 LINEORDER 表的 lo_orderdate 是否在集合中，以便排除更多数据。通过以下计算可知，布隆过滤器淘汰了约 98.7%的数据，645 494 行数据被避免返回到上一级。

```
-- 实施原 SQL 过滤条件返回到上一级的行数
SQL> SELECT COUNT(*) FROM lineorder WHERE lo_discount BETWEEN 4 AND 6 AND lo_quantity BETWEEN 26 AND 35;

  COUNT(*)
----------
    653900

-- 叠加布隆过滤器作为过滤条件返回到上一级的行数
SQL> SELECT COUNT(*) FROM lineorder WHERE lo_discount BETWEEN 4 AND 6 AND lo_quantity BETWEEN 26 AND 35 AND lo_orderdate BETWEEN 19940101 AND 19940131;

  COUNT(*)
----------
      8406

SQL> SELECT(653900 - 8406)/653900 * 100 AS skipped FROM dual;

   SKIPPED
----------
98.7144823
```

（4）第 5 行，标识布隆过滤器被使用。

（5）第 2 行，对两张表过滤后返回的结果进行哈希联结。从执行计划的 A-Rows 列可知，经过布隆过滤器过滤后的数据为 8406 行。而经过哈希联结后的数据也是 8406,实际上数据并没有发生变化。之所以仍需要哈希联结这个环节,是因为布隆过滤器可能存在误判,尽管概率很低。不过由于之前通过布隆过滤器淘汰了大部分数据,因此此时哈希计算的量已大幅减少。

(6)第 1 行,执行聚合(SUM)计算。

(7)第 0 行,返回 SQL 查询结果。

从以上测试结果可知,采用 In-Memory 联结的执行速度比未启用内存列式存储时快了 110 倍。这其中有内存扫描的原因,也有布隆过滤器的贡献。为确定它们贡献的比例,可以通过优化器提示 NO_PX_JOIN_FILTER 禁用布隆过滤器,然后查看执行的性能:

```
SELECT /* 1.2 SSB_SAMPLE_SQL */ /*+ NO_PX_JOIN_FILTER(LINEORDER) */
    SUM(lo_extendedprice * lo_discount) AS revenue
FROM
    lineorder,
    date_dim
WHERE
        lo_orderdate = d_datekey
    AND d_yearmonthnum = 199401
    AND lo_discount BETWEEN 4 AND 6
    AND lo_quantity BETWEEN 26 AND 35;

Elapsed: 00:00:00.06

Plan hash value: 2963256899

--------------------------------------------------------------------------
| Id  | Operation                   | Name     | Rows  | Bytes | Cost (%CPU)|
--------------------------------------------------------------------------
|   0 | SELECT STATEMENT            |          |       |       | 1941 (100)|
|   1 |  SORT AGGREGATE             |          |     1 |    29 |           |
|*  2 |   HASH JOIN                 |          |  8434 |  238K | 1941  (14)|
|*  3 |    TABLE ACCESS INMEMORY FULL| DATE_DIM |    31 |   341 |    1   (0)|
|*  4 |    TABLE ACCESS INMEMORY FULL| LINEORDER|  654K |   11M | 1938  (14)|
--------------------------------------------------------------------------
```

从执行计划中已经看不到布隆过滤器的信息,执行时间由之前的 0.01 秒增加到 0.06 秒。也就是说,仅启用内存列式存储可将执行速度提升 19 倍,叠加布隆过滤器可将性能再提升 6 倍。

实际上,在表扫描时可以同时应用多个布隆过滤器。使用 SSB 示例 SQL 3_1 来进行演示。在 SQL 执行时,需要将隐含参数_optimizer_vector_transformation 设置为 FALSE 以禁用向量转换。这是由于 In-Memory 联结可能选择布隆过滤器和向量转换之一作为优化手段,通过此设置可以确保优化器采用布隆过滤器。6.7 节会对向量转换进行介绍。

```
SQL> ALTER SESSION SET "_optimizer_vector_transformation" = FALSE;
SQL> @3_1
```

其执行计划如图 6-16 所示。在扫描 DATE_DIM、SUPPLIER 和 CUSTOMER 表时分别创建了 BF0000、BF0001 和 BF0002 共 3 个布隆过滤器。之后在扫描事实表 LINEORDER 时,这 3 个布隆过滤器被同时加以应用。

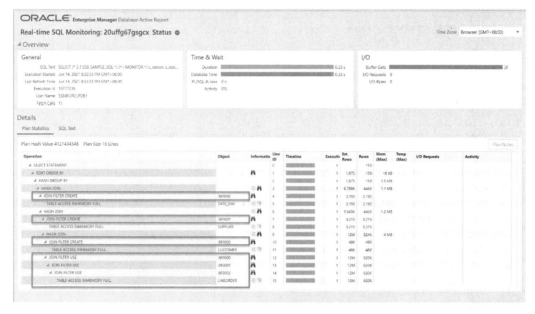

图 6-16　在扫描事实表时应用多个布隆过滤器

在典型的星形联结中，Oracle 数据库支持左深树（left-deep tree）和右深树（right-deep tree）两种联结执行方式。左深树方式为传统方式，也是排序合并和嵌套循环联结唯一支持的执行方式。在 Oracle 11g 后，哈希联结开始支持右深树执行方式，这对一个行数较多的事实表与多个行数较少的维度表之间的星型联结查询非常有效。基于图 6-16 可以绘制出树状的执行计划拓扑图，如图 6-17（b）所示，这是典型的右深树方式。

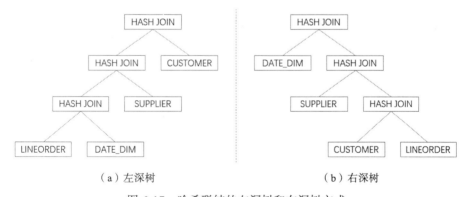

（a）左深树　　　　　　　　　　　（b）右深树

图 6-17　哈希联结的左深树和右深树方式

对于采用左深树方式的星型联结查询，事实表需要首先与一个维度表进行联结，然后再逐一与其他维度表进行联结。第一个联结将产生非常大的中间结果，并且会伴随后续整个联结过程。图 6-18 为 SSB 示例 SQL 3_1 采用左深树执行方式时的执行计划，其执行计划拓扑图如图 6-17（a）。与采用右深树方式的图 6-16 相比，中间过程中产生了大量 I/O 请求，执行时间也更长。

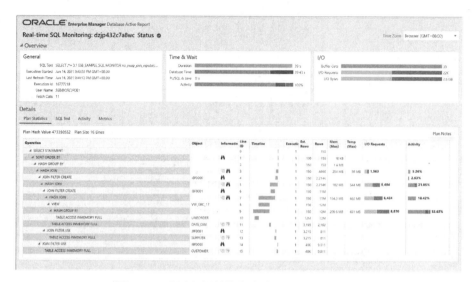

图 6-18 采用左深树执行方式的星型联结查询

实际上，优化器会基于成本自动在左深树和右深树联结方式中做出选择。从以下执行计划概要数据中可以看到，其中两个 SWAP_JOIN_INPUTS 交换了联结双方的顺序，使得较大数据量的事实表最后才处理，避免产生大量中间过程 I/O，并可以在最终对事实表一次性应用多个过滤条件，从而提高执行效率。

```
SQL> SELECT * FROM TABLE(dbms_xplan.display_cursor(FORMAT=>'TYPICAL +outline'));
…
Outline Data
-------------

  /*+
      BEGIN_OUTLINE_DATA
      IGNORE_OPTIM_EMBEDDED_HINTS
      OPTIMIZER_FEATURES_ENABLE('19.1.0')
      DB_VERSION('19.1.0')
      ALL_ROWS
      OUTLINE_LEAF(@"SEL$1")
      FULL(@"SEL$1" "CUSTOMER"@"SEL$1")
      FULL(@"SEL$1" "LINEORDER"@"SEL$1")
      FULL(@"SEL$1" "SUPPLIER"@"SEL$1")
      FULL(@"SEL$1" "DATE_DIM"@"SEL$1")
      LEADING(@"SEL$1" "CUSTOMER"@"SEL$1" "LINEORDER"@"SEL$1" "SUPPLIER"@"SEL$1"
              "DATE_DIM"@"SEL$1")
      USE_HASH(@"SEL$1" "LINEORDER"@"SEL$1")
      USE_HASH(@"SEL$1" "SUPPLIER"@"SEL$1")
      USE_HASH(@"SEL$1" "DATE_DIM"@"SEL$1")
      SWAP_JOIN_INPUTS(@"SEL$1" "SUPPLIER"@"SEL$1")
      SWAP_JOIN_INPUTS(@"SEL$1" "DATE_DIM"@"SEL$1")
      USE_HASH_AGGREGATION(@"SEL$1")
      END_OUTLINE_DATA
  */
```

6.6.3 部分表发布时的 In-Memory 联结

如果进行联结的两张表，只有一个位于内存列式存储中，是否还可以有效地利用内存列式存储？下面通过两个示例来测试一下。测试基准为：当 SSB 示例 SQL 1_2 的两张表全部发布时，执行时间为 0.01 秒，执行成本为 1941（%CPU）。

第一个示例只在内存列式存储中保留较大的事实表 LINEORDER，执行输出如下：

```
SQL> ALTER TABLE date_dim NO INMEMORY;

SQL> @1_2

     REVENUE
----------------
 192,370,252,105

Elapsed: 00:00:00.04

Execution Plan
----------------------------------------------------------
Plan hash value: 2403472142

-----------------------------------------------------------------------------------
| Id  | Operation                    | Name     | Rows  | Bytes | Cost (%CPU)|
-----------------------------------------------------------------------------------
|   0 | SELECT STATEMENT             |          |     1 |    29 |  1953  (14)|
|   1 |  SORT AGGREGATE              |          |     1 |    29 |            |
|*  2 |   HASH JOIN                  |          |  8434 |  238K |  1953  (14)|
|   3 |    JOIN FILTER CREATE        | :BF0000  |    31 |   341 |    13   (0)|
|*  4 |     TABLE ACCESS FULL        | DATE_DIM |    31 |   341 |    13   (0)|
|   5 |    JOIN FILTER USE           | :BF0000  |  654K |   11M |  1938  (14)|
|*  6 |     TABLE ACCESS INMEMORY FULL| LINEORDER|  654K |   11M |  1938  (14)|
-----------------------------------------------------------------------------------
```

从执行计划可知，虽然 DATE_DIM 表不在内存列式存储中，但由于此表较小，只有 2556 行，因此对整体性能影响不大。执行时间为 0.04 秒；执行成本为 1953，仅上升了 0.6%。

第二个示例仅将较小的维度表 DATE_DIM 发布到内存列式存储中，再次执行查询：

```
SQL> ALTER TABLE lineorder NO INMEMORY;

SQL> @1_2

   REVENUE
----------
1.9237E+11

Elapsed: 00:00:01.28

Execution Plan
----------------------------------------------------------
Plan hash value: 2963256899

-----------------------------------------------------------------------------------
| Id  | Operation                    | Name     | Rows  | Bytes | Cost (%CPU)|
-----------------------------------------------------------------------------------
|   0 | SELECT STATEMENT             |          |     1 |    29 | 45519   (1)|
|   1 |  SORT AGGREGATE              |          |     1 |    29 |            |
```

```
|* 2  |  HASH JOIN                     |           |  8434 |  238K|  45519 | (1)|
|* 3  |  TABLE ACCESS INMEMORY FULL    | DATE_DIM  |    31 |   341|      1 | (0)|
|* 4  |  TABLE ACCESS FULL             | LINEORDER |  654K |   11M|  45516 | (1)|
```

由于事实表 LINEORDER 较大，约 1200 万行，因此对性能影响较大。其执行时间为 1.28 秒；执行成本为 45 519，增加了近 24 倍。

因此，当一个较小的表和一个较大的表进行联结时，至少应将较大的表发布到内存列式存储中。需要指出，发布部分表不同于单表的不完整发布，后者是不建议的。

6.7　In-Memory 聚合

6.7.1　In-Memory 聚合基本概念

6.5.2 节介绍的聚合下推可以对简单的聚合实现性能优化，对于复杂的聚合查询，如基于分组列的聚合，以及此基础上多个维度表与事实表之间的联结，可以通过 In-Memory 聚合实现性能优化。

In-Memory 聚合实现了两类优化，首先是通过一种称为 KEY VECTOR 的数据结构将事实表与维度表之间的联结操作转换为过滤操作，从而避免了昂贵的哈希联结处理。其次，通过 VECTOR GROUP BY 替换常规的 GROUP BY 操作，实现了在扫描时同步构建报表框架和填充数据，从而实现更快的计算。这两类优化的名称中均有 VECTOR 关键字，表示它们都可以与 SIMD 向量操作相结合，因此这两类优化也被统称为向量转换。

In-Memory 聚合的概念相对复杂，通过图 6-19 加以说明。图中的 SQL 查询包含了联结操作和分组（GROUP BY）操作。联结操作涉及 LINEORDER 事实表和 DATE_DIM、SUPPLIER 两张维度表，查询条件为年份大于 1995 并且区域位于亚洲或美洲。分组键为 DATE_DIM 表的 d_year 列和 SUPPLIER 表的 s_region 列。In-Memory 聚合的整个过程如下。

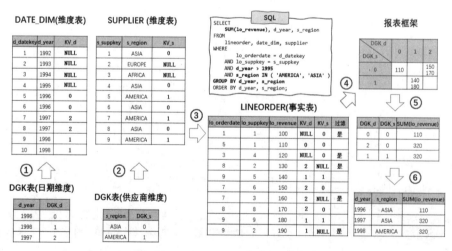

图 6-19　In-Memory 聚合原理说明

1. 建立 DATE_DIM 表的 KEY VECTOR

KEY VECTOR 是一种用于快速查询的数据结构，如图 6-19 左侧的 KV_d 和 KV_s 列。KEY VECTOR 基于 DGK（Dense Group Key）表构建，DGK 表示密集分组键，之所以称为密集是因为其通过谓词过滤掉了不符合条件的分组键。

DATE_DIM 的 DGK 表可以通过以下 SQL 语句生成。其中，DGK_d 为满足条件的分组键。为便于理解，我们将 KV_d 和 KV_s 两个 KEY VECTOR 附加在两张原始表的最右侧。

```sql
SQL>
SELECT d_year, rownum-1 AS "DGK_d"
FROM(
    SELECT d_year FROM d
    WHERE d_year > 1995
    GROUP BY d_year
);

    D_YEAR      DGK_d
---------- ----------
      1996          0
      1998          1
      1997          2
```

根据此 DGK 表即可建立 DATE_DIM 表的 KEY VECTOR，及 KV_d 列。例如 DATE_DIM 表前 4 行中的 d_year 在 DGK 表中检索不到，则在 KEY VECTOR 中填写 NULL；又如最后两行 d_year 的值为 1998，在 DGK 表对应的值为 2，则在 KEY VECTOR 中填写为 2。

2. 建立 SUPPLIER 表的 KEY VECTOR

原理和过程与第一步类似。首先根据以下 SQL 构建 DGK 表：

```sql
SQL>
SELECT s_region, rownum-1 AS "DGK_s"
FROM(
    SELECT s_region FROM s
    WHERE s_region IN('AMERICA', 'ASIA')
    GROUP BY s_region
)

S_REGION        DGK_s
------------ ----------
ASIA                 0
AMERICA              1
```

然后根据 DGK 表，可以建立 SUPPLIER 表的 KEY VECTOR，即 KV_s 列。

3. 在扫描事实表 LINEORDER 时，将以上两个 KEY VECTOR 作为过滤条件

当有更多维度表时，可以构建和应用更多的 KEY VECTOR 作为过滤条件。事实表 LINEORDER 的联结键为 lo_orderdate 列和 lo_suppkey 列，分别对应 DATE_DIM 表的 d_datekey 列和 SUPPLIER 表的 s_suppkey 列。LINEORDER 表第 1 行 lo_orderdate 的值为 1，对应 DATE_DIM 表 KV_d 列的值为 NULL；lo_suppkey 的值为 1，对应 SUPPLIER 表 KV_s 列的值为 0。KV_d 和 KV_s 中只要有一个为 NULL，事实表中的这一行就会被过滤掉。本例中，共 50% 的数据被

过滤掉。

对于结合了聚合和分组的 SQL 查询，KEY VECTOR 通常比哈希联结更高效，因为其无需逐个表进行哈希联结，可以一次性应用过滤条件，并且可以同步构建和填充报表框架。

4. 构建和填充报表框架

为加快计算，在扫描的同时构建了报表框架，满足条件的数据被填充到其中。这个报表框架是按照 SQL 查询的要求（GROUP BY）建立的。由于分组键只有两个，因此报表框架为二维表，横坐标为 DGK_d，纵坐标为 DGK_s。然后在 LINEORDER 表中，分别以 KV_d 和 KV_s 作为索引，将对应的值填入到报表框架中。例如 LINEORDER 表的第 6 行的（KV_d, KV_s）的值为（2, 0），因此在报表框架中的第 0 行第 2 列填入 150。

5. 执行聚合操作

本例中的聚合操作为求和（SUM），因此将报表框架每个单元中的值进行累加。

6. 将报表框架转换为最终的查询结果

借助于最初建立的两个 DGK 表，即可转换为最终查询结果。

In-Memory 聚合充分贯彻了之前提到的两种优化原则，即操作下推和延迟物化。通过下推 KEY VECTOR 和 VECTOR GROUP BY，在扫描的同时实现了过滤和聚合操作。延迟物化体现在以数值表示的 DGK 替代实际的分组键，并在查询过程中使用，直到最后一步才转换回原始列值。

满足以下条件时，优化器会选择 In-Memory 聚合。

（1）Database In-Memory 已启用。

（2）SQL 中包含事实表和维度表之间的联合。

（3）SQL 中包含简单形式的 GROUP BY（不支持扩展形式的 GROUP BY，如 GROUPING SETS、CUBE 和 ROLLUP）。

（4）事实表至少为 1000 万行。

KEY VECTOR 和 VECTOR GROUP BY 这两种向量转换优化并非专属于 In-Memory 聚合，在没有启用 Database In-Memory 时也可以使用。可以通过以下隐含参数控制向量转换的行为。

（1）_optimizer_vector_min_fact_rows，启用 In-Memory 聚合时，事实表至少应具有的行数。默认为 1000 万行。

（2）_optimizer_vector_transformation，是否允许 In-Memory 聚合，默认为允许。当设为否时，相当于 SQL 提示 NO_VECTOR_TRANSFORM。

（3）_always_vector_transformation，总是启用 In-Memory 聚合，默认为否。当设为是时，相当于 SQL 提示 VECTOR_TRANSFORM。

6.7.2　In-Memory 聚合性能比较

通过 4 个示例来分别比较在启用和禁用内存列式存储时，以及在启用和禁用向量转换时的性能表现。测试 SQL 如下：

```
SELECT
    SUM(lo_revenue),
    d_year,
    s_region
FROM
    lineorder,
    date_dim,
    supplier
WHERE
        lo_orderdate = d_datekey
    AND lo_suppkey = s_suppkey
    AND d_year > 1995
    AND s_region IN('AMERICA', 'ASIA')
GROUP BY
    d_year,
    s_region
ORDER BY
    d_year,
    s_region;

SUM(LO_REVENUE)     D_YEAR S_REGION
---------------     ------ ------------
     1.3573E+12       1996 AMERICA
     1.3441E+12       1996 ASIA
     1.3547E+12       1997 AMERICA
     1.3378E+12       1997 ASIA
     7.9418E+11       1998 AMERICA
     7.8566E+11       1998 ASIA

6 rows selected.
```

第一个示例将所有表发布到内存列式存储中，优化器自动选择了向量转换，执行计划如下：

```
Elapsed: 00:00:00.08

Plan hash value: 1768711973

---------------------------------------------------------------------------------
| Id | Operation                                     | Name                     |
---------------------------------------------------------------------------------
|  0 | SELECT STATEMENT                              |                          |
|  1 |  TEMP TABLE TRANSFORMATION                    |                          |
|  2 |   LOAD AS SELECT(CURSOR DURATION MEMORY)      | SYS_TEMP_0FD9D668D_AC9EA0|
|  3 |    HASH GROUP BY                              |                          |
|  4 |     KEY VECTOR CREATE BUFFERED                | :KV0000                  |
|* 5 |      TABLE ACCESS INMEMORY FULL               | DATE_DIM                 |
|  6 |   LOAD AS SELECT(CURSOR DURATION MEMORY)      | SYS_TEMP_0FD9D668E_AC9EA0|
|  7 |    HASH GROUP BY                              |                          |
|  8 |     KEY VECTOR CREATE BUFFERED                | :KV0001                  |
|* 9 |      TABLE ACCESS INMEMORY FULL               | SUPPLIER                 |
| 10 |   SORT GROUP BY                               |                          |
|* 11|    HASH JOIN                                  |                          |
```

```
|*  12 |      HASH JOIN                         |                              |
|   13 |       TABLE ACCESS FULL                | SYS_TEMP_0FD9D668E_AC9EA0    |
|   14 |       VIEW                             | VW_VT_846B3E5D               |
|   15 |        VECTOR GROUP BY                 |                              |
|   16 |         HASH GROUP BY                  |                              |
|   17 |          KEY VECTOR USE                | :KV0001                      |
|   18 |           KEY VECTOR USE               | :KV0000                      |
|*  19 |            TABLE ACCESS INMEMORY FULL  | LINEORDER                    |
|   20 |        TABLE ACCESS FULL               | SYS_TEMP_0FD9D668D_AC9EA0    |
---------------------------------------------------------------------------------

Predicate Information(identified by operation id):
---------------------------------------------------

   5 - inmemory("D_YEAR">1995)
       filter("D_YEAR">1995)
   9 - inmemory(("S_REGION"='AMERICA' OR "S_REGION"='ASIA'))
       filter(("S_REGION"='AMERICA' OR "S_REGION"='ASIA'))
  11 - access("ITEM_9"=INTERNAL_FUNCTION("C0"))
  12 - access("ITEM_8"=INTERNAL_FUNCTION("C0"))
  19 - inmemory((SYS_OP_KEY_VECTOR_FILTER("LO_ORDERDATE",:KV0000) AND
            SYS_OP_KEY_VECTOR_FILTER("LO_SUPPKEY",:KV0001)))
       filter((SYS_OP_KEY_VECTOR_FILTER("LO_ORDERDATE",:KV0000) AND
            SYS_OP_KEY_VECTOR_FILTER("LO_SUPPKEY",:KV0001)))

Note
-----
   - vector transformation used for this statement
```

在执行计划的主体部分,第 4 和第 8 行表明为两个维度表分别建立了 KEY VECTOR:KV0000 和 KV0001。之后第 17 和 18 行表明使用了这两个 KEY VECTOR,也就是在扫描事实表 LINEORDER 时将其作为过滤条件。在 Predicate Information 部分的注释 19,可看到过滤条件的详情。使用数值形式的密集分组键,原有的 SQL 从概念上被替换为以下形式的查询:

```
SELECT
    SUM(lo_revenue),
    :KV0000,
    :KV0001
FROM
    lineorder, date_dim, supplier
WHERE
        SYS_OP_KEY_VECTOR_FILTER("LO_ORDERDATE",:KV0000) IS NOT NULL
    AND SYS_OP_KEY_VECTOR_FILTER("LO_SUPPKEY",:KV0001) IS NOT NULL
GROUP BY
    :KV0000,
    :KV0001
ORDER BY
    :KV0000,
    :KV0001;
```

最后,在执行计划的 Note 部分,也表明此 SQL 使用了向量转换。

在第 2 个示例中，仍将所有表保持在内存列式存储中，但在 SQL 中嵌入提示 /*+ NO_VECTOR_TRANSFORM */ 来禁用向量转换。在执行计划中，优化器选择了布隆过滤器作为优化手段，执行时间增加到 0.49 秒。布隆过滤器和 VECTOR GROUP BY 是互斥的，优化器会自动根据成本估算选择其中一种。

```
Elapsed: 00:00:00.49

Plan hash value: 2064805782

---------------------------------------------------------------------------------
| Id  | Operation                      | Name      | Rows  | Bytes | Cost (%CPU)|
---------------------------------------------------------------------------------
|   0 | SELECT STATEMENT               |           |       |       | 2150 (100)|
|   1 |  SORT GROUP BY                 |           |     5 |   195 | 2150  (18)|
|*  2 |   HASH JOIN                    |           |  8486 |  323K | 2149  (18)|
|*  3 |    TABLE ACCESS INMEMORY FULL  | SUPPLIER  |  6479 |  113K |   68   (0)|
|   4 |    VIEW                        | VW_GBC_10 |  8486 |  174K | 2080  (19)|
|   5 |     HASH GROUP BY              |           |  8486 |  215K | 2080  (19)|
|*  6 |      HASH JOIN                 |           | 5457K |  135M | 1894  (11)|
|   7 |       JOIN FILTER CREATE       | :BF0000   |  1095 | 10950 |   13   (0)|
|*  8 |        TABLE ACCESS INMEMORY FULL| DATE_DIM|  1095 | 10950 |   13   (0)|
|   9 |       JOIN FILTER USE          | :BF0000   |   11M |  183M | 1840   (9)|
|* 10 |        TABLE ACCESS INMEMORY FULL| LINEORDER|  11M |  183M | 1840   (9)|
---------------------------------------------------------------------------------

Predicate Information(identified by operation id):
---------------------------------------------------

   2 - access("ITEM_1"="S_SUPPKEY")
   3 - inmemory(("S_REGION"='AMERICA' OR "S_REGION"='ASIA'))
       filter(("S_REGION"='AMERICA' OR "S_REGION"='ASIA'))
   6 - access("LO_ORDERDATE"="D_DATEKEY")
   8 - inmemory("D_YEAR">1995)
       filter("D_YEAR">1995)
  10 - inmemory(SYS_OP_BLOOM_FILTER(:BF0000,"LO_ORDERDATE"))
       filter(SYS_OP_BLOOM_FILTER(:BF0000,"LO_ORDERDATE"))
```

在第 3 个示例中，所有的表均不发布，优化器自动选择了向量转换。除了没有内存中全表扫描，执行计划和示例 1 非常类似。由于无法应用内存列式存储中的多种优化手段，执行时间和成本也大幅增加。

```
Elapsed: 00:00:01.23

--------------------------------------------------------------------------------
| Id  | Operation                              | Name                         |
--------------------------------------------------------------------------------
|   0 | SELECT STATEMENT                       |                              |
|   1 |  TEMP TABLE TRANSFORMATION             |                              |
|   2 |   LOAD AS SELECT(CURSOR DURATION MEMORY)| SYS_TEMP_0FD9D66A0_AC9EA0   |
|   3 |    HASH GROUP BY                       |                              |
|   4 |     KEY VECTOR CREATE BUFFERED         | :KV0000                      |
|*  5 |      TABLE ACCESS FULL                 | DATE_DIM                     |
```

```
|   6 |    LOAD AS SELECT(CURSOR DURATION MEMORY) | SYS_TEMP_0FD9D66A1_AC9EA0 |
|   7 |     HASH GROUP BY                         |                           |
|   8 |      KEY VECTOR CREATE BUFFERED           | :KV0001                   |
|*  9 |       TABLE ACCESS FULL                   | SUPPLIER                  |
|  10 |  SORT GROUP BY                            |                           |
|* 11 |   HASH JOIN                               |                           |
|* 12 |    HASH JOIN                              |                           |
|  13 |     TABLE ACCESS FULL                     | SYS_TEMP_0FD9D66A0_AC9EA0 |
|  14 |     VIEW                                  | VW_VT_846B3E5D            |
|  15 |      VECTOR GROUP BY                      |                           |
|  16 |       HASH GROUP BY                       |                           |
|  17 |        KEY VECTOR USE                     | :KV0001                   |
|  18 |         KEY VECTOR USE                    | :KV0000                   |
|* 19 |          TABLE ACCESS FULL                | LINEORDER                 |
|  20 |    TABLE ACCESS FULL                      | SYS_TEMP_0FD9D66A1_AC9EA0 |
-----------------------------------------------------------------------------

Predicate Information(identified by operation id):
---------------------------------------------------

   5 - filter("D_YEAR">1995)
   9 - filter(("S_REGION"='AMERICA' OR "S_REGION"='ASIA'))
  11 - access("ITEM_8"=INTERNAL_FUNCTION("C0"))
  12 - access("ITEM_9"=INTERNAL_FUNCTION("C0"))
  19 - filter(SYS_OP_KEY_VECTOR_FILTER("LO_ORDERDATE",:KV0000))

Note
-----
   - statistics feedback used for this statement
   - vector transformation used for this statement
```

在示例 3 的基础上,第 4 个示例利用 SQL 提示 /*+ NO_VECTOR_TRANSFORM */禁用了向量转换。执行计划显示使用了哈希分组和哈希联结,这是 4 个示例中成本最高和最慢的方式。

```
Elapsed: 00:00:01.36
Plan hash value: 2736736070

-----------------------------------------------------------------------------
| Id | Operation              | Name       | Rows  | Bytes | Cost (%CPU)|
-----------------------------------------------------------------------------
|  0 | SELECT STATEMENT       |            |       |       | 45793 (100)|
|  1 |  SORT GROUP BY         |            |     5 |   195 | 45793   (1)|
|* 2 |   HASH JOIN            |            |  8486 |  323K | 45792   (1)|
|* 3 |    TABLE ACCESS FULL   | SUPPLIER   |  6479 |  113K |    68   (0)|
|  4 |    VIEW                | VW_GBC_10  |  8486 |  174K | 45724   (1)|
|  5 |     HASH GROUP BY      |            |  8486 |  215K | 45724   (1)|
|* 6 |      HASH JOIN         |            | 5457K |  135M | 45537   (1)|
|* 7 |       TABLE ACCESS FULL| DATE_DIM   |  1095 | 10950 |    13   (0)|
|  8 |       TABLE ACCESS FULL| LINEORDER |   11M |  183M | 45483   (1)|
-----------------------------------------------------------------------------

Predicate Information(identified by operation id):
---------------------------------------------------

   2 - access("ITEM_1"="S_SUPPKEY")
   3 - filter(("S_REGION"='AMERICA' OR "S_REGION"='ASIA'))
```

```
6 - access("LO_ORDERDATE"="D_DATEKEY")
7 - filter("D_YEAR">1995)
```

将 4 个示例的结果汇总到表 6-5 中，以执行时间作为比较依据。在启用内存列式存储时，使用向量转换较未使用时可将性能提升 6 倍。在向量转换已启用前提下，启用内存列式存储较未启用时可带来 15 倍的性能提升。

表 6-5　不同优化手段组合时的聚合性能比较

测试号	内存列式存储	向量转换	执行时间（秒）	执行成本
1	启用	启用	0.08	1854
2	启用	禁用	0.49	2150
3	禁用	启用	1.23	45574
4	禁用	禁用	1.36	45793

6.8　索引优化

索引是提升查询速度最常用的手段之一，对于模式已知的查询非常有效。索引需要额外的存储空间，对 DML 操作的性能也会有一定的影响，因为执行 INSERT、UPDATE 等 DML 操作时需要更新相应的索引。索引并不适用于不可预知的分析查询，因为无法提前设计和创建索引。

Database In-Memory 可以弥补以上索引的缺陷，并可以在分析型查询中提供与索引相当或更高性能。在图 6-20 典型应用程序的时间组成摘要一图中，Database In-Memory 可以在两个部分大幅减少应用程序时间，最有效的是分析和报表应用的数据访问，其次是分析型索引的维护时间。也就是说，当查询涉及的表被置入内存列式存储时，可以考虑删除专为分析优化建立的索引。这不仅可以减少 DBA 设计和建立索引的时间，同时可以节省索引所需存储空间，并提升 OLTP 应用中插入、更新和删除操作的性能。由于内存列式存储无须持久化，因此更新成本要比索引低得多。

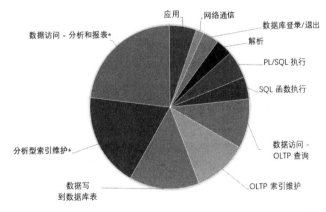

图 6-20　典型应用程序的时间组成摘要（*-Database In-Memory 可能会减少其时间）[①]

[①] 此图引用自：https://www.oracle.com/technetwork/database/in-memory/overview/twp-dbim-usage-2441076.html

这并不表示内存列式存储可以完全取代索引。对于纯交易类型工作负载，特别是涉及更新一行或少数行的 SQL，索引仍然是首选的优化手段。即便对于专为分析型查询建立的索引，也必须在仔细评估后才能进行删除。建议先将索引设为不可见状态，使 SQL 避免使用此索引。观察一段时间后，确认其不会引起各方面的性能回退后再行删除。

如果表具有索引，同时启用了 INMEMORY，优化器会根据执行成本自动选择索引优化或是内存中扫描。下面通过一个示例进行说明。测试 SQL 根据订单号查询，只涉及少数几行，因此非常适合通过索引进行优化：

```
SELECT SUM(lo_quantity) FROM lineorder WHERE lo_orderkey = 1;
```

首先根据查询条件建立索引，索引占用空间约 224MB。然后将表 LINEORDER 发布到内存列式存储中：

```
SQL> CREATE INDEX idx1 ON lineorder(lo_orderkey);
SQL> EXEC dbms_stats.gather_table_stats('SSB', 'LINEORDER');

SQL> SELECT BYTES/1024/1024 FROM user_segments WHERE segment_name = 'IDX1';

BYTES/1024/1024
---------------
            224

SQL> ALTER TABLE lineorder INMEMORY;
SQL> EXEC DBMS_INMEMORY.POPULATE('SSB', 'LINEORDER');
```

执行 SQL，优化器确实选择了索引来完成查询。如果禁用索引，优化器将转而选择内存中扫描，但其执行成本高出 433 倍：

```
SQL> SELECT SUM(lo_quantity) FROM lineorder WHERE lo_orderkey = 1;

SUM(LO_QUANTITY)
----------------
             145

SQL> @showplan

---------------------------------------------------------------------------------
| Id  | Operation                            | Name     | Rows  | Cost (%CPU)|
---------------------------------------------------------------------------------
|   0 | SELECT STATEMENT                     |          |       |     4 (100)|
|   1 |  SORT AGGREGATE                      |          |     1 |            |
|   2 |   TABLE ACCESS BY INDEX ROWID BATCHED| LINEORDER|     4 |     4   (0)|
|*  3 |    INDEX RANGE SCAN                  | IDX1     |     4 |     3   (0)|
---------------------------------------------------------------------------------

Predicate Information(identified by operation id):
---------------------------------------------------

   3 - access("LO_ORDERKEY"=1)

SQL>   SELECT /*+ NO_INDEX(lineorder idx1) */  SUM(lo_quantity) FROM lineorder WHERE
```

```
lo_orderkey = 1;

SUM(LO_QUANTITY)
----------------
             145

SQL> @showplan


---------------------------------------------------------------------------------
| Id  | Operation                   | Name     | Rows  | Bytes | Cost (%CPU)|
---------------------------------------------------------------------------------
|   0 | SELECT STATEMENT            |          |       |       | 1732  (100)|
|   1 |  SORT AGGREGATE             |          |    1  |   10  |            |
|*  2 |   TABLE ACCESS INMEMORY FULL| LINEORDER|    4  |   40  | 1732    (3)|
---------------------------------------------------------------------------------

Predicate Information(identified by operation id):
---------------------------------------------------

   2 - inmemory("LO_ORDERKEY"=1)
       filter("LO_ORDERKEY"=1)
```

第7章

Database In-Memory 高级性能优化

7.1 In-Memory 表达式

在算法中，预计算是指通过最初只计算一次并存储结果，后续可以重复利用此结果的过程。预计算可以避免重复计算，从而节省计算资源和加速后续查询时间。例如，如果旅客需要购买北京到上海的一等座火车票，最初计算一次后可以将票价结果存放到查询表中，从而省去后续旅客购买同种车票的票价计算时间。

Oracle 数据库中也支持预计算技术，如物化视图。Oracle 在 12.2 版本推出了 In-Memory 表达式，本质上也属于预计算技术，只不过是将预计算的结果存放在了内存列式存储中。在 Database In-Memory 中，In-Memory 表达式包括两类技术，即 In-Memory 虚拟列和 In-Memory Expression。前者是用户定义的、静态的表达式；后者是系统自动捕捉的、动态的表达式。无论采用哪种技术，最终表达式都会发布到 IMEU 中，IMEU 是 IMCU 的逻辑扩展，与 IMCU 一一对应。In-Memory 表达式不仅通过预计算技术提高了分析的性能，同时也节省了宝贵的计算资源，可以供交易型工作负载使用。

7.1.1 In-Memory 虚拟列

虚拟列是 Oracle 11gR1 版推出的技术。和其他普通列一样，虚拟列是表的一部分。但是虚拟列存放的是表达式，表达式由常数、相同表中的其他列、操作符、SQL 函数和用户定义 PL/SQL 函数组合而成。表达式中的函数必须是确定性的，即当输入参数固定时，输出的值不变。例如，随机函数不能包含在表达式中。

表达式可以出现在 SELECT 列表或者 WHERE 条件中，例如在以下 SQL 语句中，粗体下画线部分均为表达式：

SELECT **UPPER(CUST_ID)** FROM sales WHERE **QUANTITY_SOLD * AMOUNT_SOLD** > 1000;

In-Memory 虚拟列由初始化参数 INMEMORY_VIRTUAL_COLUMNS 控制，此参数的默认

值为 MANUAL，表示当表存在虚拟列时，除非手工指定，这些列不会被发布到内存列式存储中。将其设为 ENABLE 将启用 In-Memory 虚拟列功能，也就是在发布时会自动包含虚拟列。此参数只能在系统一级设置。

```
SQL> SHOW PARAMETER virtual_column

NAME                                 TYPE        VALUE
------------------------------------ ----------- ------------------------------
inmemory_virtual_columns             string      MANUAL

SQL> ALTER SYSTEM SET inmemory_virtual_columns='ENABLE';
System altered.
```

以下通过一个简单示例来了解如何使用 In-Memory 虚拟列。首先为 LINEORDER 表添加一列 vcol01，虚拟列的类型和表达式的结果类型一致。在后续的 SQL 负载中使用此表达式。另外，在发布到内存列式存储之前，此虚拟列不会实际占用物理空间。

```
SQL> ALTER TABLE lineorder ADD vcol01 AS (lo_extendedprice * (1 - lo_discount/100));

SQL> SELECT column_name, data_type, data_default FROM user_tab_cols WHERE table_name
= 'LINEORDER' AND column_name = 'VCOL01';

COLUMN_NAME  DATA_TYPE   DATA_DEFAULT
------------ ----------- ----------------------------------------
VCOL01       NUMBER      "LO_EXTENDEDPRICE"*(1-"LO_DISCOUNT"/100)
```

发布到内存列式存储后，可以从 V$IM_COLUMN_LEVEL 中查看到此列，UNSPECIFIED 表示此列使用默认的压缩级：

```
SQL> ALTER TABLE lineorder INMEMORY;
SQL> ALTER SYSTEM SET inmemory_virtual_columns='ENABLE';
SQL> EXEC dbms_inmemory.populate('SSB', 'LINEORDER');

SQL> SELECT table_name, column_name, inmemory_compression FROM v$im_column_level;

TABLE_NAME      COLUMN_NAME              INMEMORY_COMPRESSION
--------------- ------------------------ --------------------------
LINEORDER       VCOL01                   UNSPECIFIED
LINEORDER       LO_ORDERKEY              DEFAULT
LINEORDER       LO_LINENUMBER            DEFAULT
LINEORDER       LO_CUSTKEY               DEFAULT
...
```

从视图 V$IM_IMECOL_CU 可以查询发布到 IMEU 中列的信息。可以看到，此虚拟列为 LINEORDER 表的第 18 列。

```
SQL> SELECT column_name, internal_column_number, sql_expression FROM v$im_imecol_cu;

COLUMN_NAME         INTERNAL_COLUMN_NUMBER SQL_EXPRESSION
------------------- ---------------------- ----------------------------------------
VCOL01                                  18 "LO_EXTENDEDPRICE"*(1-"LO_DISCOUNT"/100)
VCOL01                                  18 "LO_EXTENDEDPRICE"*(1-"LO_DISCOUNT"/100)
...
```

IMEU 和 IMCU 是一一对应的，因为虚拟列实际也是表的一部分。从以下输出可以看到，IMEU 的数量是 24，和 IMCU 的数量是一致的。另外一个有趣的现象是，虽然 IMEU 中只有一列，但其发布时间几乎占到 IMCU 发布所有 17 列所用时间的 30%。这是由于发布 In-Memory 虚拟列时，除需要进行行列转换和压缩外，还需要进行表达式计算。

```
SQL> SELECT num_vircols, num_rows, time_to_populate FROM v$imeu_header
 WHERE num_rows != 0;

       NUM_VIRCOLS           NUM_ROWS   TIME_TO_POPULATE
------------------ ------------------ ------------------
                 1            570,198                433
                 1            515,507                347
                 1            534,568                394
                 1            531,983                381
                 1            504,992                355
                 1            504,780                347
                 1            514,063                357
                 1            513,998                358
                 1            516,080                344
                 1            515,501                356
                 1            513,172                369
                 1            516,584                361
                 1            514,070                358
                 1            514,064                373
                 1            514,067                363
                 1            514,053                357
                 1            514,073                353
                 1            514,061                360
                 1            514,060                356
                 1            514,060                357
                 1            514,062                356
                 1            514,066                357
                 1            514,062                357
                 1             91,872                 51

24 rows selected.

SQL> SELECT num_rows, num_cols, time_to_populate FROM v$im_header
 WHERE num_rows != 0;

          NUM_ROWS           NUM_COLS   TIME_TO_POPULATE
------------------ ------------------ ------------------
           570,198                 17              1,515
           515,507                 17              1,306
           534,568                 17              1,395
           531,983                 17              1,380
           504,992                 17              1,301
           504,780                 17              1,306
           514,063                 17              1,333
           513,998                 17              1,332
```

```
    516,080                    17              1,316
    515,501                    17              1,307
    513,172                    17              1,378
    516,584                    17              1,320
    514,070                    17              1,364
    514,064                    17              1,321
    514,067                    17              1,325
    514,053                    17              1,330
    514,073                    17              1,317
    514,061                    17              1,306
    514,060                    17              1,317
    514,060                    17              1,319
    514,062                    17              1,327
    514,066                    17              1,320
    514,062                    17              1,328
     91,872                    17                204

24 rows selected.
```

使用以下示例 SQL 进行测试，未启用内存列式存储时耗时 2.44 秒，启用内存列式存储但未启用 In-Memory 虚拟列时耗时 1.23 秒。启用 In-Memory 虚拟列后耗时为 0.31 秒，比未启用 In-Memory 虚拟列时快了近 4 倍。

```
SQL> SELECT lo_orderpriority, SUM(lo_extendedprice * (1 - lo_discount/100)) AS
  discount_price FROM lineorder GROUP BY lo_orderpriority ORDER BY lo_orderpriority;

LO_ORDERPRIORIT      DISCOUNT_PRICE
---------------      --------------------
1-URGENT             8,725,773,155,172
2-HIGH               8,735,041,593,624
3-MEDIUM             8,692,247,365,821
4-NOT SPECI          8,716,008,117,557
5-LOW                8,725,203,842,070

Elapsed: 00:00:00.31
```

从执行统计信息中也可以看到和 IMEU 相关的信息：

```
NAME                                                              VALUE
---------------------------------------------------------------   ----------
CPU used by this session                                          408
IM scan CUs columns accessed                                      144
IM scan CUs columns theoretical max                               816
IM scan CUs current                                                48
IM scan CUs memcompress for query low                              48
IM scan CUs no cleanout                                            48
IM scan CUs pcode aggregation IME                                  24
IM scan CUs pcode aggregation pushdown                             48
IM scan CUs readlist creation accumulated time                     38
IM scan CUs readlist creation number                               48
IM scan CUs split pieces                                           90
```

```
IM scan EU bytes in-memory                                    140,499,372
IM scan EU bytes uncompressed                                  80,098,802
IM scan EU rows                                                11,997,996
IM scan EUs columns accessed                                           24
IM scan EUs columns theoretical max                                   432
IM scan EUs memcompress for query low                                  24
IM scan EUs split pieces                                               22
```

纵观整个过程，In-Memory 虚拟列的配置非常简单。如果不想更改初始化参数 INMEMORY_VIRTUAL_COLUMNS，可以手工设置为虚拟列开启 INMEMORY 属性，最终效果是一样的。

```
ALTER TABLE lineorder INMEMORY INMEMORY(vcol01);
```

最后，删除虚拟列，将 LINEORDER 表还原，恢复初始化参数设置。

```
ALTER SYSTEM SET inmemory_virtual_columns = MANUAL;

ALTER TABLE lineorder DROP COLUMN vcol01;
```

7.1.2　In-Memory Expression

IME（In-Memory Expression）和 In-Memory 虚拟列基本原理一样。IME 同样也是发布到 IMEU 中，但不需要用户手工定义虚拟列，而是由系统自动捕捉表达式并生成隐含的虚拟列，随后发布到内存列式存储中。因此，In-Memory Expression 也可以认为是自动化的 In-Memory 虚拟列。

初始化参数 INMEMORY_EXPRESSIONS_USAGE 控制 IME 的使用，其默认值为 ENABLE。当其设置为 ENABLE 或 DYNAMIC_ONLY 时，均支持 IME 功能。

```
SQL> SHOW PARAMETER expression

NAME                                 TYPE        VALUE
------------------------------------ ----------- ------------------------------
inmemory_expressions_usage           string      ENABLE
```

Oracle 数据库在 12.2 版本新增了一项表达式跟踪（Expression Tracking）的功能，此功能可以捕获 SQL 负载中的热点表达式并存放于数据字典。这些信息可以供优化器和 IME 使用。

优化器将表达式统计信息存放于 ESS（Expression Statistics Store）。ESS 是 SGA 的一部分，但独立于内存列式存储。ESS 中存放了具体的表达式、表达式的执行次数和执行成本。通过 DBA_EXPRESSION_STATISTICS 可以查询 ESS 的信息。

```
SQL> SELECT expression_text, fixed_cost, evaluation_count, snapshot
FROM dba_expression_statistics WHERE table_name = 'LINEORDER'
ORDER BY fixed_cost DESC;

EXPRESSION_TEXT                                  FIXED_COST EVALUATION_COUNT SNAPSHOT
------------------------------------------------ ---------- ---------------- --------
SYS_OP_BLOOM_FILTER(:BF0000,"LO_ORDERDATE")      .000008444              440 LATEST
"LO_ORDERDATE"                                   .000008444             9240 LATEST
"LO_PARTKEY"                                     .000008444             9240 LATEST
"LO_SUPPKEY"                                     .000008444             9240 LATEST
"LO_CUSTKEY"                                     .000008444             9240 LATEST
```

```
SYS_OP_BLOOM_FILTER(:BF0002,"LO_SUPPKEY")        .000008444       20524413 LATEST
"LO_EXTENDEDPRICE"*(1-"LO_DISCOUNT"/100)         6.7552E-06      191970120 LATEST
"LO_EXTENDEDPRICE"*(1-"LO_DISCOUNT")             3.3776E-06             48 LATEST
"LO_ORDTOTALPRICE"*(1-"LO_DISCOUNT")             3.3776E-06             96 LATEST
"LO_REVENUE"                                     3.3776E-08       20533653 CUMULATIVE
...
```

在以上输出中，SNAPSHOT 列表示表达式捕获的时间段，取值 CUMULATIVE 表示自数据库创建起，LATEST 表示过去 24 小时。

使用 DBMS_INMEMORY_ADMIN PL/SQL 包中的 IME_CAPTURE_EXPRESSIONS 过程，可以将 ESS 中的排名前 20 的热点表达式以隐藏虚拟列的形式添加到对应表中，这些表达式的列名均以 SYS_IME 开始。注意，此过程只会考虑已发布到内存列式存储中的表。

可能存在一种情况，即系统自动添加的 SYS_IME 虚拟列并非全部是应用所需要的。这不但增加了后续发布的时间，同时也浪费了内存空间。在以下过程中，通过指定 CURRENT 参数捕获过去 24 小时的热点表达式并自动创建虚拟列。从视图 DBA_IM_EXPRESSIONS 或 USER_TAB_COLS 的输出可知，共有 3 列添加到 LINEORDER 表中。这些列既是虚拟列，也是隐藏列。除非显式指定，SELECT *和 DESCRIBE 命令中不会显示隐藏列。另外，虽然不是必需，In-Memory 虚拟列也可以建成隐藏列。

```
SQL> EXEC dbms_inmemory_admin.ime_capture_expressions('CURRENT');

SQL> SELECT table_name, column_name, sql_expression FROM dba_im_expressions;

TABLE_NAME    COLUMN_NAME                SQL_EXPRESSION
------------  -------------------------  ----------------------------------------
LINEORDER     SYS_IME0001000000E83D88    "LO_EXTENDEDPRICE"*(1-"LO_DISCOUNT")
LINEORDER     SYS_IME0001000000E83D89    "LO_ORDTOTALPRICE"*(1-"LO_DISCOUNT")
LINEORDER     SYS_IME0001000000E83D8A    "LO_EXTENDEDPRICE"*(1-"LO_DISCOUNT"/100)

SQL> SELECT column_id, column_name, data_default, hidden_column, virtual_column FROM
user_tab_cols WHERE table_name ='LINEORDER';

COLUMN_ID COLUMN_NAME              DATA_DEFAULT                             HID VIR
--------- ------------------------ ---------------------------------------- --- ---
        1 LO_ORDERKEY                                                       NO  NO
        2 LO_LINENUMBER                                                     NO  NO
        3 LO_CUSTKEY                                                        NO  NO
        4 LO_PARTKEY                                                        NO  NO
        5 LO_SUPPKEY                                                        NO  NO
        6 LO_ORDERDATE                                                      NO  NO
        7 LO_ORDERPRIORITY                                                  NO  NO
        8 LO_SHIPPRIORITY                                                   NO  NO
        9 LO_QUANTITY                                                       NO  NO
       10 LO_EXTENDEDPRICE                                                  NO  NO
       11 LO_ORDTOTALPRICE                                                  NO  NO
       12 LO_DISCOUNT                                                       NO  NO
       13 LO_REVENUE                                                        NO  NO
```

```
            14 LO_SUPPLYCOST                                                              NO    NO
            15 LO_TAX                                                                     NO    NO
            16 LO_COMMITDATE                                                               NO    NO
            17 LO_SHIPMODE                                                                 NO    NO
               SYS_IME0001000000E83D8A    "LO_EXTENDEDPRICE"*(1-"LO_DISCOUNT"/100)   YES  YES
               SYS_IME0001000000E83D89    "LO_ORDTOTALPRICE"*(1-"LO_DISCOUNT")       YES  YES
               SYS_IME0001000000E83D88    "LO_EXTENDEDPRICE"*(1-"LO_DISCOUNT")       YES  YES

20 rows selected.
```

但其中只有 SYS_IME0001000000E83D8A 列是所需要的。此时可以用两种方法去除不需要的表达式：直接使用 DDL 语句删除或调用 DBMS_INMEMORY PL/SQL 包中的 IME_DROP_EXPRESSIONS 过程。同时使用两种方法，最终留下所需的表达式。

```
SQL> EXEC dbms_inmemory.ime_drop_expressions(
'SSB', 'LINEORDER', 'SYS_IME0001000000E83D89');
PL/SQL procedure successfully completed.

SQL> ALTER TABLE lineorder DROP COLUMN SYS_IME0001000000E83D88;
Table altered.

SQL> SELECT table_name, column_name, sql_expression FROM dba_im_expressions;

TABLE_NAME    COLUMN_NAME                 SQL_EXPRESSION
------------  --------------------------  ----------------------------------------
LINEORDER     SYS_IME0001000000E83D8A     "LO_EXTENDEDPRICE"*(1-"LO_DISCOUNT"/100)
```

注意，此时这些新增列尚未发布到内存列式存储中，可以使用以下两种方法之一发布。其中，IME_POPULATE_EXPRESSIONS 发布所有已启用 INMEMORY 并带内存表达式的表，而 REPOPULATE 只发布指定的表。

```
-- 法1
EXEC DBMS_INMEMORY_ADMIN.IME_POPULATE_EXPRESSIONS();
-- 法2
EXEC DBMS_INMEMORY.REPOPULATE('SSB', 'LINEORDER');
```

通过以下查询可以确认这些表达式已发布到内存列式存储中的 IMEU，并且与 IMCU 数量一致：

```
SQL> SELECT COUNT(*) FROM v$im_header WHERE num_rows != 0;

  COUNT(*)
----------
        24

SQL> SELECT COUNT(*) FROM v$imeu_header WHERE num_rows != 0;

  COUNT(*)
----------
        24
```

这些新增的表达式约占用 134MB 内存空间。

```
SQL> SELECT SUM(LENGTH)/1024/1024 FROM v$im_imecol_cu;

  SIZE(MB)
----------
133.986507
```

最后，如果需要将这些表达式列从所有表定义中移除，可以运行以下过程：

```
SQL> EXEC dbms_inmemory_admin.ime_drop_all_expressions();
PL/SQL procedure successfully completed.
```

在 Oracle 18c 版本前，表达式捕获窗口只能设定为从数据库创建时开始或最近 24 小时，时间范围颗粒度较粗，可能导致不能准确捕获所需的表达式。从 Oracle 18c 版本开始，In-Memory Expression 支持动态窗口捕获，用户可以灵活指定时间范围，更精确地捕捉工作负载中的表达式。以下为使用动态窗口捕获表达式的示例，其中关键的一步是在过程 IME_CAPTURE_EXPRESSIONS 中指定新的 WINDOW 参数来实现动态窗口捕获：

```
SQL> SET SERVEROUTPUT ON

-- 开启捕获窗口
SQL> EXEC dbms_inmemory_admin.ime_open_capture_window();

-- 确认窗口状态为 OPEN
SQL>
DECLARE
   p_capture_state             varchar2(40);
   p_last_modified             timestamp;
BEGIN
   DBMS_INMEMORY_ADMIN.IME_GET_CAPTURE_STATE(p_capture_state, p_last_modified);
   DBMS_OUTPUT.PUT_LINE('State is '||p_capture_state||' on '|| p_last_modified);
END;
/

State is OPEN on 03-JUN-21 01.11.57.431943 PM

-- 执行示例 SQL 工作负载
SQL> SELECT lo_orderpriority, SUM(lo_extendedprice * (1 - lo_discount/100)) AS
discount_price FROM lineorder GROUP BY lo_orderpriority ORDER BY lo_orderpriority;

-- 关闭捕获窗口
SQL> EXEC dbms_inmemory_admin.ime_close_capture_window();

-- 确认窗口状态为 CLOSE
SQL>
DECLARE
   p_capture_state             varchar2(40);
   p_last_modified             timestamp;
BEGIN
   DBMS_INMEMORY_ADMIN.IME_GET_CAPTURE_STATE(p_capture_state, p_last_modified);
```

```
    DBMS_OUTPUT.PUT_LINE('State is '||p_capture_state||' on '|| p_last_modified);
END;
/

State is CLOSE on 03-JUN-21 01.16.42.775590 PM

-- 验证表达式已捕获
SQL> SELECT expression_text, fixed_cost, evaluation_count, snapshot
FROM  dba_expression_statistics WHERE table_name = 'LINEORDER' AND SNAPSHOT = 'WINDOW'
ORDER BY fixed_cost DESC;

EXPRESSION_TEXT                                    FIXED_COST EVALUATION_COUNT SNAPSHOT
-------------------------------------------------- ---------- ---------------- --------
"LO_EXTENDEDPRICE"*(1-"LO_DISCOUNT"/100)           6.7552E-06         11997996 WINDOW
"LO_DISCOUNT"                                      3.3776E-08         11997996 WINDOW
"LO_ORDERPRIORITY"                                 3.3776E-08         11997996 WINDOW
"LO_EXTENDEDPRICE"                                 3.3776E-08         11997996 WINDOW

-- 将捕获的表达式以虚拟列形式发布到内存列式存储
SQL> EXEC dbms_inmemory_admin.ime_capture_expressions('WINDOW');

-- 验证对象已添加虚拟列
SQL> SELECT column_id, column_name, data_default, hidden_column, virtual_column FROM
user_tab_cols WHERE table_name ='LINEORDER' AND column_name LIKE 'SYS_IME%';

 COLUMN_ID COLUMN_NAME               DATA_DEFAULT                                   HID VIR
---------- ------------------------- ---------------------------------------------- --- ---
           SYS_IME0001000000EC0E1C   "LO_EXTENDEDPRICE"*(1-"LO_DISCOUNT"/100)       YES YES

-- 查询所有 SYS_IME 虚拟列
SQL> SELECT table_name, column_name, sql_expression FROM dba_im_expressions;

TABLE_NAME    COLUMN_NAME               SQL_EXPRESSION
------------- ------------------------- ----------------------------------------
LINEORDER     SYS_IME0001000000EC0E1C   "LO_EXTENDEDPRICE"*(1-"LO_DISCOUNT"/100)

-- 发布这些 SYS_IME 虚拟列
SQL> EXEC dbms_inmemory_admin.ime_populate_expressions();
```

测试完成后，执行以下过程以清除所有 SYS_IME 虚拟列：

```
SQL> EXEC dbms_inmemory_admin.ime_drop_all_expressions();

SQL> SELECT table_name, column_name, sql_expression FROM dba_im_expressions;
no rows selected
```

和 In-Memory Expression 操作相关的过程主要由两个 PL/SQL 包提供，它们是 DBMS_INMEMORY 和 DBMS_INMEMORY_ADMIN，详见表 7-1。

表 7-1　In-Memory Expression 相关 PL/SQL 过程

PL/SQL 包	过程	描述
DBMS_INMEMORY	IME_DROP_EXPRESSIONS	删除指定表中的 SYS_IME 虚拟列
DBMS_INMEMORY_ADMIN	IME_CAPTURE_EXPRESSIONS	捕获指定时间段内排名前 20 的热点表达式。时间段可指定为 CUMULATIVE（自数据库创建起）、CURRENT（过去 24 小时）和 WINDOW（指定窗口）
DBMS_INMEMORY_ADMIN	IME_CLOSE_CAPTURE_WINDOW	关闭捕获窗口
DBMS_INMEMORY_ADMIN	IME_DROP_ALL_EXPRESSIONS	删除所有表中的 SYS_IME 虚拟列
DBMS_INMEMORY_ADMIN	IME_GET_CAPTURE_STATE	查询窗口状态，可能的值为 OPEN、CLOSED 或 DEFAULT（未使用）
DBMS_INMEMORY_ADMIN	IME_OPEN_CAPTURE_WINDOW	打开捕获窗口
DBMS_INMEMORY_ADMIN	IME_POPULATE_EXPRESSIONS	发布最近一次执行 IME_CAPTURE_EXPRESSIONS 时捕获的热点表达式

7.2　In-Memory 联结优化

7.2.1　联结组（Join Group）

联结组是 Oracle 12.2 版本新增的特性，可针对哈希联结做进一步的性能优化。在 12.2 版本以前，进行哈希联结的两列各自采用本地字典进行压缩，因此即使对于相同的列值，在两张表中的压缩编码也可能不同。这导致这两列必须解压才能进行后续的联结操作。联结组使用全局字典对来自不同表的联结键进行压缩，因此可以无须解压直接用压缩值进行比较，同时也避免了后续的哈希操作。联结组可以包括同一张表或不同表中的列，最多支持 255 列。表中的列只能属于一个联结组，Active Data Guard 的备数据库端不支持联结组，外部表也不支持联结组。

第 2 章已经给出了安装 Oracle 数据库的标准示例。以下基于 HR 用户下的 REGIONS 表和 COUNTRIES 表来说明在启用联结组时的哈希联结过程，SQL 查询如图 7-1 左上角所示。其大致过程如下。

（1）建立联结组，创建全局字典。REGION 和 COUNTRIES 两表将共享全局字典编码。为简化，假设 REGION_ID 列 1~4 的值依次编码为 0~3。在图 7-1 中，以括号中的值表示全局字典编码后的值，而括号外的值表示列的原始值。例如 2(1) 表示 REGION_ID 的原始值为 2，字典编码后的值为 1。

（2）扫描 REGIONS 表。根据过滤谓词（region_name = 'Asia'）返回满足条件的压缩格式的联结键，即 REGION_ID 3 对应的字典编码。同时构建布隆过滤器。

（3）构建联结键数组。由于 REGION_ID 列共有 4 个不同的值，因此数组共有 4 个元素，从 ARRAY[0] 到 ARRAY[3]。REGION_ID 3 对应的字典编码为 2，因此 ARRAY[2]=1，其他不匹配的元素均设为 0。

（4）扫描 COUNTRIES 表，并结合下放的布隆过滤器作为过滤条件，以压缩格式返回匹配的行。

（5）完成哈希联结。使用联结组时，完成哈希联结并不需要在检测端执行哈希计算，而是通过之前构建的联结键数组来实现。例如在上一步中，如果返回的 REGION_ID 为 0（压缩编码值），由于 ARRAY[0]=0，因此此行被过滤掉；如果返回的 REGION_ID 为 2（压缩编码值），由于 ARRAY[2]=1，因此此行被保留。然后补齐 SELECT 列表中所需的 COUNTRY_NAME 列，完成哈希联结。

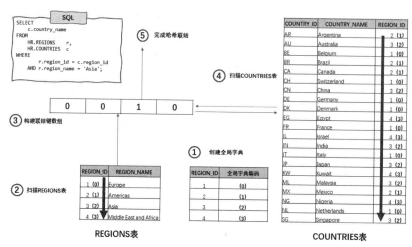

图 7-1　启用联结组时的哈希联结过程

如果没有联结组所建立的全局字典，则扫描 REGIONS 和 COUNTRIES 表后返回的数据必须解压缩，布隆过滤器也必须基于解压缩后的值建立。而且由于布隆过滤器可能存在误判，返回的数据还需要进行一次哈希计算。简而言之，使用联结组省去了解压缩和哈希计算的开销。

下面通过一个示例了解如何配置和使用联结组，测试用例使用 SSB 示例 SQL 4_1。注意在此 SQL 中额外添加了 SQL 提示 NO_VECTOR_TRANSFORM。这是由于向量转换和联结组这两种优化手段是互斥的，为确保联结组的使用，需要通过此 SQL 提示来禁用向量转换。

```
SELECT /* 4.1 SSB_SAMPLE_SQL */ /*+ NO_VECTOR_TRANSFORM MONITOR */
    d_year,
    c_nation,
    SUM(lo_revenue - lo_supplycost) AS profit
FROM
    date_dim,
    customer,
    supplier,
    part,
    lineorder
WHERE
        lo_custkey = c_custkey
    AND lo_suppkey = s_suppkey
    AND lo_partkey = p_partkey
    AND lo_orderdate = d_datekey
    AND c_region = 'AMERICA'
```

```
   AND s_region = 'AMERICA'
   AND ( p_mfgr = 'MFGR#1'
       OR p_mfgr = 'MFGR#2' )
GROUP BY
   d_year,
   c_nation
ORDER BY
   d_year,
   c_nation;
```

启用联结组前,首先将 SSB 示例的 5 张表全部发布到内存列式存储,其内存空间占用如下:

```
$ ./dbim_get_popstatus.sh

        SEGMENT         IN MEM      ON DISK   COMPRESSION  BYTES NOT  POPULATED
OWNER   NAME            SIZE(KB)    SIZE(KB)  RATIO        POPULATED  STATUS
------- --------------  ----------  --------- ------------ ---------- ---------
SSB     SUPPLIER         2,304        1,952     .85             0     COMPLETED
SSB     CUSTOMER        18,688       30,176    1.61             0     COMPLETED
SSB     LINEORDER      699,968    1,371,040    1.96             0     COMPLETED
SSB     PART            25,024      105,256    4.21             0     COMPLETED
SSB     DATE_DIM         1,280          344     .27             0     COMPLETED


TOTAL_DISK_SIZE(KB) TOTAL_IM_SIZE(KB) OVERALL_COMPRESSION_RATIO
------------------- ----------------- -------------------------
            1508768            747264                      2.02
```

在 SSB 示例 SQL 4_1 中,总共有 4 组表的联结,因此需要建立 jg1~jg4 共 4 个联结组。在创建联结组的语句中,表的顺序可以互换,但联结列必须保持一致。定义完成后,需要重新发布才能生效。

```
ALTER TABLE lineorder NO INMEMORY;
ALTER TABLE part NO INMEMORY;
ALTER TABLE customer NO INMEMORY;
ALTER TABLE supplier NO INMEMORY;
ALTER TABLE date_dim NO INMEMORY;

CREATE INMEMORY JOIN GROUP jg1 (lineorder(lo_custkey), customer(c_custkey));
CREATE INMEMORY JOIN GROUP jg2 (lineorder(lo_suppkey), supplier(s_suppkey));
CREATE INMEMORY JOIN GROUP jg3 (lineorder(lo_partkey), part(p_partkey));
CREATE INMEMORY JOIN GROUP jg4 (lineorder(lo_orderdate), date_dim(d_datekey));

ALTER TABLE lineorder INMEMORY;
ALTER TABLE part INMEMORY;
ALTER TABLE customer INMEMORY;
ALTER TABLE supplier INMEMORY;
ALTER TABLE date_dim INMEMORY;

EXEC dbms_inmemory.populate('SSB', 'LINEORDER');
EXEC dbms_inmemory.populate('SSB', 'CUSTOMER');
EXEC dbms_inmemory.populate('SSB', 'DATE_DIM');
```

```
EXEC dbms_inmemory.populate('SSB', 'PART');
EXEC dbms_inmemory.populate('SSB', 'SUPPLIER');
```

等待所有表发布完成,整体内存空间占用为 744 192 KB,较使用联结组前下降了 3072 KB,内存空间的节省全部来自事实表 LINEORDER。这说明单一全局字典比多个本地字典占用的内存空间更少。

```
$ ./dbim_get_popstatus.sh

            SEGMENT        IN MEM         ON DISK  COMPRESSION  BYTES NOT  POPULATED
OWNER       NAME           SIZE(KB)       SIZE(KB)       RATIO  POPULATED  STATUS
-----       --------       --------       --------  -----------  ---------  ---------
SSB         SUPPLIER          2,304          1,952          .85          0  COMPLETED
SSB         CUSTOMER         18,688         30,176         1.61          0  COMPLETED
SSB         LINEORDER       696,896      1,371,040         1.97          0  COMPLETED
SSB         PART             25,024        105,256         4.21          0  COMPLETED
SSB         DATE_DIM          1,280            344          .27          0  COMPLETED


TOTAL_DISK_SIZE(KB) TOTAL_IM_SIZE(KB) OVERALL_COMPRESSION_RATIO
------------------- ----------------- -------------------------
            1508768            744192                      2.03
```

从视图 USER_JOINGROUPS 中可以查询联结组的信息。GD_ADDRESS 相同表示各列共享了全局字典。

```
SQL> SELECT joingroup_name, table_name, column_name, gd_address
     FROM user_joingroups ORDER BY joingroup_name;

JOINGROUP_NAME      TABLE_NAME          COLUMN_NAME         GD_ADDRESS
---------------     -------------       ---------------     ----------------
JG1                 CUSTOMER            C_CUSTKEY           00000000863FFF40
JG1                 LINEORDER           LO_CUSTKEY          00000000863FFF40
JG2                 LINEORDER           LO_SUPPKEY          0000000086FFFF40
JG2                 SUPPLIER            S_SUPPKEY           0000000086FFFF40
JG3                 LINEORDER           LO_PARTKEY          00000000867FFF40
JG3                 PART                P_PARTKEY           00000000867FFF40
JG4                 LINEORDER           LO_ORDERDATE        00000000871FFF40
JG4                 DATE_DIM            D_DATEKEY           00000000871FFF40

8 rows selected.
```

使用另一个联合查询也可以证明这些列共享了全局字典,其中,表 V$IM_COL_CU 中的列 SEGMENT_DICTIONARY_ADDRESS 为 00 表示使用了本地字典,否则为全局字典:

```
SQL> SELECT b.table_name, b.column_name, a.segment_dictionary_address
FROM v$im_col_cu a, v$im_column_level b
WHERE a.objd = b.obj_num AND a.column_number = b.segment_column_id
AND a.segment_dictionary_address != '00';
```

```
TABLE_NAME        COLUMN_NAME         SEGMENT_DICTIONARY_ADDRESS
---------------   -----------------   --------------------------
SUPPLIER          S_SUPPKEY           0000000086FFFF40
PART              P_PARTKEY           00000000867FFF40
CUSTOMER          C_CUSTKEY           00000000863FFF40
DATE_DIM          D_DATEKEY           00000000871FFF40
LINEORDER         LO_CUSTKEY          00000000863FFF40
LINEORDER         LO_PARTKEY          00000000867FFF40
LINEORDER         LO_SUPPKEY          0000000086FFFF40
LINEORDER         LO_ORDERDATE        00000000871FFF40
...
```

和联结组相关的系统视图还包括 V$IM_GLOBALDICT、V$IM_GLOBALDICT_VERSION、V$IM_GLOBALDICT_SORTORDER 和 V$IM_GLOBALDICT_PIECEMAP。

使用联结组时的执行时间和执行计划如下，可以看到其中也使用了布隆过滤器对哈希联结进行优化：

```
SQL> @4_1
Elapsed: 00:00:00.32

SQL> @../../../showplan
Plan hash value: 1592036000

---------------------------------------------------------------------------------------
| Id  | Operation                        | Name     | Rows  |TempSpc| Cost (%CPU)|
---------------------------------------------------------------------------------------
|   0 | SELECT STATEMENT                 |          |       |       | 58772 (100)|
|   1 |  SORT GROUP BY                   |          |    35 |       | 58772   (2)|
|*  2 |   HASH JOIN                      |          | 9790K |       | 58486   (1)|
|   3 |    TABLE ACCESS INMEMORY FULL    | DATE_DIM |  2556 |       |     1   (0)|
|*  4 |    HASH JOIN                     |          | 9790K |       | 58457   (1)|
|   5 |     JOIN FILTER CREATE           | :BF0000  |  3264 |       |     3   (0)|
|*  6 |      TABLE ACCESS INMEMORY FULL  | SUPPLIER |  3264 |       |     3   (0)|
|*  7 |     HASH JOIN                    |          |   11M | 2152K | 58421   (1)|
|   8 |      JOIN FILTER CREATE          | :BF0001  | 47777 |       |    42  (10)|
|*  9 |       TABLE ACCESS INMEMORY FULL | CUSTOMER | 47777 |       |    42  (10)|
|* 10 |      HASH JOIN                   |          |   11M | 9368K | 27015   (2)|
|  11 |       JOIN FILTER CREATE         | :BF0002  |  399K |       |   152  (13)|
|* 12 |        TABLE ACCESS INMEMORY FULL| PART     |  399K |       |   152  (13)|
|  13 |       JOIN FILTER USE            | :BF0000  |   11M |       |  1954  (12)|
|  14 |       JOIN FILTER USE            | :BF0001  |   11M |       |  1954  (12)|
|  15 |       JOIN FILTER USE            | :BF0002  |   11M |       |  1954  (12)|
|* 16 |        TABLE ACCESS INMEMORY FULL| LINEORDER|   11M |       |  1954  (12)|
---------------------------------------------------------------------------------------
```

要确认联结组是否使用，只有通过实时 SQL 监控报告。如图 7-2 所示，单击执行计划第 4 行中的望远镜图标，在弹出窗口中的 Columnar Encodings Leveraged 为正数，并且没有出现 Columnar Encodings Observed 时，表示已使用联结组。

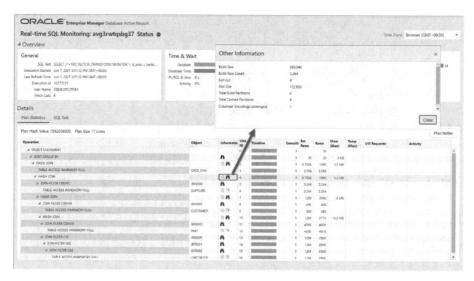

图 7-2　从实时 SQL 监控报告中确认联结组已使用

如果不再希望使用联结组，可以执行以下 SQL 语句进行删除。删除联结组同时也会将对象从内存列式存储中清除：

```
DROP INMEMORY JOIN GROUP jg1;
DROP INMEMORY JOIN GROUP jg2;
DROP INMEMORY JOIN GROUP jg3;
DROP INMEMORY JOIN GROUP jg4;
```

在前面的测试中，在 SQL 语句中通过 SQL 提示强行禁止了向量转换，使得 SQL 优化器选择了布隆过滤器和联结组。表 7-2 列举了采用其他几种优化手段时的测试结果，仅就 SSB 示例 SQL 4_1 而言，采用向量转换时的性能最好。

表 7-2　SSB 示例 SQL 4_1 使用不同优化手段时的性能测试结果

内存列式存储	优 化 手 段	CPU 时间/秒	ELAPSED 时间/秒
否	N/A	7.32	11.67
否	向量转换	6.64	11.46
是	布隆过滤器	0.39	0.403
是	布隆过滤器+联结组	0.316	0.321
是	向量转换	0.178	0.180

在实际的工作负载中，SQL 优化器会基于成本自动选择最适合的优化手段，有时可能选择联结组，有时可能向量转换效率更高。最重要的是，这一切都是自动的，无须人工干预。

需要注意的是，由于启用联结组需要重新发布对象，因此最好在最初设计时就将联结组考虑在内，或者后续启用时提前规划好时间，以最小化对业务系统的影响。

7.2.2　In-Memory 深度向量化

In-Memory 深度向量化是 Oracle 数据库 21c 版本发布的基于 SIMD 的框架，是在之前向量化操作基础上的进一步深化和优化。In-Memory 向量化联结是此框架中的核心技术，可以优化哈希联结中的哈希计算、构建、探测和收集操作。In-Memory 向量化联结可以和联结组、In-Memory 动态扫描、聚合下推以及 In-Memory 压缩等功能配合使用。

以下来看一个 In-Memory 向量化联结的示例，执行的 SQL 查询如下：

```sql
SELECT /*+ NO_VECTOR_TRANSFORM */ COUNT(*)
FROM
    customer,
    lineorder
WHERE
        c_custkey = lo_custkey
    AND c_nation = 'FRANCE';
```

In-Memory 深度向量化需配合哈希联结使用。在第 6 章已经介绍过，哈希联结的主要优化手段为布隆过滤器和 VECTOR GROUP BY，这两项技术是互斥的，只能二选一。因此需要在 SQL 语句中加入 NO_VECTOR_TRANSFORM 提示，以禁用 VECTOR GROUP BY。

在图 7-3 的实时 SQL 监控报告中，单击执行计划第 2 行的望远镜图标，输出信息中的 DeepVec Hash Joins 为 1，表明已启用 In-Memory 深度向量化。

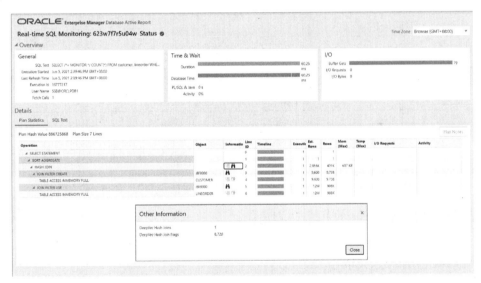

图 7-3　从实时 SQL 监控报告中确认 In-Memory 深度向量化已启用

初始化参数 INMEMORY_DEEP_VECTORIZATION 可以控制开启或关闭 In-Memory 深度向量化功能，默认值为开启。即使设为开启，In-Memory 深度向量化也并非一定会被使用，具体需要看 SQL 优化器所评估的执行成本。无论如何，In-Memory 深度向量化对于用户是透明的，也无须修改应用。

```
SQL> SHOW PARAMETER inmemory_deep_vectorization

NAME                                 TYPE        VALUE
------------------------------------ ----------- ------------------------------
inmemory_deep_vectorization          boolean     TRUE
```

In-Memory 深度向量化既可以加速普通的哈希联结，也可以加速基于联结组的哈希联结。在随书示例脚本 chap07/im_deep_vectorization.sql 中，对开启和关闭 In-Memory 深度向量化以及建立和未建立联结组时的 4 种组合做了性能比较。控制 In-Memory 深度向量化的 SQL 语句如下：

```
-- 开启 In-Memory 深度向量化
ALTER SESSION SET inmemory_deep_vectorization = TRUE;
-- 禁用 In-Memory 深度向量化
ALTER SESSION SET inmemory_deep_vectorization = FALSE;
```

建立联结组的 SQL 语句如下：

```
ALTER TABLE lineorder NO INMEMORY;
ALTER TABLE customer NO INMEMORY;
CREATE INMEMORY JOIN GROUP jg1(lineorder(lo_custkey), customer(c_custkey));
ALTER TABLE lineorder INMEMORY;
ALTER TABLE customer INMEMORY;
```

最终 4 轮测试的结果如表 7-3 所示。从测试结果可知，在未建立联结组时，In-Memory 深度向量化性能提升约为 2.4 倍；建立联结组后，In-Memory 深度向量化性能提升约为 2.33 倍；两种优化手段叠加后的总体性能提升为 6.33 倍。另外，单独使用 In-Memory 深度向量化或联结组均能加速哈希联结，且执行时间和 CPU 消耗接近。但 In-Memory 深度向量化更简单，因其为完全自动执行，无须额外的配置。

表 7-3　In-Memory 深度向量化性能测试结果

测试序号	深度向量化	联结组	执行时间（秒）	CPU 消耗
1	关闭	未建立	0.19	26
2	开启	未建立	0.08	14
3	关闭	已建立	0.07	14
4	开启	已建立	0.03	8

7.3　In-Memory 数据类型优化

7.3.1　In-Memory JSON 列

JSON (JavaScript Object Notation)是一种基于文本的轻量级数据交互格式，它易于机器解析，编码工作量少，运行速度快。JSON 可用于系统之间的数据交换和传输，将复杂数据转换为简单易懂的数据模型，存储临时数据等，是最受开发人员青睐的数据格式之一。

Oracle 数据库从 12.0.1.2 版本开始提供 JSON 支持，从此在结构化数据之外还可以存储和处理半结构化数据。和数据库中其他类型的数据一样，JSON 数据可以使用 OCI、.NET 和 JDBC 进行访问。JSON 文件可以使用 SQL Loader 导入数据库，JSON 表可以和传统的关系型表之间进行联结。此外，存储在数据库中的 JSON 数据可以获得额外的安全性和可靠性保护。在 Oracle 12.0.1.2 版本，JSON 数据以文本形式存于数据库中，数据类型仍使用标准的 SQL 数据类型，如 VARCHAR2、CLOB 和 BLOB。建议使用 BLOB，因为可以避免字符集转换。为了保证数据符合 JSON 语法，可以为 JSON 列添加 IS JSON 检查约束。Oracle 数据库 21c 版本新增了原生的 JSON 数据类型，新的 JSON 数据类型采用优化的二进制格式 OSON 存储数据，OSON 格式支持 SIMD 向量处理，可加速 JSON 数据的查询和更新。

```
-- Oracle 21c 之前，通过 IS JSON 检查约束确保 JSON 数据合法性
SQL> CREATE TABLE json_table(a BLOB CHECK(a IS JSON));
Table created.

-- Oracle 21c 开始，支持原生的 JSON 类型
SQL> CREATE TABLE json_table(a JSON);
Table created.
```

Database In-Memory 对 JSON 数据的优化从 Oracle 12.2.0.1 版本开始。加载到内存列式存储中的数据以 OSON 格式存储，可加速 JSON 路径访问操作的性能，包括 JSON_TABLE、JSON_VALUE 和 JSON_EXISTS。到 Oracle 21c 版本，全文搜索操作 JSON_TEXTCONTAINS 也可以通过 In-Memory 全文本列特性提升性能，详见 7.3.3 节。

当满足以下条件时，Oracle 数据库会自动为 JSON 列创建 In-Memory 虚拟列。

（1）包含 JSON 数据的列添加了 IS JSON 检查约束或使用原生 JSON 类型定义。

（2）IMMEMORY_EXPRESSIONS_USAGE 初始化参数设置为 STATIC_ONLY 或 ENABLE。此参数默认值为 ENABLE。

（3）如果包含 JSON 数据的列不是用原生 JSON 类型定义，还需要将初始化参数 MAX_STRING_SIZE 设置为 EXTENDED。此参数默认值为 STANDARD。

Oracle 数据库在 12c 版本之前，VARCHAR2 数据类型最多只支持 4000 字节。从 12c 版本开始，通过将 MAX_STRING_SIZE 设置为 EXTENDED，VARCHAR2 最多可支持 32 767 字节。可以为非 CDB，整个 CDB 或单个 PDB 设置 MAX_STRING_SIZE 为 EXTENDED。假设 PDB 的名字为 orclpdb1，以下为单个 PDB 的设置过程：

```
CONNECT / AS SYSDBA

-- 重启 PDB，进入 UPGRADE 模式
ALTER PLUGGABLE DATABASE orclpdb1 CLOSE;
ALTER PLUGGABLE DATABASE orclpdb1 OPEN UPGRADE;

-- 切换到 PDB
ALTER SESSION SET container = orclpdb1;

-- 修改参数为 EXTENDED
ALTER SYSTEM SET max_string_size = "EXTENDED";

-- 根据需要，调整部分视图中 VARCHAR2、NVARCHAR2 和 RAW 列的定义
@?/rdbms/admin/utl32k.sql

-- 重启 PDB
SHUTDOWN IMMEDIATE;
STARTUP;
```

```
-- 重新编译数据库中的无效对象
@?/rdbms/admin/utlrp.sql
```

以上设置过程是不可逆的,也就是说 MAX_STRING_SIZE 修改为 EXTENDED 后,不能再改回 STANDARD。

通过一个示例来了解 In-Memory JSON 列的使用过程。以 Oracle 数据库标准示例中 OE 用户下的 J_PURCHASEORDER 表为例(示例数据的基础安装参见 2.4.3 节),首先建立数据库目录,然后基于此目录下的 JSON 数据文件建立外部表:

```
CREATE OR REPLACE DIRECTORY order_entry_dir AS
    '/home/oracle/db-sample-schemas/order_entry';

CREATE TABLE json_dump_file_contents(json_document BLOB)
  ORGANIZATION EXTERNAL(TYPE ORACLE_LOADER DEFAULT DIRECTORY order_entry_dir
  ACCESS PARAMETERS(RECORDS DELIMITED BY 0x'0A'
                    DISABLE_DIRECTORY_LINK_CHECK
                    FIELDS(json_document CHAR(5000)))
  LOCATION(order_entry_dir:'PurchaseOrders.dmp'))
  PARALLEL   REJECT LIMIT UNLIMITED;
```

接下来,根据使用 IS JSON 检查约束还是 JSON 原生数据类型,测试表的语句也有所不同。先来看使用 IS JSON 检查约束的情形。按照以下语句建表,并从外部表导入数据:

```
CREATE TABLE j_purchaseorder(
    id           VARCHAR2(32) NOT NULL PRIMARY KEY,
    date_loaded  TIMESTAMP(6) WITH TIME ZONE,
    po_document  BLOB CONSTRAINT ensure_json CHECK(po_document IS JSON)
)
    LOB(po_document)STORE AS(CACHE);

INSERT INTO j_purchaseorder(id, date_loaded, po_document)
  SELECT SYS_GUID(), SYSTIMESTAMP, json_document FROM json_dump_file_contents
    WHERE json_document IS JSON;

COMMIT;
```

此表包含 10 000 条数据,其中 po_document 列为 JSON 格式数据。虽然尚未发布到内存列式存储,但 J_PURCHASEORDER 表已经添加了 OSON 格式的虚拟列,列名前缀为 SYS_IME_OSON。

```
SQL> SELECT column_id, column_name, data_type, data_length, hidden_column,
virtual_column FROM user_tab_cols WHERE table_name = 'J_PURCHASEORDER';

COLUMN_ID COLUMN_NAME                   DATA_TYPE                      DATA_LENGTH HID VIR
--------- ----------------------------- ------------------------------ ----------- --- ---
        1 ID                            VARCHAR2                                32 NO  NO
        2 DATE_LOADED                   TIMESTAMP(6) WITH TIME ZONE             13 NO  NO
        3 PO_DOCUMENT                   BLOB                                  4000 NO  NO
          SYS_IME_OSON_EC8C219757D      RAW                                  32767 YES YES
          64FC6BF5D4B236731B9E7

SQL> select * from USER_JSON_COLUMNS;
```

TABLE_NAME	OBJEC	COLUMN_NAME	FORMAT	DATA_TYPE
J_PURCHASEORDER	TABLE	PO_DOCUMENT	**TEXT**	**BLOB**

将表发布到内存列式存储,并执行测试 SQL:

```
ALTER TABLE j_purchaseorder INMEMORY;
EXEC dbms_inmemory.populate('SSB', 'J_PURCHASEORDER');

-- 等待发布完成,执行以下测试 SQL
SELECT
    po.po_document
FROM
    j_purchaseorder po
WHERE
    JSON_EXISTS ( po.po_document, '$.LineItems[*]?(@.Part.UPCCode == 85391628927 &&
@.Quantity > 3)' );
```

在执行计划中,可以看到 OSON 和 ime 关键字,说明 OSON 格式的虚拟列已发布到 IMEU 并被使用:

```
-------------------------------------------------------------------------------------
| Id  | Operation                  | Name            | Rows  | Bytes | Cost (%CPU)|
-------------------------------------------------------------------------------------
|   0 | SELECT STATEMENT           |                 |       |       |    18 (100)|
|*  1 |  TABLE ACCESS INMEMORY FULL| J_PURCHASEORDER |    93 | 1669K |    18   (6)|
-------------------------------------------------------------------------------------

Predicate Information (identified by operation id):
---------------------------------------------------

   1 - inmemory(JSON_EXISTS2("PO"."PO_DOCUMENT" FORMAT JSON,
       '$.LineItems[*]?(@.Part.UPCCode == 85391628927 && @.Quantity > 3)' FALSE ON ERROR,
       OSON("PO_DOCUMENT" FORMAT JSON, 'ime' RETURNING RAW(32767) NULL ON ERROR))=1)
       filter(JSON_EXISTS2("PO"."PO_DOCUMENT" FORMAT JSON,
       '$.LineItems[*]?(@.Part.UPCCode == 85391628927 && @.Quantity > 3)' FALSE ON ERROR,
       OSON("PO_DOCUMENT" FORMAT JSON, 'ime' RETURNING RAW(32767) NULL ON ERROR))=1)

Note
-----
   - dynamic statistics used: dynamic sampling (level=2)
```

同样,在执行统计信息中也可以看到 EU 关键字:

```
NAME                                                              VALUE
---------------------------------------------------------------- ----------
CPU used by this session                                             15
IM scan CUs columns accessed                                          2
IM scan CUs memcompress for query low                                 2
IM scan EU rows                                                   10000
IM scan EUs columns accessed                                          1
IM scan EUs memcompress for query low                                 1
IM scan rows                                                      20000
```

```
IM scan rows projected                                438
IM scan rows valid                                  10428
IM scan segments minmax eligible                        1
session logical reads                                3429
session logical reads - IM                           3260
session pga memory                               26470384
table scans (IM)                                        2

14 rows selected.
```

接下来，看一下使用原生 JSON 数据类型的情形，此时数据库版本必须为 21c 版本或以上。初始化参数 MAX_STRING_SIZE 不要求设置为 EXTENDED，可以保持其默认值 STANDARD。按以下语句建表，并从外部表导入数据：

```
CREATE TABLE j_purchaseorder(
    id              VARCHAR2(32) NOT NULL PRIMARY KEY,
    date_loaded     TIMESTAMP(6) WITH TIME ZONE,
    po_document     JSON
);

INSERT INTO j_purchaseorder(id, date_loaded, po_document)
  SELECT SYS_GUID(), SYSTIMESTAMP, json_document FROM json_dump_file_contents
    WHERE json_document IS JSON;

COMMIT;
```

同样，虚拟列在未启用内存列式存储时已经创建，列名前缀为 SYS_IME_OSON。

```
SQL> SELECT column_id, column_name, data_type, data_length, hidden_column,
virtual_column FROM user_tab_cols WHERE table_name = 'J_PURCHASEORDER';

COLUMN_ID COLUMN_NAME                DATA_TYPE                      DATA_LENGTH HID VIR
--------- -------------------------- ------------------------------ ----------- --- ---
        1 ID                         VARCHAR2                                32 NO  NO
        2 DATE_LOADED                TIMESTAMP(6) WITH TIME ZONE             13 NO  NO
        3 PO_DOCUMENT                JSON                                  8200 NO  NO
          SYS_IME_OSON_767FC01D92E   RAW                                  32767 YES YES
          54FEEBF653534DFC368D4

SQL> SELECT * FROM user_json_columns;

TABLE_NAME         OBJEC COLUMN_NAME          FORMAT    DATA_TYPE
----------------   ----- ------------------   --------  ---------------
J_PURCHASEORDER    TABLE PO_DOCUMENT          OSON      JSON
```

采用与之前相同的测试 SQL，在执行计划也可以看到 OSON 关键字：

```
---------------------------------------------------------------------------------
| Id | Operation                   | Name            | Rows | Bytes | Cost (%CPU)|
---------------------------------------------------------------------------------
|  0 | SELECT STATEMENT            |                 |      |       |   15 (100) |
|* 1 | TABLE ACCESS INMEMORY FULL  | J_PURCHASEORDER |   22 |  440K |   15   (7) |
---------------------------------------------------------------------------------
```

```
Predicate Information(identified by operation id):
---------------------------------------------------

   1 - inmemory(JSON_EXISTS2("PO"."PO_DOCUMENT" /*+ LOB_BY_VALUE */  FORMAT OSON,
             '$.LineItems[*]?(@.Part.UPCCode == 85391628927 && @.Quantity > 3)' FALSE
ON ERROR)=1)
       filter(JSON_EXISTS2("PO"."PO_DOCUMENT" /*+ LOB_BY_VALUE */  FORMAT OSON,
             '$.LineItems[*]?(@.Part.UPCCode == 85391628927 && @.Quantity > 3)' FALSE
ON ERROR)=1)

Note
-----
   - dynamic statistics used: dynamic sampling (level=2)
```

在执行统计信息中已看不到 EU 关键字,这是由于存放在原生 JSON 数据类型定义的列中的数据已经是 OSON 格式,在内存列式存储中存放在 IMCU 中。而在 IMEU 中存放的是优化的 JSON 索引结构,而非 OSON 数据。

```
NAME                                                             VALUE
---------------------------------------------------------- ----------
CPU used by this session                                            12
IM scan CUs columns accessed                                         2
IM scan CUs memcompress for query low                                2
IM scan rows                                                     20000
IM scan rows projected                                             492
IM scan rows valid                                               10482
IM scan segments minmax eligible                                     1
session logical reads                                             2983
session logical reads - IM                                        2756
session pga memory                                            22079472
table scans(IM)                                                      2

11 rows selected.
```

对比两种方式的空间占用,JSON 原生数据类型占用的磁盘和内存空间均少于采用 IS JSON 检查约束时的情形。另外,采用 JSON 原生数据类型时的 USED_LEN 仅为 740 字节,也说明其存放的并非实际 JSON 数据。

```
-- 执行以下 SQL 脚本和命令
SQL> !../chap04/dbim_get_popstatus.sh
SQL> SELECT allocated_len, used_len, num_vircols, num_rows, time_to_populate FROM
v$imeu_header WHERE num_rows != 0;

-- 采用 IS JSON 检查约束的输出
TOTAL_DISK_SIZE(KB) TOTAL_IM_SIZE(KB) OVERALL_COMPRESSION_RATIO
------------------- ----------------- -------------------------
              13040             15616                       .84

ALLOCATED_LEN   USED_LEN NUM_VIRCOLS   NUM_ROWS TIME_TO_POPULATE
------------- ---------- ----------- ---------- ----------------
      5242880    5225765           1      10000              245
```

```
-- 采用JSON原生数据类型的输出
TOTAL_DISK_SIZE(KB) TOTAL_IM_SIZE(KB) OVERALL_COMPRESSION_RATIO
------------------- ----------------- --------------------------
              11024             10496                       1.05

ALLOCATED_LEN   USED_LEN NUM_VIRCOLS   NUM_ROWS TIME_TO_POPULATE
------------- ---------- ----------- ---------- ----------------
      1048576        740           1      10000              317
```

将测试表 J_PURCHASEORDER 中的数据放大十倍，以不同的 JSON 列定义方式与是否发布到内存列式存储相组合，其性能和空间测试结果如表 7-4 所示。可以看到，采用 In-Memory JSON 列特性后，执行时间比之前快了 6 倍。另外，即使未启用 INMEMORY，采用 JSON 原生数据类型也有很好的性能表现。

表 7-4 In-Memory JSON 列性能和空间比较

序号	JSON 列定义	INMEMORY	执行时间/秒	CPU 时间 (1/100 秒)	磁盘空间/KB	内存空间/KB
1	IS JSON 检查约束	否	1.58	240	127688	N/A
2	IS JSON 检查约束	是	0.28	22	127688	155072
3	JSON 原生数据类型	否	0.28	151	113416	N/A
4	JSON 原生数据类型	是	0.24	30	113416	127424

测试完成后，可使用以下 SQL 语句清理测试环境：

```
ALTER TABLE j_purchaseorder NO INMEMORY;
DROP TABLE j_purchaseorder;
DROP TABLE json_dump_file_contents;
```

7.3.2 In-Memory 优化运算

NUMBER 是 Oracle 数据库原生的数据类型，用来存放定点和浮点数。NUMBER 的定义由精度和标度组成，精度指所有的位数，标度是小数点后的位数。例如 12.345 的精度为 5，标度为 3。NUMBER 类型的精度最多为 38。Oracle 数据库内部以变长格式存储 NUMBER 数据类型，通过 vsize 和 dump 函数可以查看其存储详情：

```
SQL> CREATE TABLE numtest(n NUMBER);

SQL> INSERT INTO numtest VALUES(-100.123);
SQL> INSERT INTO numtest VALUES(-100);
SQL> INSERT INTO numtest VALUES(0);
SQL> INSERT INTO numtest VALUES(100);
SQL> INSERT INTO numtest VALUES(100.123);

SQL> SELECT n, vsize(n), dump(n) FROM numtest;
```

```
       N   VSIZE(N)  DUMP(N)
--------- --------- ------------------------------------
 -100.123         6  Typ=2 Len=6: 61,100,101,89,71,102
     -100         3  Typ=2 Len=3: 61,100,102
        0         1  Typ=2 Len=1: 128
      100         2  Typ=2 Len=2: 194,2
  100.123         5  Typ=2 Len=5: 194,2,1,13,31
```

NUMBER 数据类型的变长特性使其不能很好地与 SIMD 向量计算结合。尽管通过压缩转换成定长格式后，可以结合 SIMD 进行等值和比较查询，但数学运算仍需要解压还原后才能进行。In-Memory 优化运算是 Oracle 18c 新增功能，使用优化的格式对 NUMBER 数据类型重新编码。当表的压缩级别为 QUERY LOW 时，这种格式可以结合硬件的 SIMD 特性来实现快速计算，此特性有助于提升简单聚合和 GROUP BY 聚合的性能。

由于并非支持所有类型的运算，优化的 NUMBER 格式并不能完全取代原生的 NUMBER 数据类型，因此内存列式存储中需要同时存储两种格式，因此会带来额外的内存消耗。

In-Memory 优化运算功能由初始化参数 INMEMORY_OPTIMIZED_ARITHMETIC 控制，其默认值为 DISABLE。

```
SQL> SHOW PARAMETER inmemory_optimized_arithmetic

NAME                                 TYPE        VALUE
------------------------------------ ----------- ------------------------------
inmemory_optimized_arithmetic        string      DISABLE
```

通过一个示例了解启用 In-Memory 优化运算对性能和内存空间的影响。以 SSB 示例 SQL 4_1 作为测试查询，其中包含 NUMBER 类型列和数学运算。在没有开启 In-Memory 优化运算时，先记录内存空间消耗和性能表现：

```
$ ./dbim_get_popstatus.sh

                     IN MEM     ON DISK  COMPRESSION  BYTES NOT  POPULATED
       SEGMENT       SIZE(KB)   SIZE(KB)      RATIO  POPULATED   STATUS
OWNER  NAME
------ ----------- ---------- ---------- ----------- ---------- ----------
SSB    LINEORDER      579,136  1,332,376        2.3          0  COMPLETED
SSB    DATE_DIM         1,280        344         .27         0  COMPLETED
SSB    SUPPLIER         2,304      1,952         .85         0  COMPLETED
SSB    CUSTOMER        18,688     30,176        1.61         0  COMPLETED
SSB    PART            21,952    105,256        4.79         0  COMPLETED


TOTAL_DISK_SIZE(KB)  TOTAL_IM_SIZE(KB)  OVERALL_COMPRESSION_RATIO
-------------------  -----------------  -------------------------
            1470104             623360                       2.36

SQL> @4_1
Elapsed: 00:00:00.12
```

然后开启 In-Memory 优化运算：

```
SQL> ALTER SYSTEM SET inmemory_optimized_arithmetic = ENABLE;
```

重新发布 SSB 示例所有对象，会发现有两张表的内存空间占用发生了变化，其中，LINEORDER 表增加了 20%，PART 表增加了 14%。

```
$ ./dbim_get_popstatus.sh
            SEGMENT          IN MEM      ON DISK  COMPRESSION   BYTES NOT  POPULATED
OWNER       NAME             SIZE(KB)    SIZE(KB)       RATIO   POPULATED  STATUS
------      ----------       --------    --------  -----------  ---------  ---------
SSB         LINEORDER         699,968   1,332,376          1.9          0  COMPLETED
SSB         DATE_DIM            1,280         344          .27          0  COMPLETED
SSB         SUPPLIER            2,304       1,952          .85          0  COMPLETED
SSB         CUSTOMER           18,688      30,176         1.61          0  COMPLETED
SSB         PART               25,024     105,256         4.21          0  COMPLETED


TOTAL_DISK_SIZE(KB)  TOTAL_IM_SIZE(KB)  OVERALL_COMPRESSION_RATIO
-------------------  -----------------  -------------------------
            1470104             747264                       1.97
```

再次运行测试负载，其执行时间为 0.08 秒：

```
SQL> @4_1
Elapsed: 00:00:00.08
```

总结以上的测试，在启用 In-Memory 优化运算后，内存占用增加了近 20%，运算速度快了 1.5 倍。因此，在实际的业务系统中，需要仔细评估新增内存的代价和性能提升的程度，然后决定是否启用 In-Memory 优化运算特性。

7.3.3 In-Memory 全文本列

Oracle Text 使用标准 SQL 来搜索和分析 Oracle 数据库内的文本和文档，并可以用纯文本、高亮关键字和 HTML 格式显示结果。Oracle 数据库并没有专门的 Text 数据类型，使用 CHAR、VARCHAR2、CLOB、BLOB 或 JSON 类型都可以存储文本，因此用这些类型定义的列也称为全文本列。

Oracle 数据库的数据类型分为标量数据类型和非标量数据类型两种。标量数据类型只存储一个值，如数字、字符串、日期等。非标量数据类型可存储多个值，如文本和文档。LOB 和 JSON 是典型的非标量数据类型。在 Oracle 21c 版本之前，内存列式存储不支持在谓词中的非标量数据类型，如 XML 和 JSON。这些类型有其特定的谓词操作符和查询模式，如 CLOB 列的 CONTAIINS 操作符和 JSON 列的 JSON_TEXTCONTAINS 操作符。对于这些列的快速查询依赖于特定类型的索引，如全文本索引、XML 搜索索引和 JSON 搜索索引。Oracle 数据库 21c 版本新增了 In-Memory 全文本列功能，可利用 INMEMORY TEXT 关键字标注全文本列并发布到内存列式存储中，谓词中出现的 CONTAINS 或 JSON_TEXTCONTAINS 操作符可利用特定的内存列式存储格式实现快速搜索。当分析型应用同时涉及标量和非标量数据类型列时，无须访问行式存储，在内存列式存储中即可实现所有列的访问，从而提升查询性能。

使用 In-Memory 全文本列功能需要设置两个初始化参数，其中，MAX_STRING_SIZE 需要设为 EXTENDED，具体方法请参照 7.3.1 节。此外，INMEMORY_VIRTUAL_COLUMNS 需要设为 ENABLE，其默认值为 MANUAL。

```
SQL> SHOW PARAMETER max_string_size
NAME                                 TYPE        VALUE
------------------------------------ ----------- ------------------------------
max_string_size                      string      EXTENDED

SQL> SHOW PARAMETER inmemory_virtual_columns
NAME                                 TYPE        VALUE
------------------------------------ ----------- ------------------------------
inmemory_virtual_columns             string      MANUAL

SQL> ALTER SYSTEM SET inmemory_virtual_columns = ENABLE;
System altered.
```

下面通过一个示例来了解如何使用 In-Memory 全文本列功能。选用 Oracle 数据库标准示例中 OE 用户下的 PRODUCT_INFORMATION 表，此表的 PRODUCT_DESCRIPTION 列为全文本列，类型为 VARCHAR2(2000)。

在没有发布到内存列式存储之前，对于全文本列的查询需要建立特定域的索引，否则报错。例如：

```
SQL> SELECT product_id, product_name FROM product_information WHERE
 contains(product_description, 'software') > 0;
*
ERROR at line 1:
ORA-20000: Oracle Text error:
DRG-10599: column is not indexed
```

暂时先不建立索引，因为内存列式存储中针对全文本列的部分操作可以不依赖这些索引。使用 INMEMORY TEXT 标注全文本列，系统会自动为表添加虚拟列，其列名前缀为 SYS_IME_IVDX：

```
SQL> ALTER TABLE product_information INMEMORY INMEMORY TEXT(PRODUCT_DESCRIPTION);
Table altered.

SQL> SELECT
    column_id, column_name, data_type, data_length, hidden_column, virtual_column
FROM
    user_tab_cols
WHERE
    table_name = 'PRODUCT_INFORMATION';

COLUMN_ID COLUMN_NAME                    DATA_TYPE                    DATA_LENGTH HID VIR
---------- ------------------------------ ---------------------------- ----------- --- ---
         1 PRODUCT_ID                     NUMBER                                22 NO  NO
         2 PRODUCT_NAME                   VARCHAR2                              50 NO  NO
         3 PRODUCT_DESCRIPTION            VARCHAR2                            2000 NO  NO
         4 CATEGORY_ID                    NUMBER                                22 NO  NO
         5 WEIGHT_CLASS                   NUMBER                                22 NO  NO
```

```
     6 WARRANTY_PERIOD            INTERVAL YEAR(2) TO MONTH          5 NO   NO
     7 SUPPLIER_ID                NUMBER                            22 NO   NO
     8 PRODUCT_STATUS             VARCHAR2                          20 NO   NO
     9 LIST_PRICE                 NUMBER                            22 NO   NO
    10 MIN_PRICE                  NUMBER                            22 NO   NO
    11 CATALOG_URL                VARCHAR2                          50 NO   NO
       SYS_IME_IVDX_3F522754274   RAW                            32767 YES  YES
       94FB2BF1A3ED922671DE4

12 rows selected.
```

发布此表,可以看到虚拟列已发布到 IMEU:

```
SQL> EXEC dbms_inmemory.populate('OE', 'PRODUCT_INFORMATION');
PL/SQL procedure successfully completed.

SQL> SELECT allocated_len, used_len, num_vircols, num_rows, time_to_populate FROM
v$imeu_header WHERE num_rows != 0;

ALLOCATED_LEN    USED_LEN NUM_VIRCOLS    NUM_ROWS TIME_TO_POPULATE
------------- ----------- ----------- ----------- ----------------
      1048576       47764           1         288              214
```

发布到内存列式存储后,虽然没有建立全文本索引,但查询可正常执行。

```
SQL> SELECT product_id, product_name FROM product_information WHERE
 contains(product_description, 'software') > 0;

PRODUCT_ID PRODUCT_NAME
---------- --------------------------------------------------
      2459 LaserPro 1200/8/BW
      1822 SPNIX4.0 - SL
...
55 rows selected.

Elapsed: 00:00:00.01
```

执行计划显示为内存中全表扫描,统计信息中也可以看到 IM scan EU 关键字,说明使用了发布到 IMEU 中的虚拟列。

```
SQL> @showplan

PLAN_TABLE_OUTPUT
-------------------------------------
SQL_ID  cgtrbvavx6m6s, child number 0
-------------------------------------
SELECT product_id, product_name from product_information WHERE
CONTAINS(product_description, 'software', 1) > 0

Plan hash value: 2715330242

--------------------------------------------------------------------------
| Id | Operation                          | Name   | Rows  | Cost (%CPU)|
--------------------------------------------------------------------------
```

```
|   0 | SELECT STATEMENT             |                     |    |  1 (100)|
|*  1 |  TABLE ACCESS INMEMORY FULL| PRODUCT_INFORMATION | 14 |  1   (0)|

Predicate Information(identified by operation id):
---------------------------------------------------

   1 - inmemory(SYS_CTX_CONTAINS2("PRODUCT_DESCRIPTION",'software',
             SYS_CTX_MKIVIDX("PRODUCT_DESCRIPTION" RETURNING RAW(32767)))>0)
       filter(SYS_CTX_CONTAINS2("PRODUCT_DESCRIPTION",'software',
             SYS_CTX_MKIVIDX("PRODUCT_DESCRIPTION" RETURNING RAW(32767)))>0)


22 rows selected.

SQL> @imstats

NAME                                                          VALUE
------------------------------------------------------------  ----------
CPU used by this session                                               4
IM scan CUs columns accessed                                           3
IM scan CUs memcompress for query low                                  1
IM scan EU rows                                                      288
IM scan EUs columns accessed                                           1
IM scan EUs memcompress for query low                                  1
IM scan rows                                                         288
IM scan rows projected                                                55
IM scan rows valid                                                   288
IM scan segments minmax eligible                                       1
session logical reads                                                510
session logical reads - IM                                            13
session pga memory                                               3074032
table scans(IM)                                                        1

14 rows selected.
```

但是，In-Memory 全文本列并不能支持所有全文本操作，一些操作仍需要使用全文本索引。例如 SCORE 操作符，SCORE 操作符可以对文本匹配评分，出现多个匹配时比仅有一个匹配时的评分高：

```
SQL> SELECT score(1), product_id, product_name FROM product_information WHERE
CONTAINS(product_description, 'software', 1) > 0;
*
ERROR at line 1:
ORA-20000: Oracle Text error:
DRG-10599: column is not indexed
```

如果需要关闭 In-Memory 全文本列功能，表和全文本列都需要指定 NO INMEMORY 属性。只有这样，虚拟列才能从表中删除：

```
ALTER TABLE product_information NO INMEMORY NO INMEMORY TEXT(product_description);
```

为表创建全文本索引以支持 SCORE 操作。创建索引时会自动生成符号表，表名格式为 DR$indexname$I。此时匹配和评分操作均可正常执行，执行计划也显示使用了全文本索引：

```
SQL> CREATE INDEX demoindex ON product_information(product_description)
     INDEXTYPE IS ctxsys.context;

SQL> SELECT * FROM dr$demoindex$i WHERE token_text = 'SOFTWARE';
TOKEN_TEXT      TOKEN_TYPE TOKEN_FIRST TOKEN_LAST TOKEN_COUNT
-----------     ---------- ----------- ---------- -----------
TOKEN_INFO
--------------------------------------------------------------------------------
SOFTWARE                 0          71        288          55
008815508803018803018803018803018803018803018803018803018803018803018803018803
0188030188030188030288030288030188030188030B880B01880D0188010188010190030401

SQL> SELECT score(1), product_id, product_name FROM product_information WHERE
CONTAINS(product_description, 'software', 1) > 0;

  SCORE(1) PRODUCT_ID PRODUCT_NAME
---------- ---------- --------------------
         5       2459 LaserPro 1200/8/BW
         5       1822 SPNIX4.0 - SL
...
55 rows selected.

Elapsed: 00:00:00.01

SQL> @showplan
--------------------------------------------------------------------------------
| Id  | Operation                    | Name                | Rows | Cost (%CPU)|
--------------------------------------------------------------------------------
|   0 | SELECT STATEMENT             |                     |      |   4  (100)|
|   1 |  TABLE ACCESS BY INDEX ROWID | PRODUCT_INFORMATION |    1 |   4    (0)|
|*  2 |   DOMAIN INDEX               | DEMOINDEX           |      |   4    (0)|
--------------------------------------------------------------------------------

Predicate Information(identified by operation id):
---------------------------------------------------

  2 - access("CTXSYS"."CONTAINS"("PRODUCT_DESCRIPTION",'software',1)>0)
```

从以上示例结果可知，In-Memory 全文本列功能支持将全文本列发布到内存列式存储中，对于部分操作可避免访问行式存储，从而提升性能。但是其并不能完全取代全文本索引，因此不要轻易删除全文本索引，SQL 优化器也会自动判断使用全文本索引或是 In-Memory 全文本列。

7.3.4　In-Memory Spatial 支持

In-Memory Spatial 是 Oracle 21c 版本的新增特性，当空间表发布到内存列式存储后，可以无须借助空间索引直接进行空间过滤操作。

Oracle Spatial 最初属于 Oracle Spatial and Graph 数据库选件，可用于管理和分析地理空间数据、基于位置的数据。2019 年底，此选件开放给所有数据库版本使用，以支持其多模融合数据库的发展战略。

下面通过一个简单的示例了解 In-Memory Spatial 功能。首先建立一个 100 万行的表，此表总共两列，其中一列的数据类型为 SDO_GEOMETRY。

```
CREATE TABLE t2(
    id         NUMBER,
    geometry   SDO_GEOMETRY
);

BEGIN
    FOR i IN 1..1000000 LOOP
        INSERT INTO t2 VALUES(
            i,
            sdo_geometry(2001, 4326, sdo_point_type(dbms_random.value(1, 360) - 180,
dbms_random.value(1, 360) - 90, NULL),
                         NULL,
                         NULL)
        );

        IF MOD(i, 5000) = 0 THEN
            COMMIT;
        END IF;
    END LOOP;

    COMMIT;
END;
/
```

SDO_GEOMETRY 为复合数据类型，可以构建 Spatial 对象。

```
CREATE TYPE sdo_geometry AS OBJECT(
  SDO_GTYPE NUMBER,                        -- 几何图形类型 ID，如点、线和多边形等
  SDO_SRID NUMBER,                         -- 坐标系 ID
  SDO_POINT SDO_POINT_TYPE,                -- 只适用于点
  SDO_ELEM_INFO SDO_ELEM_INFO_ARRAY,       -- 只适用于线、多边形等
  SDO_ORDINATES SDO_ORDINATE_ARRAY         -- 只适用于线、多边形等
);
```

在以上随机生成的表中，SDO_GTYPE 指定为 2001，表示是二维空间的点信息，其他值如 2002、2003 表示线和多边形；坐标系指定为 4326，表示经纬度，其他值如 3857 表示墨卡托坐标系。

此时，空间表 T2 各列的定义如下，其中有多个隐藏列：

```
SQL> SELECT column_id, qualified_col_name, data_type, hidden_column, virtual_column
  FROM user_tab_cols WHERE table_name = 'T2' ORDER BY column_id, column_name;

COLUMN_ID QUALIFIED_COL_NAME              DATA_TYPE              HID VIR
--------- ------------------------------  --------------------   --- ---
        1 ID                              NUMBER                 NO  NO
```

```
    2 GEOMETRY                          SDO_GEOMETRY            NO   NO
    2 "GEOMETRY"."SDO_GTYPE"            NUMBER                  YES  NO
    2 "GEOMETRY"."SDO_SRID"             NUMBER                  YES  NO
    2 "GEOMETRY"."SDO_POINT"."X"        NUMBER                  YES  NO
    2 "GEOMETRY"."SDO_POINT"."Y"        NUMBER                  YES  NO
    2 "GEOMETRY"."SDO_POINT"."Z"        NUMBER                  YES  NO
    2 "GEOMETRY"."SDO_ELEM_INFO"        SDO_ELEM_INFO_ARRAY     YES  NO
    2 "GEOMETRY"."SDO_ORDINATES"        SDO_ORDINATE_ARRAY      YES  NO
```

接下来添加 Spatial 元数据，这是建立 Spatial 索引的前提，然后建立 Spatial 索引。

```
INSERT INTO user_sdo_geom_metadata VALUES(
   'T2',
   'GEOMETRY',
   sdo_dim_array(sdo_dim_element('x', - 180, 180, 0.05), sdo_dim_element('y', - 90, 90, 0.05)),
   4326
);

COMMIT;

CREATE INDEX t2_sidx ON t2 ( geometry ) INDEXTYPE IS mdsys.spatial_index;
```

Spatial 索引对于高性能的地理空间操作非常关键，例如在以下的 SQL 查询中，SDO_FILTER 操作判断两个几何图形是否相交。当有 Spatial 索引时执行时间为 0.01 秒，无索引时则长达 3 分钟。

```
SQL> SELECT id FROM t2 WHERE
    sdo_filter(geometry,   sdo_geometry(2001,   4326,   sdo_point_type(-72.432246,
156.378981, NULL), NULL, NULL)) = 'TRUE';


---------------------------------------------------------------------------
| Id | Operation                          | Name    | Rows | Bytes | Cost (%CPU)|
---------------------------------------------------------------------------
|  0 | SELECT STATEMENT                   |         |      |       |  1 (100)|
|  1 |  TABLE ACCESS BY INDEX ROWID       | T2      |   1  | 3848  |  1   (0)|
|* 2 |   DOMAIN INDEX (SEL: 0.000000 %)   | T2_SIDX |      |       |  1   (0)|
---------------------------------------------------------------------------

Predicate Information(identified by operation id):
---------------------------------------------------

   2 - access("MDSYS"."SDO_FILTER"("GEOMETRY","MDSYS"."SDO_GEOMETRY"(2001,4326,"SDO
            _POINT_TYPE"((-72.432246),156.378981,NULL),NULL,NULL))='TRUE')

Note
-----
   - dynamic statistics used: dynamic sampling (level=2)
```

需要为 SDO_GEOMETRY 类型的列指定 INMEMORY SPATIAL 属性，然后将表 T2 发布到内存列式存储：

```
SQL> ALTER TABLE t2 INMEMORY PRIORITY HIGH INMEMORY spatial(geometry);
```

```
SQL> EXEC dbms_inmemory.populate('SSB','T2');
```

发布完成后，可以看到表 T2 新增了 6 列，列名均已 SYS_IME 开头。这表明其用到了在 7.1.2 节介绍的 In-Memory Expression 特性。新增的 6 个虚拟列为在 3 个维度上的最小和最大值。

```
SQL> SELECT table_name, column_name, data_type, data_default
  FROM user_tab_cols WHERE column_name LIKE 'SYS_IME%';

TABLE_NAME  COLUMN_NAME                          DATA_TYPE        DATA_DEFAULT
----------  -----------------------------------  ---------------  ------------------------------
T2          SYS_IME_SDO_1F611E4D949B4            BINARY_DOUBLE    SDO_GEOM_MAX_Y(SYS_OP_NOEXPAND
            FD7BFB57E8EC8D9DB29                                   ("GEOMETRY"))
T2          SYS_IME_SDO_26554C6AA7924            BINARY_DOUBLE    SDO_GEOM_MIN_Z(SYS_OP_NOEXPAND
            F0BBFE80F6C6D46C349                                   ("GEOMETRY"))
T2          SYS_IME_SDO_8D903E50ED1B4            BINARY_DOUBLE    SDO_GEOM_MAX_Z(SYS_OP_NOEXPAND
            F11BFB8A65A35BFC9AC                                   ("GEOMETRY"))
T2          SYS_IME_SDO_EFA7D642A1634            BINARY_DOUBLE    SDO_GEOM_MIN_X(SYS_OP_NOEXPAND
            FA2BF814C87EC42E8D6                                   ("GEOMETRY"))
T2          SYS_IME_SDO_43D5ADF040374            BINARY_DOUBLE    SDO_GEOM_MAX_X(SYS_OP_NOEXPAND
            FFDBF2EBA1DDC9ED9EF                                   ("GEOMETRY"))
T2          SYS_IME_SDO_28770B865C8F4            BINARY_DOUBLE    SDO_GEOM_MIN_Y(SYS_OP_NOEXPAND
            F19BF825412F5B93868                                   ("GEOMETRY"))

6 rows selected.
```

从执行计划中可以看到，发布到内存列式存储后，SDO_FILTER 操作可以不借助 Spatial 索引，直接使用发布到内存列式存储中的虚拟列完成操作。

```
SQL> @../showplan

PLAN_TABLE_OUTPUT
---------------------------------------------------------------------------
SQL_ID  8rfg0grvw3845, child number 0
---------------------------------------
SELECT     id FROM     t2 WHERE    sdo_filter(geometry,
sdo_geometry(2001, 4326, sdo_point_type(- 72.432246, 156.378981, NULL),
NULL, NULL)) = 'TRUE'

Plan hash value: 1513984157

---------------------------------------------------------------------------
| Id  | Operation                   | Name | Rows  | Bytes | Cost (%CPU)| Time     |
---------------------------------------------------------------------------
|   0 | SELECT STATEMENT            |      |       |       | 2669K(100)|          |
|*  1 |  TABLE ACCESS INMEMORY FULL | T2   |  9594 |   35M | 2669K(100)| 00:01:45 |
---------------------------------------------------------------------------

Predicate Information (identified by operation id):
---------------------------------------------------

   1 - filter((SDO_GEOM_MAX_X("GEOMETRY")>=SDO_GEOM_MIN_X("MDSYS"."SDO_GEOM
       ETRY"(2001,4326,"SDO_POINT_TYPE"((-72.432246),156.378981,NULL),NULL,NULL))-
```

```
              7.8480526674024172E-009D AND SDO_GEOM_MIN_X("GEOMETRY")<=SDO_GEOM_MAX_X
              ("MDSYS"."SDO_GEOMETRY"(2001,4326,"SDO_POINT_TYPE"((-72.432246),156.378981,
              NULL),NULL,NULL))+7.8480526674024172E-009D AND

              SDO_GEOM_MAX_Y("GEOMETRY")>=SDO_GEOM_MIN_Y("MDSYS"."SDO_GEOMETRY"(2001,4326,
              "SDO_POINT_TYPE"((-72.432246),156.378981,NULL),NULL,NULL))-7.8480526674024
              172E-009D AND SDO_GEOM_MIN_Y("GEOMETRY")<=SDO_GEOM_MAX_Y("MDSYS"."SDO_GEOME
              TRY"(2001,4326,"SDO_POINT_TYPE"((-72.432246),156.378981,NULL),NULL,NULL))+
              7.8480526674024172E-009D AND SDO_GEOM_MAX_Z("GEOMETRY")>=SDO_GEOM_MIN_Z
              ("MDSYS"."SDO_GEOMETRY"(2001,4326,"SDO_POINT_TYPE"((-72.432246),156.378981,
              NULL),NULL,NULL))-7.8480526674024172E-009D AND

              SDO_GEOM_MIN_Z("GEOMETRY")<=SDO_GEOM_MAX_Z("MDSYS"."SDO_GEOMETRY"(2001,4326,
              "SDO_POINT_TYPE"((-72.432246),156.378981,NULL),NULL,NULL))+7.8480526674024
              172E-009D))

   Note
   -----
      - dynamic statistics used: dynamic sampling (level=2)
```
实验完成后，在最终清除环境时，除删除表外，还需删除对应的元数据：
```
SQL> DROP TABLE t2;
SQL> DELETE FROM user_sdo_geom_metadata WHERE table_name = 'T2';
```
总的来说，In-Memory Spatial 支持是一项非常小的改进，使用的场景非常有限。例如新增虚拟列只适用于 SDO_FILTER 操作，而对于 SDO_WITHIN_DISTANCE 等操作并不支持。因此使用此功能前需仔细评估，并且不要轻易删除 Spatial 索引。

7.4 In-Memory 扫描优化

7.4.1 In-Memory 动态扫描

In-Memory 动态扫描也称为 IMDS，是 Oracle 数据库 18c 版本新增的功能。这项功能对于应用透明，可以根据 CPU 利用率，自动启动多线程对 In-Memory 对象进行并行扫描。传统的 Oracle 并行执行仍然支持，和 IMDS 可以配合使用。但 IMDS 更加灵活，开销更小，并且可以基于 CPU 利用率动态调整并行线程的数量。IMDS 可以跨 IMCU 执行，即可以同时扫描多个 IMCU；也可以在 IMCU 内部启用，即可以同时扫描多个列。

实际上，性能提升技术无非从两方面同时着手。一方面是减少或消除对资源的消耗，如 Database In-Memory 通过将对象置入内存列式存储消除了磁盘 I/O，又如下推技术通过就地处理减少了数据的往复传递；另一方面是充分利用资源，例如 Oracle 并行处理和 In-Memory 动态扫描，通过将闲置的资源加以利用来加快处理速度。

在 Oracle 18c 版本中，In-Memory 动态扫描功能依赖于数据库资源计划（Resource Manager），启用资源计划需要设置初始化参数 RESOURCE_MANAGER_PLAN。Oracle 19c 版本以后不再需要设置此初始化参数，只需 INMEMORY_SIZE 大于 0 即可自动启动资源计划。以下为 Oracle 18c 时设置数据库资源计划的示例：

```
SQL> SHOW PARAMETER resource_manager_plan
NAME                                 TYPE        VALUE
------------------------------------ ----------- ------------------------------
resource_manager_plan                string

SQL> SELECT PLAN FROM dba_rsrc_plans;
PLAN
--------------------------------------------------------------------------------
MIXED_WORKLOAD_PLAN
DEFAULT_MAINTENANCE_PLAN
DEFAULT_PLAN
INTERNAL_QUIESCE
INTERNAL_PLAN
APPQOS_PLAN
ETL_CRITICAL_PLAN
ORA$AUTOTASK_PLAN
ORA$ROOT_PLAN
ORA$QOS_PLAN
DSS_PLAN

11 rows selected.

SQL> ALTER SYSTEM SET resource_manager_plan = default_plan;
System altered.
```

下面看一个 In-Memory 动态扫描的例子，使用的 SQL 查询为 SSB 示例 SQL 1_1，为体现 IMDS 使用前后的差异，将 LINEORDER 表放大到近 6000 万行。在没有启用 IMDS 之前，此查询耗时 0.24 秒：

```
SQL> SELECT COUNT(*) FROM lineorder;
  COUNT(*)
----------
  59986052

SQL> @1_1
   REVENUE
----------
4.4728E+12

Elapsed: 00:00:00.24
```

如果数据库版本为 18c，可以通过将资源计划设为空来禁用 IMDS。如果版本是 19c，可以通过将隐含参数_inmemory_dynamic_scans 设置为 DISABLE 禁用，实验完毕后请修改回默认值 AUTO。

```
-- Oracle 18c
SQL> ALTER SYSTEM SET resource_manager_plan = '';
System altered.
```

```
-- Oracle 19c
ALTER SESSION SET "_inmemory_dynamic_scans" = DISABLE;
```

图 7-4 为未启用 IMDS 时的实时 SQL 监控报告。从其中的 Time & Wait 部分或者执行计划的第 13 行都可以看到，会话中 100%的活动都用于 CPU: IN MEMORY，也就是内存中列扫描。

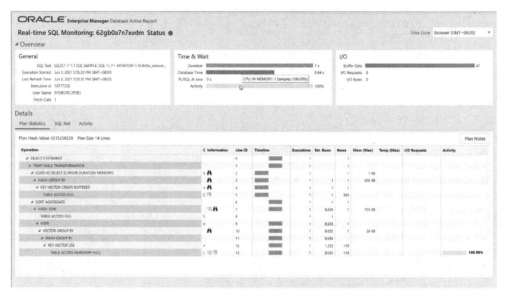

图 7-4 未启用 IMDS 时的实时 SQL 监控报告

启用 IMDS 后，SQL 查询的速度提升了 8 倍，为 0.03 秒。

```
SQL> @1_1
  REVENUE
----------
4.4728E+12


Elapsed: 00:00:00.03
```

在执行统计信息中，IM scan (dynamic) multi-threaded scans 非零表示已启用 IMDS。

```
SQL> @../../../imstats

NAME                                                             VALUE
---------------------------------------------------------------- ----------
CPU used by this session                                         100
IM scan(dynamic)executing tasks                                  1491
IM scan(dynamic)max degree                                       16
IM scan(dynamic)multi-threaded scans                             1
IM scan(dynamic)pending tasks                                    4945
IM scan(dynamic)rows                                             59986052
IM scan(dynamic)rs2 rowsets                                      472
IM scan(dynamic)task execution time                              1019728
IM scan(dynamic)task reap time                                   72820
IM scan(dynamic)task submission time                             21444
IM scan(dynamic)tasks processed by thread                        117
IM scan CUs columns accessed                                     472
```

```
IM scan CUs memcompress for query low                    118
IM scan CUs pcode aggregation pushdown                   118
IM scan CUs predicates applied                           354
IM scan bytes in-memory                             3127271848
IM scan bytes uncompressed                          5748587884
IM scan delta - only base scan                           118
IM scan dict engine results reused                       118
IM scan rows                                        59986052
table scans(IM)                                            1
```

IM scan (dynamic) tasks processed by thread 表示使用 IMDS 并行处理的 IMCU 数量,其值 117 小于 LINEORDER 表实际的 IMCU 数量 118。这是由于 IMDS 对第一个 IMCU 实施非并行处理,以评估并行处理是否适合,如果可行才会对余下的 IMCU 进行并行处理。

```
SQL> SELECT COUNT(*) FROM v$im_header WHERE num_rows !=0;

  COUNT(*)
----------
       118
```

在图 7-5 所示的实时 SQL 监控报告中,单击执行计划第 13 行的望远镜图标,Dynamic Scan Tasks on Thread 为 118,与执行统计信息中的值一致,也说明启用了 IMDS。

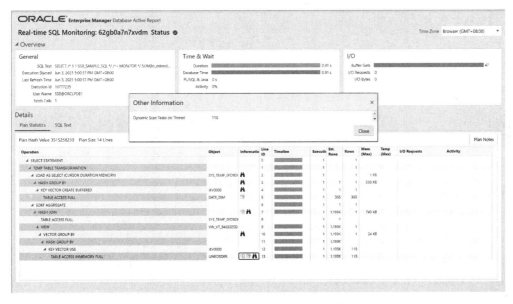

图 7-5 启用 IMDS 时的实时 SQL 监控报告

7.4.2 In-Memory 混合扫描

在 Oracle 数据库 21c 版本之前,如果 SQL 查询涉及到的任意一列未发布到内存列式存储中,此查询将会整个转移到行式存储中执行。从 21c 版本开始,Database In-Memory 支持 In-Memory 混合扫描,对于满足特定条件的 SQL,可以同时利用行式存储和内存列式存储来查询数据。

在满足以下条件时,优化器会考虑 In-Memory 混合扫描。

（1）WHERE 条件中仅包括 INMEMORY 列。

（2）SELECT 列表中同时包括 INMEMORY 列和非 INMEMORY 列。

从以上条件也可以看出，In-Memory 混合扫描使用效果最佳时，可以通过 INMEMORY 列过滤掉绝大部分数据，而只有少量数据需要从行式存储中访问。例如以下示例 SQL：

```
SELECT
    lo_orderpriority,
    SUM(lo_extendedprice *(1 - lo_discount / 100)) AS discount_price
FROM
    lineorder
WHERE
        lo_shipmode = 'AIR'
    AND lo_discount > 8
    AND lo_quantity > 30
    AND lo_tax > 5
    AND lo_supplycost > 90000
GROUP BY
    lo_orderpriority;
```

此 SQL 的 WHERE 条件可以将数据从近 1200 万行过滤到仅 2 万行，相当于淘汰掉约 99.8% 的数据。

```
SQL> SELECT COUNT(*) FROM lineorder;

  COUNT(*)
----------
  11997996

SELECT
    COUNT(*)
FROM
    lineorder
WHERE
        lo_shipmode = 'AIR'
    AND lo_discount > 8
    AND lo_quantity > 30
    AND lo_tax > 5
    AND lo_supplycost > 90000;

  COUNT(*)
----------
     21042
```

为评估 IMDS 带来的性能提升，首先建立性能基线。当此表未发布到内存列式存储时，执行时间为 10.7 秒；所有列完整发布到内存列式存储后，执行时间为 0.04 秒。

将 SELECT 列表中的 lo_extendedprice 列设置为 NO INMEMORY，再发布 LINEORDER 表：

```
ALTER TABLE lineorder INMEMORY PRIORITY HIGH NO INMEMORY(lo_extendedprice);
```

等待发布完成后，再次执行之前的查询。在执行计划中的第 2 行，TABLE ACCESS

INMEMORY FULL (HYBRID)说明 In-Memory 混合扫描已启用。此时的执行时间为 0.09 秒，相对于未启用 In-Memory 混合扫描时有 100 多倍的性能提升。

```
SQL> @../../../showplan

PLAN_TABLE_OUTPUT
-------------------------------------------------------------------------------
SQL_ID  dbt3qna8pyadb, child number 0
-------------------------------------------------------------------------------
select lo_orderpriority, sum(lo_extendedprice * (1 - lo_discount/100))
as discount_price from lineorder where lo_shipmode = 'AIR' and
lo_discount > 8 and lo_quantity > 30 and lo_tax > 5 and lo_supplycost >
90000 group by lo_orderpriority

Plan hash value: 3675673598

PLAN_TABLE_OUTPUT
-------------------------------------------------------------------------------
| Id  | Operation                            | Name     | Rows  | Cost (%CPU)|
-------------------------------------------------------------------------------
|   0 | SELECT STATEMENT                     |          |       | 46632 (100)|
|   1 |  HASH GROUP BY                       |          |     5 | 46632   (1)|
|*  2 |   TABLE ACCESS INMEMORY FULL (HYBRID)| LINEORDER| 26234 | 46631   (1)|
-------------------------------------------------------------------------------

Predicate Information (identified by operation id):
-------------------------------------------------------------------------------

   2 - filter(("LO_DISCOUNT">8 AND "LO_TAX">5 AND "LO_QUANTITY">30 AND
              "LO_SUPPLYCOST">90000 AND "LO_SHIPMODE"='AIR'))
```

不过需要注意，In-Memory 混合扫描并非在任何时候都能提升性能。在某些时候，其执行效率甚至会比未启用内存列式存储时更低。例如对于 SSB 示例 SQL 1_1，在完整发布时，其执行时间为 0.06 秒。如果仅将 lo_extendedprice 列设置为 NO INMEMORY，其执行时间为 1 分 38 秒。从执行计划中的 HYBRID 关键字可以确认 In-Memory 混合扫描已启用：

```
PLAN_TABLE_OUTPUT
-------------------------------------------------------------------------------
SQL_ID  62gb0a7n7xvdm, child number 0
-------------------------------------------------------------------------------

-------------------------------------------------------------------------------
| Id  | Operation                            | Name     | Rows  | Cost (%CPU)|
-------------------------------------------------------------------------------
|   0 | SELECT STATEMENT                     |          |       | 46608 (100)|
|   1 |  SORT AGGREGATE                      |          |     1 |            |
|*  2 |   HASH JOIN                          |          |  340K | 46608   (1)|
|*  3 |    TABLE ACCESS INMEMORY FULL        | DATE_DIM |   365 |     1   (0)|
|*  4 |    TABLE ACCESS INMEMORY FULL (HYBRID)| LINEORDER| 2243K | 46601   (1)|
-------------------------------------------------------------------------------
```

但是，如果不启用 Database In-Memory，其执行时间为 11 秒。也就是说，启用 In-Memory 混合扫描后执行速度反而下降。因此，启用 Database In-Memory 前需要仔细评估其对性能是否有提升，必要时可锁定执行计划，以避免采用 In-Memory 混合扫描。另外，在绝大多数情况下，都不建议使用非完整发布，应尽可能保证查询涉及的表和列均发布到内存列式存储中。

第8章

Database In-Memory 与高可用性

8.1 Oracle 最高可用性架构

Oracle 数据库高可用性基于最高可用性架构（MAA）。Oracle 最高可用性架构是 Oracle 的最佳实践蓝图，它基于成熟的 Oracle 高可用性技术、端到端验证、专家建议和客户体验。MAA 总结了 30 多年来为企业级客户解决最复杂和最困难的高可用问题时的经验和教训，是基于 Oracle 技术提供业务连续性的最佳实践和参考架构。MAA 可减少或消除计划内和计划外停机，适用于本地数据中心和公有云环境。

Oracle 最高可用性架构中的关键技术如图 8-1 所示，包括数据保护、本地高可用、数据复制和应用连续性四个方面。

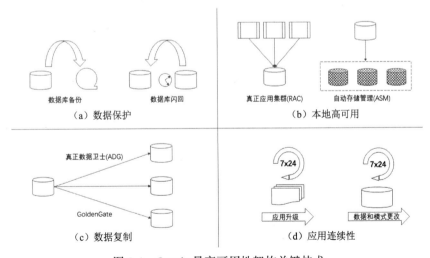

图 8-1 Oracle 最高可用性架构关键技术

数据是用户业务运营的基础，因此数据保护也是最高可用性架构中最基础和最重要的保护手段。无论是手工还是通过备份软件自动实施 Oracle 数据库备份，底层都会调用 RMAN（Oracle

Recovery Manager）来执行。RMAN 管理整个数据库的备份、还原和恢复过程，备份目标可以是磁盘、磁带或公有云。其他一些高级备份功能包括备份压缩和加密、跨模式和跨平台的备份和恢复、多租户备份和从备份中恢复单个数据库表等。传统数据库备份的恢复点目标（RTO）通常为几小时到 1 天，Oracle 提供零数据丢失恢复一体机（ZDLRA），使用成熟的 Data Guard 技术传输重做日志，使得 RTO 降低至亚秒级，最大限度地减少了数据的丢失。数据备份主要防止异常中断引起的数据丢失，而 Oracle 数据库的闪回功能则主要是用于恢复人为错误。例如当数据被错误的修改或删除时，闪回可以快速将数据恢复到错误操作前的状态。闪回比数据库恢复的时间更短，操作也更为简便。此外，闪回可以将某行数据恢复至指定时间点，恢复指定的事务或恢复整库。

在数据保护的基础上，下一步需要实现本地的高可用。在这一领域，Oracle 提供 RAC（Real Application Clusters）实现服务器高可用，以及 ASM（Automatic Storage Management）实现存储高可用。RAC 是全共享的数据库集群技术，支持在集群中所有服务器上运行多个数据库实例，而在应用程序看来，它是一个统一的数据库。RAC 是 Oracle 独有的技术，也是用户广为采用的数据库选件。相对于传统的主备集群，RAC 集群中的所有节点都可以执行读写操作。因此在服务器故障时，应用可以快速切换到其他正常节点继续业务处理，服务器宕机引发的业务中断被最小化。RAC 还支持业务负载均衡，以及在数据库整合方案中作为构建数据库资源池的技术组件。ASM 是由 Oracle 提供的卷管理技术，可用于数据库存储或构建集群文件系统。ASM 支持两重或三重镜像，如果由于硬件损坏导致镜像副本减少，ASM 可以利用余下的空间重新建立一份镜像，以防止二次故障。通过 ASM 还可以实现通常说的"双存储"方案，即使两台相互镜像的存储中的一台出现问题，数据不会丢失，应用也不会中断，从而消除了存储端的单点故障。

本地高可用可以防止绝大部分的计划内和计划外停机引发的业务中断。如果整个集群或站点失效，通过备份恢复通常无法满足恢复时间目标（RTO）的要求，此时就需要通过数据复制技术在另一站点提供数据库副本，这就是通常所说的灾难恢复。Oracle 提供 ADG（Active Data Guard）和 GoldenGate 两种技术来实现灾难恢复，ADG 提供物理数据库副本，GoldenGate 提供逻辑数据库副本。在绝大多数情况下，会推荐采用 ADG 来实现 Oracle 数据库的灾难恢复。这是由于 ADG 是 Oracle 数据库的一部分，支持所有数据类型，切换和回切操作简单，可实现 Oracle 数据库的完整和深度保护。GoldenGate 比 ADG 更灵活，例如源和目标数据库可以是不同的操作系统、不同的版本，甚至不同类型的数据库。但是 GoldenGate 的管理比 ADG 复杂，并非支持所有的数据类型，不支持同步复制。和 ADG 目标端只读不同，GoldenGate 的目标端允许写，这使得 GoldenGate 常用于负载分担的场景，即通过复制提供另一份数据库副本来实现负载转移，以减轻源数据库的负担。GoldenGate 还可以实现双活数据中心，不过这常用于在两个数据中心分别执行不同的业务。如果两端执行相同的业务，还需要提供防冲突机制来避免写覆盖现象。写覆盖现象可能导致业务逻辑错误，例如一张火车票被来自两个数据中心的用户同时购买到，从而导致座席重号。ADG 和 GoldenGate 还可以实现近乎零停机的数据库迁移。

在应用运行过程中，当 RAC 集群中部分实例发生故障时，应用程序连续性可以在余下正常实例重放应用请求；如果整个 RAC 集群发生故障，还可以在备用站点的数据库中重放请求。应用感知不到底层故障的发生，从而可以提供良好的用户体验。此外，Oracle 还支持一系列联机操作，允许在数据和模式重组时，应用仍可以访问数据库，从而提高整体数据库的可用性，减少了计划停机时间。例如，可利用联机数据文件移动实现将不常访问的数据文件移至具有较低成本的存储，或通过联机分区移动实现在线分区重定义。联机表重定义可在业务不中断的前提下对表的物理结构进行变更，例如在线加密、在线压缩等。基于版本的重定义（EBR）特性允许在不中断应用程序的前提下进行联机应用程序升级。简单来说，EBR 允许为一个表定义新旧两个版本，分别用于升级前后的应用。应用通过视图来引用这两个版本，因此新旧应用可以同时运行。新的表定义可能具有新增列，或合并列，或不同的列数据类型。当旧应用继续运行时，数据的改变通过触发器传递到新版本的表定义，因此后续可以实现零停机的应用迁移。

以上四个方面是最高可用性架构中最为主要的业务连续性保障技术，Database In-Memory 可以与架构中涉及的技术和功能无缝集成，其中比较重要的是 RAC 和 ADG。

8.2　Database In-Memory 与 RAC

8.2.1　利用 OCI 搭建 RAC 实验环境

搭建 RAC 实验环境最快、最简单的方式是利用 Oracle 公有云，即 Oracle Cloud Infrastructure，简称 OCI。在 OCI 中可以创建基于虚拟机、物理机和 Exadata 的 Oracle 数据库服务，RAC 数据库服务可以在虚拟机或 Exadata 环境提供支持。

以下简单介绍在 OCI 上创建两节点 RAC 虚拟机环境的过程。登录 Oracle 公有云，进入 OCI 控制台，首先单击左上角的"三明治"菜单，选择"Oracle 数据库" > "裸金属、VM 和 Exadata"，然后单击"创建数据库系统"按钮，进入如图 8-2 所示界面。

图 8-2　创建数据库系统 1/2：数据库系统配置

在此步骤中，需要指定数据库系统的基本配置，其中的关键配置信息如下。

（1）配置类型，由于裸金属类型不支持 RAC，因此只能选择虚拟机或 Exadata，前者支持两节点 RAC，后者支持最多 8 节点 RAC。由于是实验环境，此处选择虚拟机。

（2）选择配置，RAC 数据库服务要求每个数据库节点至少两个 OCPU，因此此处选择 VM.Standard2.2，这也是支持 RAC 的最小配置。使用此配置的每个 RAC 节点将具有 4 vCPU 和 30GB 内存。

（3）数据库系统节点总数，此处只能选择 2，这也是虚拟机配置类型唯一支持的 RAC 选项。

（4）Oracle 数据库软件版本，只能选择 Enterprise Edition Extreme Performance，这也是唯一支持 RAC 和 Database In-Memory 的版本。

（5）存储管理软件，对于 RAC，只能选择 Oracle Grid Infrastructure。如果是单实例，还可以选择 Logical Volume Manager。Logical Volume Manager 的创建速度会更快，但功能较简单。

（6）SSH 密钥指定 SSH 密钥对中的公钥，后续可以使用对应私钥登录数据库节点。

（7）网络信息指定虚拟云网络和客户端子网。

（8）主机名前缀必须指定。本例此处输入 dbim，最终两个数据库节点的主机名将为 dbim1 和 dbim2。

完成数据库系统基本配置后，单击"下一步"，进入数据库配置页面，如图 8-3 所示。

图 8-3 创建数据库系统 2/2：数据库配置

此步骤中的关键配置信息如下：

（1）数据库名称，系统自动生成的名称格式为 DB 加日期，如 DB0601。此处输入 DBRAC。

（2）数据库唯一名称后缀，输入 21c，与数据库名称共同组成数据库唯一名称，即 DBRAC_21c。

（3）数据库映像，即数据库版本，可以选择从 11.2 到 21c 共 6 个版本，本例选择 21c。

（4）PDB 名称，输入 orclpdb1。

（5）管理员密码，输入满足复杂要求的密码，必须为 9~30 个字符，并且必须至少包含 2 个大写字符、2 个小写字符、2 个特殊字符和 2 个数字字符。

（6）工作负载类型，可以是事务处理或数据仓库，此处选择事务处理。

所有信息输入完毕后，单击左下方"创建数据库系统"按钮，开始实际的数据库系统创建过程，大约 1.5 小时后，数据库系统就绪。如图 8-4 所示，查看"工作请求"页面下关联的资源，可以查看到工作请求进度、任务状态详情和关联资源的状态。

图 8-4　创建数据库系统：任务详情

至此，基于 OCI 的 RAC 环境创建完毕。使用与之前指定公钥配对的私钥，即可登录任一数据库节点进行实验。

8.2.2　利用 Vagrant 搭建 RAC 实验环境

通过 OCI 搭建 RAC 实验环境，虽然比较快捷，但需要拥有 Oracle 公有云账户。而通过 Oracle 在 GitHub 上的 Vagrant 项目来搭建，只需要一台宿主机即可。第 2 章已经演示了通过 Vagrant 项目搭建了 Database In-Memory 环境。实际上这个项目也可以用于搭建容器、GoldenGate、RAC 和 ADG 等更复杂的环境。

通过 Vagrant 创建两节点 RAC 环境的简明过程如下。

（1）创建一台宿主主机，Oracle Vagrant RAC 项目会在宿主主机上创建两个虚拟机和共享存储。每个虚拟机默认配置为 8GB 内存，共享存储约为 100GB。如果你的笔记本电脑具有至少 20GB 内存，那么可以直接将其作为宿主主机。如果不满足条件，比较简单的方法是在公有云中创建宿主主机。

（2）在宿主主机上安装虚拟化引擎，Vagrant 支持多种虚拟机引擎，Oracle Vagrant 项目支持 VirtualBox 和 libVirt。

（3）在宿主主机上安装 Vagrant 软件。

（4）下载 Oracle 在 GitHub 上的 Vagrant 项目。

（5）进入 RAC 所在目录，运行 vagrant up。

下面通过一个示例了解通过 Vagrant 创建两节点 RAC 环境的详细过程。首先创建一台宿主主机，操作系统为 Oracle Linux 7，内存为 30GB，磁盘容量为 512GB。

1. 安装 Vagrant 软件

第一步是安装 Vagrant 软件，可以从官网[①]获取不同操作系统 Vagrant 软件的下载地址。安装过程如下：

```
# 使用 root 用户进行安装
sudo -s

# 下载 Vagrant 软件，约 13MB。压缩包中实际只包含一个可执行文件
wget https://releases.hashicorp.com/vagrant/2.2.14/vagrant_2.2.14_linux_amd64.zip

# 解压到可执行文件目录
unzip -d /usr/bin vagrant_2.2.14_linux_amd64.zip

# Vagrant 需要在用户空间实现文件系统，因此需要安装 FUSE(Filesystem in Userspace)模块
yum install fuse -y

# 如果普通用户需要使用 Vagrant，需要将其加入 fuse 用户组
# 如果是 Oracle 公有云环境，请使用 opc；如果是本地环境，建议使用 oracle
sudo usermod -a -G fuse opc

# 安装 vagrant 依赖的 BSD tar 软件
yum install bsdtar -y
```

通过以下命令确认 Vagrant 已成功安装：

```
$ vagrant --version
Vagrant 2.2.14

$ which vagrant
/bin/vagrant
$ ls -ld /bin
lrwxrwxrwx. 1 root root 7 Jul 22 17:10 /bin -> usr/bin

$ ls -l /usr/bin/vagrant
-rwxr-xr-x. 1 root root 13787176 Nov 20 18:36 /usr/bin/vagrant
```

2. 安装 Oracle VirtualBox

接下来安装 Oracle VirtualBox，安装过程如下：

```
# 使用 root 用户进行安装
sudo -s

# 下载包含 VirtualBox 的资料库文件
cd /etc/yum.repos.d
wget https://download.virtualbox.org/virtualbox/rpm/el/virtualbox.repo

# 安装 VirtualBox
yum install VirtualBox-6.1 -y
```

[①] https://www.vagrantup.com/downloads

```
# 安装 VirtualBox 扩展包(Extension Pack)
wget
https://download.virtualbox.org/virtualbox/6.1.18/Oracle_VM_VirtualBox_Extension_Pack-6.1.18.vbox-extpack
vboxmanage extpack install Oracle_VM_VirtualBox_Extension_Pack-6.1.18.vbox-extpack
```

通过以下命令确认 VirtualBox 已安装成功:

```
$ yum list installed|grep -i virtualbox
VirtualBox-6.1.x86_64              6.1.22_144080_el7-1        @virtualbox

$ vboxmanage --version
6.1.22r144080
```

3. 创建 RAC 环境的 Vagrant Box

最后一步也是时间最长的一步,是创建 RAC 环境的 Vagrant Box。首先需要克隆 GitHub 上的 Oracle Vagrant 项目:

```
sudo yum install git -y
sudo su - opc
git clone https://github.com/oracle/vagrant-projects.git
```

然后从 Oracle 官网[①]下载数据库和 Grid 软件,放置到 RAC 对应的目录下。这两个软件合计约 5.6GB。

```
$ cd ~/vagrant-projects/OracleRAC/ORCL_software

$ ls -l
total 5809472
-rw-rw-r--. 1 opc opc 3059705302 Feb  9 03:56 LINUX.X64_193000_db_home.zip
-rw-rw-r--. 1 opc opc 2889184573 Feb  9 04:03 LINUX.X64_193000_grid_home.zip
-rw-rw-r--. 1 opc opc        198 May  2 07:06 put_Oracle-software_here.txt
```

在建立 RAC Vagrant Box 前,可以预先查看每个 RAC 节点的配置。从配置文件可知,这两个 RAC 节点的主机名分别为 node1 和 node2。每个 RAC 节点配置 8GB 内存,两个虚拟 CPU。4 块 20GB 的磁盘被用来建立 ASM 磁盘组,作为两个 RAC 节点的共享存储。

```
$ cd ~/vagrant-projects/OracleRAC/config
$ ls
vagrant.yml

$ cat vagrant.yml
# ----------------------------------------------
# vagrant.yml for VirtualBox
# ----------------------------------------------
node1:
  vm_name: node1
  mem_size: 8192
  cpus: 2
  public_ip:   192.168.56.111
```

[①] https://www.oracle.com/database/technologies/oracle-database-software-downloads.html

```
    vip_ip:      192.168.56.112
    private_ip: 192.168.200.111
    u01_disk: ./node1_u01.vdi

  node2:
    vm_name: node2
    mem_size: 8192
    cpus: 2
    public_ip: 192.168.56.121
    vip_ip:      192.168.56.122
    private_ip: 192.168.200.122
    u01_disk: ./node2_u01.vdi

shared:
  prefix_name:   vgtol7-rac
  # ----------------------------------------
  domain: localdomain
  scan_ip1: 192.168.56.115
  scan_ip2: 192.168.56.116
  scan_ip3: 192.168.56.117
  # ----------------------------------------
  non_rotational: 'on'
  asm_disk_path:
  asm_disk_num:    4
  asm_disk_size: 20
  p1_ratio:       80

env:
  provider: virtualbox
  # ----------------------------------------
  gi_software:    LINUX.X64_193000_grid_home.zip
  db_software:    LINUX.X64_193000_db_home.zip
  # ----------------------------------------
  root_password:    welcome1
  grid_password:    welcome1
  oracle_password: welcome1
  sys_password:    welcome1
  pdb_password:    welcome1
  # ----------------------------------------
  ora_languages:  en,en_GB
  # ----------------------------------------
  nomgmtdb:       true
  orestart:       false
  # ----------------------------------------
  db_name:        DB193H1
  pdb_name:       PDB1
  db_type:        RAC
  cdb:            false
```

在备注文件的末尾,可以设置数据库名,是否使用多租户架构以及 PDB 名称。为保持一致,建议将 cdb 设为 true,pdb_name 修改为 ORCLPDB1。

进入 RAC 项目所在目录，开始创建两节点 RAC Vagrant Box，目前的数据库版本为 Oracle 19.3。取决于宿主机的性能和网速，安装时间可能持续数小时，整个过程无须人工介入：

```
$ cd ~/vagrant-projects/OracleRAC
$ vagrant up
```

如果安装成功，最终会显示类似于如下的信息：

```
node1: Instance DB193H11 is running on node node1
node1: Instance DB193H12 is running on node node2
node1: -------------------------------------------------------------
node1: SUCCESS: 2021-05-02 14:37:05: Oracle RAC on Vagrant has been created successfully!
```

也可以使用 vagrant 命令确认两个 RAC 节点的运行状态：

```
$ vagrant status

Oracle RAC(Real Application Cluster)Vagrant box for KVM/libVirt or VirtualBox
Copyright(c)1982-2020 Oracle and/or its affiliates

Author: Ruggero Citton <ruggero.citton@oracle.com>
        RAC Pack, Cloud Innovation and Solution Engineering Team


--------------------
Detected virtualbox
--------------------
getting Proxy Configuration from Host…
Current machine states:

node2                     running(virtualbox)
node1                     running(virtualbox)

This environment represents multiple VMs. The VMs are all listed
above with their current state. For more information about a specific
VM, run `vagrant status NAME`.
```

两个 RAC 节点的 Vagrant 主机显示名称分别为 node1 和 node2，使用以下命令登录：

```
$ cd ~/vagrant-projects/OracleRAC
$ vagrant ssh node1
$ vagrant ssh node2
```

在 RAC 节点内部，默认不能连接互联网，此时可通过 /vagrant 共享目录实现与宿主机的文件交换。

```
$ df /vagrant
Filesystem        1K-blocks       Used Available Use% Mounted on
vagrant           524062508  129648936 394413572  25% /vagrant
```

也可以补充 DNS 设置后，通过互联网获取和交换数据：

```
sudo bash -c 'echo "nameserver 8.8.8.8" >> /etc/resolv.conf'
git clone https://github.com/XiaoYu-HN/dbimbook
```

使用 Virtual Box 的命令也可以查询这两个 RAC 虚拟机的状态：

```
$ vboxmanage list vms
"vgtol7-rac-node2" {4f40ec20-b51f-4f85-b001-de35558525f6}
"vgtol7-rac-node1" {c28ee06f-ef59-4060-afdb-a11d7b5a0cc5}

$ vboxmanage showvminfo vgtol7-rac-node1
Name:                    vgtol7-rac-node1
Groups:                  /vgtol7-rac
Guest OS:                Oracle(64-bit)
...
Memory size:             8192MB
...
SATA Controller (0, 0): /home/opc/VirtualBox VMs/vgtol7-rac/vgtol7-rac-node1/box-disk001.vmdk (UUID: 5d56ae09-1433-455f-8bb4-8d4a98daba5b)
SATA Controller (1, 0): /home/opc/vagrant-projects/OracleRAC/node1_u01.vdi (UUID: 6608c5c6-ebd2-4b25-a39a-054e635ef3d6)
SATA Controller (2, 0): /home/opc/vagrant-projects/OracleRAC/asm_disk0.vdi (UUID: 65d1e319-6d33-40ea-9812-875ce53eeae0)
SATA Controller (3, 0): /home/opc/vagrant-projects/OracleRAC/asm_disk1.vdi (UUID: 37cb96c2-4048-4a99-9b5f-e85f73de1d04)
SATA Controller (4, 0): /home/opc/vagrant-projects/OracleRAC/asm_disk2.vdi (UUID: 323fa7fd-d566-42e7-896b-d37d08dc8094)
SATA Controller (5, 0): /home/opc/vagrant-projects/OracleRAC/asm_disk3.vdi (UUID: cb2bd56f-fc65-4e28-bd50-a77aadac29a3)
...

$ vboxmanage list hdds
...
UUID:             4840f79e-f203-40a1-955a-6dd1b389d5af
Parent UUID:      base
State:            locked write
Type:             normal(base)
Location:         /home/opc/vagrant-projects/OracleRAC/node2_u01.vdi
Storage format:   VDI
Capacity:         102400 MBytes
Encryption:       disabled

UUID:             65d1e319-6d33-40ea-9812-875ce53eeae0
Parent UUID:      base
State:            locked read
Type:             shareable
Location:         /home/opc/vagrant-projects/OracleRAC/asm_disk0.vdi
Storage format:   VDI
Capacity:         20480 MBytes
Encryption:       disabled
...
```

最后，在宿主机上可以安装 Oracle Instant Client，这样方便使用 SQL Plus 分别连接 RAC 节点进行管理。具体步骤如下：

```
$ hostname
vagrant-rac

$ sudo yum install -y oracle-instantclient18.3-sqlplus.x86_64

$ yum list installed | grep instantclient
oracle-instantclient18.3-basic.x86_64
oracle-instantclient18.3-sqlplus.x86_64

# 将 sqlplus 加到 PATH 路径中
$ export PATH=$PATH:/usr/lib/oracle/18.3/client64/bin/

# 将 Oracle Instant Client 使用的动态链接库加到路径中
$ sudo sh -c "echo /usr/lib/oracle/18.3/client64/lib > /etc/ld.so.conf.d/oracle-instantclient.conf"
$ sudo ldconfig
```

在宿主机上，使用以下命令分别连接到两个 RAC 实例：

```
$ sqlplus sys/welcome1@192.168.56.111/DB193H1 as sysdba    # 登录 RAC 实例 1
$ sqlplus sys/welcome1@192.168.56.121/DB193H1 as sysdba    # 登录 RAC 实例 2
```

8.2.3　In-Memory 数据分布

在 RAC 环境下，每个 RAC 实例都拥有独立的内存列式存储。默认情况下，对象会分布到集群中所有的节点，也就是说，每个实例会发布此数据库对象的一部分数据。In-Memory 对象支持以下 4 种分布策略，可以用 INMEMORY DISTRIBUTE 子句指定，此选项仅适用于 RAC。

（1）BY ROWID RANGE，数据以 IMCU 为单位被分布到不同的 RAC 实例。此策略基于 IMCU 中第一行的 ROWID 列哈希运算的结果，将 IMCU 发布到 RAC 集群中不同的实例。此策略可以基本保证数据在所有实例间均匀分布。

（2）BY PARTITION，数据以分区为单位分布到不同的 RAC 实例。此策略最适合与哈希分区结合，基本可以保证数据在所有实例间的均匀分布和均衡的访问，其他分区类型则未必能保证。例如对于基于时间的范围分区，最接近当前时间的分区可能访问更频繁。此策略的另一个好处是可以实现智能分区联结（Partition-Wise Join），假设事实表和维度表均基于同一列做哈希分区，这两个表的分区对将位于同一实例，因此它们之间的联结可以在本地进行，从而提升查询的性能。

（3）BY SUBPARTITION，数据以子分区为单位分布到不同 RAC 实例。此策略同样非常适合于哈希分区。例如当父分区类型导致数据分布不均时，子分区采用哈希分区可以纠正这种偏差。

（4）AUTO，根据表的分区类型和优化器统计信息，自动为数据库对象选择以上三种分布策略之一。

基于之前创建的两节点 RAC 集群，通过下面的实例对分布策略做进一步的了解。首先需要为 RAC 集群中每个实例启用 Database In-Memory，在任一示例所在主机执行以下脚本，然后逐一重启其他实例使配置生效：

```
cd ~/dbimbook/chap04
./dbim_enable.sh
```

为便于演示和观察，先只为最大的表 LINEORDER 开启 INMEMORY 属性。可以看到，其默认的分布策略为 AUTO。

```
SQL> ALTER TABLE lineorder INMEMORY;
SQL> SELECT inmemory_distribute FROM user_segments WHERE segment_name = 'LINEORDER';

INMEMORY_DISTRIBUTE
-------------------
AUTO
```

将表 LINEORDER 发布到内存列式存储。由于 LINEORDER 表没有分区，因此 AUTO 对应的实际分布策略为 BY ROWID RANGE。实际的数据分布如下：

```
SQL> SELECT COUNT(*) FROM lineorder;

  COUNT(*)
----------
  11997996

SQL> SELECT inst_id, SUM(num_rows), SUM(num_rows)/11997996, COUNT(*) FROM gv$im_header
WHERE is_head_piece=1 GROUP BY inst_id;

   INST_ID SUM(NUM_ROWS) SUM(NUM_ROWS)/11997996   COUNT(*)
---------- ------------- ---------------------- ----------
         1       6682119             .556936258         13
         2       5315877             .443063742         10

SQL> SELECT inst_id, segment_name, inmemory_size/1024/1024 im_mb, BYTES/1024/1024
disk_mb, bytes_not_populated, populate_status AS pop_status FROM gv$im_segments;

   INST_ID SEGMENT_NAME      IM_MB    DISK_MB BYTES_NOT_POPULATED POP_STATUS
---------- ------------ ---------- ---------- ------------------- ----------
         1 LINEORDER         314.5 1301.14844           604848128 COMPLETED
         2 LINEORDER       249.9375 1301.14844           759504896 COMPLETED
```

从以上输出可知，LINEORDER 表在内存列式存储中共分配了 23 个 IMCU。在实例 1 上有 13 个 IMCU，约 56% 的行，在实例 2 上有 10 个 IMCU，约 44% 的行，数据分布基本保持均匀。另外，注意输出中的 BYTES_NOT_POPULATED 并不为 0，这对 RAC 集群来说是正常的，并不代表不完整发布。实际上，BYTES_NOT_POPULATED 所表示的数据会在其他实例的内存列式存储中发布。

接下来测试 BY PARTITION 分布策略，为此需要先将 LINEORDER 表改造为分区表，并使用 LO_ORDERKEY 列作为范围分区键。可以使用第 6 章介绍的在线重定义实现分区表的迁移，这里使用另一种方法，即分区交换加分区拆解来实现分区表的改造。

```
-- 创建空的临时范围分区表 lineorder_p，只有一个分区 p1
CREATE TABLE lineorder_p PARTITION BY RANGE(lo_orderkey)
```

```sql
(PARTITION p1 VALUES LESS THAN(MAXVALUE)) AS SELECT * FROM lineorder WHERE 1=2;

-- 将 lineorder 和 lineorder_p 实施分区交换，此时 lineorder 中所有数据迁移到 lineorder_p
ALTER TABLE lineorder_p EXCHANGE PARTITION p1 WITH TABLE lineorder WITHOUT VALIDATION;

-- 删除 lineorder 表，将 lineorder_p 重命名为 lineorder，此时 lineorder 变为分区表
DROP TABLE lineorder;
RENAME lineorder_p TO lineorder;

-- 由于 lineorder 表有近 1200 万行，并且分区键 LO_ORDERKEY 基本均匀分布
-- 因此以 300 万为界，将 lineorder 表拆分为 4 个范围分区：p1，p2，p3 和 p4
ALTER TABLE lineorder
  SPLIT PARTITION p1 AT(3000000)INTO(PARTITION p1, PARTITION p2);

ALTER TABLE lineorder
  SPLIT PARTITION p2 AT(6000000)INTO(PARTITION p2, PARTITION p3);

ALTER TABLE lineorder
  SPLIT PARTITION p3 AT(9000000)INTO(PARTITION p3, PARTITION p4);

-- 重新生成优化器统计信息
EXEC DBMS_STATS.gather_table_stats(USER, 'LINEORDER', CASCADE => TRUE);
```

至此，LINEORDER 表已改造为具有 4 个分区的范围分区表，每个分区中包含的数据量接近，为 300 万行左右。

```
SQL> SELECT partition_name, num_rows, high_value FROM  user_tab_partitions
 WHERE table_name = 'LINEORDER';

PARTITION_NAME           NUM_ROWS HIGH_VALUE
-------------------- ---------- --------------------
P1                    2999666 3000000
P2                    3001547 6000000
P3                    2998481 9000000
P4                    2998302 MAXVALUE

SQL> SELECT table_name, partitioning_type, partition_count FROM all_part_tables WHERE
table_name = 'LINEORDER';

TABLE_NAME           PARTITION PARTITION_COUNT
-------------------- --------- ---------------
LINEORDER            RANGE                   4
```

将分布策略改为按分区发布，发布表并等待发布完成。从结果来看，基于范围分区的数据分布并不均衡。实例 1 上只有 1 个分区，其他 3 个分区均位于实例 2。

```
SQL> ALTER TABLE lineorder INMEMORY DISTRIBUTE BY PARTITION;
SQL> EXEC DBMS_INMEMORY.POPULATE('SSB', 'LINEORDER');

SQL> SELECT segment_name, partition_name, inmemory_distribute FROM user_segments WHERE
segment_name = 'LINEORDER';
```

```
SEGMENT_NAME          PARTITION_NAME        INMEMORY_DISTRIBUTE
--------------------  --------------------  -------------------
LINEORDER             P1                    BY PARTITION
LINEORDER             P2                    BY PARTITION
LINEORDER             P3                    BY PARTITION
LINEORDER             P4                    BY PARTITION

SQL>
SELECT inst_id, segment_name, inmemory_size/1024/1024 im_mb, bytes/1024/1024
disk_mb, bytes_not_populated, populate_status AS pop_status FROM gv$im_segments;

   INST_ID SEGMENT_NAME      IM_MB     DISK_MB BYTES_NOT_POPULATED POP_STATUS
---------- ------------ ---------- ---------- ------------------- -------------
         1 LINEORDER     142.1875  325.382813                   0 COMPLETED
         2 LINEORDER     142.1875  324.960938                   0 COMPLETED
         2 LINEORDER     141.1875  325.734375                   0 COMPLETED
         2 LINEORDER     142.1875   325.40625                   0 COMPLETED
```

使用以下命令将分区表还原为普通表。至此，第一个基于范围分区的测试结束。

```
ALTER TABLE lineorder MERGE PARTITIONS p1, p2, p3, p4 INTO PARTITION p4;

CREATE TABLE lineorder_nop AS SELECT * FROM lineorder WHERE 1=2;

ALTER TABLE lineorder EXCHANGE PARTITION p4 WITH TABLE lineorder_nop WITHOUT VALIDATION;

DROP TABLE lineorder;

RENAME lineorder_nop TO lineorder;
```

接下来，将表 LINEORDER 改为哈希分区，分区数量为 4，过程如下：

```
SQL> CREATE TABLE lineorder_p PARTITION BY HASH(lo_orderkey)
PARTITIONS 4 AS SELECT * FROM lineorder;

SQL> DROP TABLE lineorder;

SQL> RENAME lineorder_p TO lineorder;

SQL> EXEC DBMS_STATS.gather_table_stats(USER, 'LINEORDER', CASCADE => TRUE);

SQL> SELECT partition_name, num_rows, high_value FROM  user_tab_partitions
 WHERE table_name = 'LINEORDER';

PARTITION_NAME           NUM_ROWS HIGH_VALUE
--------------------  ---------- --------------------
SYS_P684                 2995606
SYS_P685                 2999880
SYS_P686                 3004104
SYS_P687                 2998406

SQL> SELECT table_name, partitioning_type, partition_count FROM all_part_tables WHERE
table_name = 'LINEORDER';
```

```
TABLE_NAME            PARTITION PARTITION_COUNT
--------------------  --------- ---------------
LINEORDER             HASH                    4
```

将分布策略再次改为按分区发布，发布表并等待发布完成。最终的分布非常均衡，实例 1 和实例 2 上各有 2 个分区。

```
SQL> ALTER TABLE lineorder INMEMORY DISTRIBUTE BY PARTITION;
SQL> EXEC DBMS_INMEMORY.POPULATE('SSB', 'LINEORDER');

SQL> SELECT inst_id, segment_name, inmemory_size/1024/1024 im_mb, bytes/1024/1024
disk_mb, bytes_not_populated, populate_status AS pop_status FROM gv$im_segments;

   INST_ID SEGMENT_NAME      IM_MB     DISK_MB BYTES_NOT_POPULATED POP_STATUS
---------- ------------ ---------- ---------- ------------------- -------------
         2 LINEORDER      142.1875  325.414063                   0 COMPLETED
         2 LINEORDER      141.1875      325.25                   0 COMPLETED
         1 LINEORDER      141.1875    324.9375                   0 COMPLETED
         1 LINEORDER      141.1875     325.875                   0 COMPLETED
```

按分区分布的测试结束，使用以下命令将哈希分区表还原为普通表：

```
CREATE TABLE lineorder_tmp AS SELECT * FROM lineorder NOLOGGING;

DROP TABLE lineorder;

RENAME lineorder_tmp TO lineorder;
```

最后，测试按子分区发布策略。通过查询 LO_ORDERDATE 列的最小值和最大值，可知 LINEORDER 表包含了从 1992 年到 1998 年共 7 年的数据。因此将 LINEORDER 建立复合分区，父分区创建为范围分区，分区键为 LO_ORDERDATE 列，子分区创建为哈希分区，分区键为 LO_ORDERDATE 列，每一个父分区创建 4 个子分区。以下为创建复合分区的过程，最终一共有 28 个子分区，每一个子分区中包含的行数接近。

```
SQL> SELECT MIN(lo_orderdate), MAX(lo_orderdate) FROM lineorder;

MIN(LO_ORDERDATE) MAX(LO_ORDERDATE)
----------------- -----------------
         19920101          19980802

SQL> CREATE TABLE lineorder_p PARTITION BY RANGE (lo_orderdate) INTERVAL (10000)
SUBPARTITION BY HASH (lo_orderkey) SUBPARTITIONS 4
(PARTITION p1 VALUES LESS THAN (19930101))
AS SELECT * FROM lineorder NOLOGGING;

SQL> DROP TABLE lineorder;
SQL> RENAME lineorder_p TO lineorder;

SQL> EXEC dbms_stats.gather_table_stats(
NULL,'LINEORDER', granularity=>'SUBPARTITION');

SQL> SELECT partition_name, subpartition_name, num_rows FROM  all_tab_subpartitions
```

```
  WHERE table_name = 'LINEORDER';

PARTITION_NAME        SUBPARTITION_NAME       NUM_ROWS
--------------------  --------------------  ----------
P1                    SYS_SUBP764               454731
P1                    SYS_SUBP765               454147
P1                    SYS_SUBP766               455173
P1                    SYS_SUBP767               454080
SYS_P792              SYS_SUBP788               455922
SYS_P792              SYS_SUBP789               457763
SYS_P792              SYS_SUBP790               454699
SYS_P792              SYS_SUBP791               454443
SYS_P797              SYS_SUBP793               453275
SYS_P797              SYS_SUBP794               456435
SYS_P797              SYS_SUBP795               456898
SYS_P797              SYS_SUBP796               455948
SYS_P782              SYS_SUBP778               455799
SYS_P782              SYS_SUBP779               452909
SYS_P782              SYS_SUBP780               456488
SYS_P782              SYS_SUBP781               453410
SYS_P772              SYS_SUBP768               456719
SYS_P772              SYS_SUBP769               455221
SYS_P772              SYS_SUBP770               457110
SYS_P772              SYS_SUBP771               457778
SYS_P777              SYS_SUBP773               453200
SYS_P777              SYS_SUBP774               454482
SYS_P777              SYS_SUBP775               456553
SYS_P777              SYS_SUBP776               454845
SYS_P787              SYS_SUBP783               265960
SYS_P787              SYS_SUBP784               268923
SYS_P787              SYS_SUBP785               267183
SYS_P787              SYS_SUBP786               267902

28 rows selected.
```

修改分布策略为按子分区，发布表并等待发布完成。查询系统视图，结果显示所有数据均匀的分布在两个 RAC 实例，每个实例包含 14 个 IMCU。

```
SQL> ALTER TABLE lineorder INMEMORY DISTRIBUTE BY SUBPARTITION;
SQL> EXEC DBMS_INMEMORY.POPULATE('SSB', 'LINEORDER');

SQL> SELECT inst_id, segment_name, inmemory_size/1024/1024 im_mb, bytes/1024/1024
disk_mb, bytes_not_populated, populate_status AS pop_status FROM gv$im_segments;

   INST_ID SEGMENT_NAME          IM_MB    DISK_MB BYTES_NOT_POPULATED POP_STATUS
---------- ---------------- ---------- ---------- ------------------- ------------
         1 LINEORDER             21.25 49.1640625                   0 COMPLETED
         1 LINEORDER             21.25 49.4453125                   0 COMPLETED
         1 LINEORDER             21.25 49.5859375                   0 COMPLETED
         1 LINEORDER             13.25 28.8515625                   0 COMPLETED
         1 LINEORDER             21.25    49.53125                  0 COMPLETED
         1 LINEORDER             21.25 49.4609375                   0 COMPLETED
         1 LINEORDER             21.25   49.515625                  0 COMPLETED
```

```
         1 LINEORDER          21.25   49.171875                    0 COMPLETED
         1 LINEORDER          13.25   28.984375                    0 COMPLETED
         1 LINEORDER          21.25   49.328125                    0 COMPLETED
         1 LINEORDER          21.25   49.328125                    0 COMPLETED
         1 LINEORDER          21.25     49.5625                    0 COMPLETED
         1 LINEORDER          21.25      49.375                    0 COMPLETED
         1 LINEORDER          21.25   49.546875                    0 COMPLETED
         2 LINEORDER          21.25     49.1875                    0 COMPLETED
         2 LINEORDER          21.25   49.515625                    0 COMPLETED
         2 LINEORDER          13.25     29.0625                    0 COMPLETED
         2 LINEORDER          21.25   49.265625                    0 COMPLETED
         2 LINEORDER          21.25   49.296875                    0 COMPLETED
         2 LINEORDER          21.25  49.4609375                    0 COMPLETED
         2 LINEORDER          21.25  49.2578125                    0 COMPLETED
         2 LINEORDER          21.25  49.3828125                    0 COMPLETED
         2 LINEORDER          21.25  49.3046875                    0 COMPLETED
         2 LINEORDER          21.25  49.1328125                    0 COMPLETED
         2 LINEORDER          21.25  49.6640625                    0 COMPLETED
         2 LINEORDER          13.25   29.171875                    0 COMPLETED
         2 LINEORDER          21.25     49.34375                   0 COMPLETED
         2 LINEORDER          21.25     49.65625                   0 COMPLETED

28 rows selected.

SQL> SELECT inst_id, COUNT(*) num_imcu, SUM(inmemory_size)/1024/1024 im_mb,
SUM(bytes)/1024/1024 disk_mb, SUM(bytes_not_populated) FROM gv$im_segments WHERE
segment_name = 'LINEORDER' GROUP BY inst_id;

   INST_ID   NUM_IMCU        IM_MB    DISK_MB SUM(BYTES_NOT_POPULATED)
---------- ---------- ---------- ---------- ------------------------
         1         14        281.5 650.851563                        0
         2         14        281.5 650.703125                        0
```

实验结束，使用以下命令将混合分区表还原为普通表：

```
CREATE TABLE lineorder_tmp AS SELECT * FROM lineorder NOLOGGING;

DROP TABLE lineorder;

RENAME lineorder_tmp TO lineorder;
```

从以上测试结果可以得出结论，基于分区或子分区的分布策略最适合于哈希分区表，可以保证数据最均衡的分布；基于 ROWID 的分布策略可以保证数据分布基本均衡，当原表没有分区或分区类型不是哈希分区时，按 ROWID 的分布策略就是最佳选择。

最后，对于没有分区的小表，如果其大小只有 1 个 IMCU，此表将仅在一个实例中发布。在以下示例中，除 PART 外的所有维度表均为这种情形。

```
SQL> SELECT inst_id, segment_name, inmemory_size/1024/1024 im_mb, BYTES/1024/1024
  disk_mb, bytes_not_populated, populate_status AS pop_status FROM gv$im_segments;

   INST_ID SEGMENT_NAME        IM_MB    DISK_MB BYTES_NOT_POPULATED POP_STATUS
---------- ------------ ---------- ---------- ------------------- --------------
         1 PART            21.4375  98.890625                   0 COMPLETED
         1 DATE_DIM           1.25    .296875                   0 COMPLETED
```

```
         1 SUPPLIER              2.25      1.875                 0 COMPLETED
         1 CUSTOMER             17.25     29.234375              0 COMPLETED

SQL> SELECT inst_id, object_name, SUM(num_rows), COUNT(*) num_imcu FROM gv$im_header,
user_objects WHERE is_head_piece=1 AND object_id = table_objn GROUP BY inst_id,
object_name;

   INST_ID OBJECT_NAME   SUM(NUM_ROWS)   NUM_IMCU
---------- ------------- --------------- ----------
         1 SUPPLIER             16000          1
         1 PART               1000000          2
         1 CUSTOMER            240000          1
         1 DATE_DIM              2556          1
```

8.2.4 In-Memory 复制

8.2.3 节介绍的 In-Memory 数据分布解决了内存容量和性能扩展的问题。在此基础上，随之而来的另一个重要的需求就是系统的容错性。如果 RAC 集群中的部分实例失效，虽然查询并不会出错，但是原本位于内存列式存储中的数据需要从 Buffer Cache 或磁盘中读取，这可能会对性能产生负面影响。为此，Oracle 提供了 In-Memory 复制功能，此功能也称为 In-Memory 容错。

In-Memory 复制可以跨 RAC 实例复制内存列式存储中的数据。当某一实例失效时，可以保证数据副本存在于一个或多个实例的内存列式存储中，从而保证数据访问的性能。为保证数据复制的性能，目前此功能只在基于 Exadata 的 RAC 环境提供。In-Memory 复制通过在 DDL 语句中加入 DUPLICATE 子句启用，共提供两种选项，如图 8-5 所示。

图 8-5 In-Memory 复制的两种选项

（1）DUPLICATE，数据复制到另一个 RAC 实例上，因此总共有两份数据，位于两个不同的 RAC 实例。

（2）DUPLICATE ALL，数据复制到所有 RAC 实例上。如果指定了 DISTRIBUTE FOR SERVICE，则复制到服务所对应的所有 RAC 实例上。此时的 In-Memory 分布策略被强制设定为 DISTRIBUTE AUTO。

通过示例来了解 In-Memory 复制的两种复制选项，测试环境为两节点 RAC 集群。为便于演示,只基于 LINEORDER 一张表进行测试,此表总共有 11 997 996 行数据。先来看一下默认设置,也就是没有复制时的数据发布情况。此时的数据分布策略为 AUTO,数据近乎均匀的分布到两个实例,总共 23 个 IMCU。其中 11 个 IMCU,近 49%的行分布到实例 1,占用内存约 275MB;其余 12 个 IMCU,超过 51%的行分布到实例 2,约占用内存 287MB。BYTES_NOT_POPULATED 不为 0,表示对象有一部分数据发布在其他的实例,这也可以说明其没有启用 In-Memory 复制。

```
SQL> SELECT inmemory_distribute FROM user_segments WHERE segment_name = 'LINEORDER';

INMEMORY_DISTRIBUTE
-------------------
AUTO

SQL> SELECT inst_id, SUM(num_rows), SUM(num_rows)/11997996, COUNT(*) "#IMCU" FROM
gv$im_header WHERE is_head_piece=1 GROUP BY inst_id;

   INST_ID SUM(NUM_ROWS) SUM(NUM_ROWS)/11997996      #IMCU
---------- ------------- ----------------------- ----------
         1       5858298                    .488         11
         2       6139698                    .512         12

SQL> SELECT inst_id, segment_name, inmemory_size/1024/1024 im_mb, bytes/1024/1024
disk_mb, bytes_not_populated, populate_status AS pop_status FROM gv$im_segments;

   INST_ID SEGMENT_NAME          IM_MB    DISK_MB BYTES_NOT_POPULATED POP_STATUS
---------- ------------- ------------- ---------- ------------------- -----------
         1 LINEORDER           275.125 1301.14844            697786368 COMPLETED
         2 LINEORDER          287.3125 1301.14844            666566656 COMPLETED

SQL> SELECT inst_id, pool, round(alloc_bytes/1024) "ALLOC(KB)", round(used_bytes/1024)
"USED(KB)", populate_status, con_id FROM gv$inmemory_area;

   INST_ID POOL         ALLOC(KB)  USED(KB) POPULATE_STATUS       CON_ID
---------- ---------- ----------- --------- --------------- ----------
         2 1MB POOL       2128896    291840 DONE                     3
         2 64KB POOL       916928      2368 DONE                     3
         1 1MB POOL       2128896    279552 DONE                     3
         1 64KB POOL       916928      2176 DONE                     3
```

将 In-Memory 复制策略设置为 DUPLICATE 并发布:

```
ALTER TABLE lineorder INMEMORY DUPLICATE;
EXEC dbms_inmemory.populate('SSB', 'LINEORDER');
```

待发布完成后,可以从系统视图中查询其分布情况。首先确认 In-Memory 复制策略为 DUPLICATE,这表示每一个 IMCU 都会在另一 RAC 实例的内存列式存储中建立副本。由于 RAC 集群只有两个节点,所以此时的 DUPLICATE 设置和 DUPLICATE ALL 是等效的。这表示从一个 RAC 实例的内存列式存储中可以访问到对象的所有数据。这可以从以下输出中的多处得到印证,包括两个实例拥有相同数量的 IMCU,相同的内存占用,以及 BYTES_NOT_POPULATED 为 0。

```
SQL> SELECT inmemory_duplicate FROM user_tables WHERE table_name = 'LINEORDER';
```

```
INMEMORY_DUPLICATE
------------------
DUPLICATE

SQL> SELECT inst_id, SUM(num_rows), SUM(num_rows)/11997996, COUNT(*) "#IMCU" FROM
gv$im_header WHERE is_head_piece=1 GROUP BY inst_id;

   INST_ID SUM(NUM_ROWS) SUM(NUM_ROWS)/11997996      #IMCU
---------- ------------- ----------------------- ----------
         1      11997996                       1         23
         2      11997996                       1         23

SQL> SELECT inst_id, segment_name, inmemory_size/1024/1024 IM_MB, bytes/1024/1024
DISK_MB, bytes_not_populated, populate_status AS pop_status FROM gv$im_segments;

   INST_ID SEGMENT_NAME        IM_MB    DISK_MB BYTES_NOT_POPULATED POP_STATUS
---------- ------------- ----------- ---------- ------------------- ----------
         2 LINEORDER         562.375 1301.14844                   0 COMPLETED
         1 LINEORDER         562.375 1301.14844                   0 COMPLETED

SQL> SELECT inst_id, pool, round(alloc_bytes/1024) "ALLOC(KB)", round(used_bytes/1024)
"USED(KB)", populate_status, con_id FROM gv$inmemory_area;

   INST_ID POOL       ALLOC(KB)   USED(KB) POPULATE_STATUS     CON_ID
---------- --------- ---------- ---------- --------------- ----------
         2 1MB POOL     2128896     571392 DONE                     3
         2 64KB POOL     916928       4480 DONE                     3
         1 1MB POOL     2128896     571392 DONE                     3
         1 64KB POOL     916928       4480 DONE                     3
```

另一个显著的不同是内存占用增加了一倍，这也是采用 In-Memory 复制时最大的开销。对于 N 节点的 RAC 集群，如果采用 DUPLICATE，内存占用会增加 1 倍；如果采用 DUPLICATE ALL，因为有 N 个副本，内存占用也会增加 N 倍。更多的副本消耗更多的内存，但同时也增强了系统的容错性和性能。N 越大，损失一个 RAC 实例对整体性能的影响越小。由于内存列式存储中数据可用性的保证，RAC 集群可以实施滚动升级和打补丁，同时对应用访问性能的下降没有太大影响。又例如对于星形模型，可以让事实表按分区或子分区分布，而维表使用 DUPLICATE 发布到所有实例。这样星形联结可以通过本地访问实现，从而避免了跨实例的访问。

当 RAC 节点或实例失效时，经过一段时间，受影响的 IMCU 会在其他实例重新发布，详情请参见 8.2.7 节。

8.2.5　In-Memory 与并行执行

Oracle RAC 是共享架构，数据通过 RAC 节点间的私有网络传递。而基于 RAC 的 Database In-Memory 是无共享架构，实例只能访问位于本地内存列式存储中的数据，IMCU 不会通过 RAC 私有网络进行传递。如果数据不在本实例的内存列式存储中，这些数据只能从 Buffer Cache 或磁盘中读取。因此，当一个表的数据被分布在多个 RAC 实例时，必须启用并行，才能保证所有的数据均从内存列式存储中获取。

初始化参数 PARALLEL_DEGREE_POLICY 可以控制是否启用自动并行控制，此参数的默认值为 MANUAL，即不启用自动并行。

```
SQL> SHOW PARAMETER parallel_degree_policy;

NAME                                 TYPE        VALUE
------------------------------------ ----------- ------------------------------
parallel_degree_policy               string      MANUAL
```

当自动并行控制未启用时，先将 LINEORDER 表发布到内存列式存储，然后在实例 1 中执行查询，执行时间为 3.2 秒。

```
SQL> SELECT COUNT(*) FROM lineorder;

  COUNT(*)
----------
  11997996


Elapsed: 00:00:03.20

SQL> @xplan

PLAN_TABLE_OUTPUT
--------------------------------------------------------------------------------
SQL_ID  0g5kz6486txsb, child number 0
-------------------------------------
SELECT COUNT(*) FROM lineorder

Plan hash value: 2267213921

--------------------------------------------------------------------------------
| Id  | Operation                   | Name     | Rows  | Cost (%CPU)| Time     |
--------------------------------------------------------------------------------
|   0 | SELECT STATEMENT            |          |       |  1713 (100)|          |
|   1 |  SORT AGGREGATE             |          |     1 |            |          |
|   2 |   TABLE ACCESS INMEMORY FULL| LINEORDER|   11M |  1713   (2)| 00:00:01 |
--------------------------------------------------------------------------------

14 rows selected.

SQL> @imstats

NAME                                                                    VALUE
----------------------------------------------------------------------- ----------
CPU used by this session                                                   306
IM scan CUs memcompress for query low                                       15
IM scan rows                                                           7323340
```

```
IM scan rows projected                              7138861
IM scan rows valid                                  7138861
physical reads                                        67466
physical reads direct                                 67466
session logical reads                                172583
session logical reads - IM                           101681
session pga memory                                 23910208
table scan disk IMC fallback                        4859135
table scans(IM)                                           1
table scans(long tables)                                  1

13 rows selected.
```

执行计划显示 TABLE ACCESS INMEMORY FULL，表示内存中的全表扫描，但在执行统计信息中，仍存在大量的物理读，这表明部分数据是从磁盘中读取的。

从以下查询得知，7 323 340 行数据位于实例 1 的内存列式存储中，此值与统计信息中的 IM scan rows 一致。这也侧面说明了其余 4 674 656 行数据是从磁盘中读取的。

```
SQL> SELECT inst_id, object_name, SUM(num_rows), COUNT(*) num_imcu FROM gv$im_header,
user_objects WHERE is_head_piece=1 AND object_id = table_objn GROUP BY inst_id,
object_name;

   INST_ID OBJECT_NAME          SUM(NUM_ROWS)   NUM_IMCU
---------- -------------------- ------------- ----------
         1 LINEORDER                  7323340         15
         2 LINEORDER                  4674656          9
```

将 PARALLEL_DEGREE_POLICY 设为 AUTO，此时 Oracle 会启用自动并行，即根据查询的成本估算决定是否并行运行查询。启动自动并行后，运行与之前相同的查询，查询时间为 0.18 秒，比之前快了近 18 倍。

```
-- 由于是系统一级的改变，因此只需要在 RAC 集群中任一实例中执行
SQL> ALTER SYSTEM SET parallel_degree_policy = AUTO;

SQL> SELECT COUNT(*) FROM lineorder;

  COUNT(*)
----------
  11997996

Elapsed: 00:00:00.18

SQL> @xplan

PLAN_TABLE_OUTPUT
--------------------------------------------------------------------------------
SQL_ID  0g5kz6486txsb, child number 3
-------------------------------------
select count(*) from lineorder
```

```
Plan hash value: 396151021

---------------------------------------------------------------------------------
| Id  | Operation                    | Name      | TQ    |IN-OUT| PQ Distrib |
---------------------------------------------------------------------------------
|   0 | SELECT STATEMENT             |           |       |      |            |
|   1 |  SORT AGGREGATE              |           |       |      |            |
|   2 |   PX COORDINATOR             |           |       |      |            |
|   3 |    PX SEND QC (RANDOM)       | :TQ10000  | Q1,00 | P->S | QC (RAND)  |
|   4 |     SORT AGGREGATE           |           | Q1,00 | PCWP |            |
|   5 |      PX BLOCK ITERATOR       |           | Q1,00 | PCWC |            |
|*  6 |       TABLE ACCESS INMEMORY FULL| LINEORDER | Q1,00 | PCWP |            |
---------------------------------------------------------------------------------

Predicate Information (identified by operation id):
---------------------------------------------------

   6 - inmemory(:Z>=:Z AND :Z<=:Z)

Note
-----
   - automatic DOP: Computed Degree of Parallelism is 8
   - parallel scans affinitized for inmemory

28 rows selected.

SQL> @imstats

NAME                                              VALUE
-----------------------------------------------   ----------
CPU used by this session                          91
IM scan CUs memcompress for query low             24
IM scan rows                                      11997996
IM scan rows projected                            11997996
IM scan rows valid                                11997996
session logical reads                             177857
session logical reads - IM                        166587
session pga memory                                25286464
table scans(IM)                                   24
table scans(long tables)                          24

10 rows selected.
```

在执行计划的 Note 部分，automatic DOP 表示自动并行已启用，并且并行度被设置为 8。在执行统计信息部分，没有发现任何与物理读相关的信息。IM scan rows 为 11 997 996，这正好是 LINEORDER 表的总行数，这表示所有的数据都是从内存列式存储中读取的。

如图 8-6 所示,通过实时 SQL 监控报告也可以确认并行执行已生效。在该图左上角 Execution Plan 一行的两个图标,8 表示并行度,2 表示并行执行跨两个 RAC 实例。

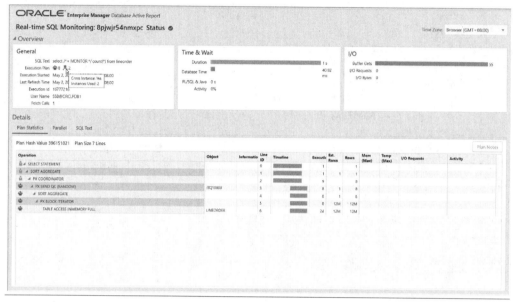

图 8-6　通过实时 SQL 监控报告确认并行执行

也可以在 SQL 语句中加入 PARALLEL 提示来启用并行,指定的并行度必须大于启用了内存列式存储的实例的数量。由于在实际生产应用中未必能对 SQL 进行修改,因此建议通过初始化参数来启用自动并行。

```
SQL> SHOW PARAMETER parallel_degree_policy

NAME                                 TYPE        VALUE
------------------------------------ ----------- ------------------------------
parallel_degree_policy               string      MANUAL

SQL> SELECT /*+ PARALLEL(4) */ COUNT(*) FROM lineorder;

  COUNT(*)
----------
  11997996

Elapsed: 00:00:00.02
```

另一个与并行执行相关的初始化参数是 PARALLEL_FORCE_LOCAL,默认值为 FALSE,表示允许并行执行。如果设为 TRUE,则表示 SQL 只能在发起查询的节点上执行,这将导致只能在本实例内存列式存储中获取部分数据,余下的数据需要从 Buffer Cache 或磁盘中查询。在 Oracle 电子商务套件(E-Business Suite,EBS)中,PARALLEL_FORCE_LOCAL 的推荐值为 TRUE,目的是为了减少节点间的通信。因此当 EBS 与 Database In-Memory 配合使用时,需要做特殊的考虑,其最佳实践和操作指南请参见 Oracle 技术白皮书:如何在 Oracle EBS 中使用

Database In-Memory[①]。

8.2.6　In-Memory 与实例子集发布

实例子集发布是指仅将数据库对象发布到 RAC 集群中的部分实例。有两种实现方式，第一种方式是通过禁止部分实例的内存列式存储，使得数据库对象只发布到 RAC 集群中其余的实例；第二种方式是通过配置 RAC 服务，数据库对象只发布到 RAC 服务所关联的实例。

先来看第一种方式。假设 RAC 集群包括 2 个实例，通过禁用实例 1 的内存列式存储，所有的数据库对象将只发布到实例 2 中。首先在实例 1 中将 INMEMORY_SIZE 设为 0，从而禁用内存列式存储，然后重启使其生效。

```
-- 查询实例 ID, 确认为实例 1
SQL> SELECT sys_context('userenv','instance_name') AS sid FROM dual;

SID
--------------------------------------------------------------------------------
DBRAC1

-- 将实例 1 的 INMEMORY_SIZE 设为 0, 从而禁用内存列式存储
SQL> ALTER SYSTEM SET inmemory_size=0 SID='DBRAC1' SCOPE=SPFILE;
System altered.

-- 重启实例使变更生效
SQL> SHUTDOWN IMMEDIATE;
Database closed.
Database dismounted.
ORACLE instance shut down.

SQL> STARTUP
ORACLE instance started.
Total System Global Area 2634021168 bytes
Fixed Size                  9689392 bytes
Variable Size             922746880 bytes
Database Buffers         1677721600 bytes
Redo Buffers               23863296 bytes
Database mounted.
Database opened.
```

通过启用并行的全表扫描发布表 LINEORDER。由于实例 1 中的内存列式存储被禁止，此时 lineorder 表只被发布到实例 2 中：

```
SQL> ALTER TABLE lineorder INMEMORY;

SQL> SELECT /*+ FULL(p) PARALLEL(4) */ COUNT(*) FROM lineorder p;

SQL> SELECT inst_id, segment_name, inmemory_size, bytes_not_populated,
```

[①] https://support.oracle.com/epmos/faces/DocumentDisplay?id=2025309.1

```
populate_status FROM gv$im_segments;

  INST_ID SEGMENT_NAME INMEMORY_SIZE BYTES_NOT_POPULATED POPULATE_STAT
---------- ------------ ------------- ------------------- -------------
        2 LINEORDER        590938112                    0 COMPLETED
```

需要特别注意，这里所说的禁用内存列式存储是指在 CDB 一级，也就是整个实例层面。如果仅在 PDB 一级禁用，在对象会被不完整发布，也就是虽然实例 1 上不会发布任何数据，但实例 2 中也只会发布部分数据：

```
SQL> SELECT inst_id, segment_name, inmemory_size, bytes_not_populated,
 populate_status FROM gv$im_segments;

  INST_ID SEGMENT_NAME INMEMORY_SIZE BYTES_NOT_POPULATED POPULATE_STAT
---------- ------------ ------------- ------------------- -------------
        2 LINEORDER        230424576           832970752 COMPLETED
```

实例子集发布的第二种方式需要配合数据库服务来使用，这也是 Database In-Memory 推荐的方式。数据库服务是 Oracle 非常重要的基本概念，每一个数据库都可以用一个或多个服务表示，例如默认的数据库服务名使用初始化参数 DB_NAME 和 DB_DOMAIN 组合而成：

```
SQL> SHOW PARAMETER service_names

NAME          TYPE   VALUE
------------- ------ --------------------------------------------------
service_names string DBRAC_21c.sub12092311540.training.oraclevcn.com
```

客户端使用服务名来确定需要访问的数据库。一个服务名可以与多个数据库实例关联，典型的如 RAC 集群中的实例，或 ADG 中的主备实例；一个实例也可以关联多个服务名，也就是被多个服务引用。监听作为客户端和数据库实例间的中介，并将请求路由到适合的实例。由于客户端连接数据库只需要指定服务名，无须指定具体的实例名，因此服务可以认为是数据库负载的逻辑抽象。特别是在 RAC 集群中，可以利用服务来实现应用分区。例如在一个两节点的 RAC 集群中，节点 1 负责应用负载 A，节点 2 负责应用负载 B，而节点 1 和节点 2 同时负责应用负载 C。

通过服务来实现 In-Memory 对象的实例子集发布可以有两种方式，如图 8-7 所示。例如在一个 3 节点 RAC 集群中，共定义了两个数据库服务 S1 和 S2。第 1 种方式是通过实例层面的初始化参数 PARALLEL_INSTANCE_GROUP 来控制。此参数将并行查询限定在了其指定服务所运行的实例，In-Memory 对象也将只发布到这些实例的内存列式存储中。由于一个实例只能通过此参数指定一个服务，因此这种方式适合于实例可以与服务一一对应的情形。也就是说，一个实例不会被多个服务共同使用。显然，这种方式还需要与网络服务名配合，以限定客户端只能连接到相应的服务。否则，以图 8-7（a）中的 Table A 为例，当连接到实例 1（I1）时，发布目标将仅限于实例 1；当连接到实例 2（I2）或实例 3（I3）时，发布目标仅限于实例 2 和实例 3。

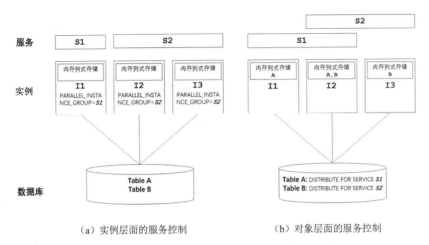

(a)实例层面的服务控制　　　　　　　(b)对象层面的服务控制

图 8-7　通过服务控制 In-Memory 对象的发布

第 2 种方式是在数据库对象层面实现,可实现更精确的控制。数据库对象可以通过 **DISTRIBUTE FOR SERVICE** 子句与服务绑定。也就是说,这些对象只会发布到启动此服务所有实例的内存列式存储中。如果服务迁移到新的实例,对象也会跟随服务迁移,即从原内存列式存储中移除,并发布到新示例的内存对象存储中。如果服务终止,与其绑定的对象也将从内存列式存储中移除。图 8-7(b)中,这表示 Table A 会被发布到实例 1 和 2 中,表 B 会被发布到实例 2 和实例 3 中。

通过对象层面的服务控制来实现 In-Memory 对象的实例子集发布是推荐的方式,先介绍其具体的配置过程。仍然使用 SSB 模型中的 5 个表。假设应用 A 需要 SUPPLIER 和 PART 表,仅通过 RAC 节点 1 的内存列式存储访问;应用 B 需要 CUSTOMER 和 DATE_DIM 表,仅通过 RAC 节点 2 上的内存列式存储访问;应用 C 需要 LINEORDER 表,可通过 RAC 节点 1 和 2 上的内存列式存储访问。首先获取数据库名和两个实例名,并定义环境变量。

```
$ srvctl config database                    # 获取数据库名
DBRAC_21c

$ export DBNAME=$(srvctl config database)

$ srvctl status database -d $DBNAME -v      # 获取 RAC 集群中的两个实例名
Instance DBRAC1 is running on node dbim1. Instance status: Open.
Instance DBRAC2 is running on node dbim2. Instance status: Open.

$ export INST1=DBRAC1
$ export INST2=DBRAC2
```

使用 SQL 语句也可以查询数据库名和实例状态:

```
SQL> SHOW PARAMETER db_unique_name

NAME                                 TYPE        VALUE
------------------------------------ ----------- ------------------------------
db_unique_name                       string      DBRAC_21c
```

```
SQL> SELECT instance_name, host_name, instance_role, database_type, status FROM
gv$instance;

INSTANCE_NAME     HOST_NAME     INSTANCE_ROLE        DATABASE_TYPE     STATUS
---------------   ----------    -----------------    -------------     --------
DBRAC1            dbim1         PRIMARY_INSTANCE     RAC               OPEN
DBRAC2            dbim2         PRIMARY_INSTANCE     RAC               OPEN
```

然后定义 3 个服务，IM1、IM2 和 IM_ALL。服务 IM1 只运行于 1 个实例，主实例为实例 1，备实例为实例 2；服务 IM2 只运行于 1 个实例，主实例为实例 2，备实例为实例 1；服务 IM_ALL 运行于 RAC 集群中的所有实例，即实例 1 和实例 2：

```
$ srvctl status service -db $DBNAME -v
Database DBRAC_21c does not have services

$ srvctl add service -db $DBNAME -service IM1 -preferred $INST1 -available $INST2
$ srvctl add service -db $DBNAME -service IM2 -preferred $INST2 -available $INST1
$ srvctl add service -db $DBNAME -service IM_ALL -preferred "$INST1,$INST2"
```

启动所有服务，确认服务均在运行：

```
$ srvctl status service -db $DBNAME
Service im1 is not running.
Service im2 is not running.
Service im_all is not running.

$ srvctl start service -db $DBNAME -service "IM1,IM2,IM_ALL"

$ srvctl status service -db $DBNAME
Service im1 is running on instance(s) DBRAC1
Service im2 is running on instance(s) DBRAC2
Service im_all is running on instance(s) DBRAC1,DBRAC2
```

此时，在 RAC 节点 1 上可以看到 IM1 和 IM_ALL 服务，在 RAC 节点 2 上可以看到 IM2 和 IM_ALL 服务：

```
$ srvctl status database -d $DBNAME -v
Instance DBRAC1 is running on node dbim1 with online services IM1,IM_ALL. Instance
status: Open.
Instance DBRAC1 is running on node dbim2 with online services IM2,IM_ALL. Instance
status: Open.

$ hostname
dbim1

$ srvctl status listener -v
Listener LISTENER is enabled
Listener LISTENER is running on node(s): dbim1,dbim2

$ lsnrctl status
...
Service "im1.sub12092311540.training.oraclevcn.com" has 1 instance(s).
  Instance "DBRAC1", status READY, has 1 handler(s) for this service…
```

```
Service "im_all.sub12092311540.training.oraclevcn.com" has 1 instance(s).
  Instance "DBRAC1", status READY, has 1 handler(s) for this service…

$ hostname
dbim2

$ lsnrctl status
…
Service "im2.sub12092311540.training.oraclevcn.com" has 1 instance(s).
  Instance "DBRAC2", status READY, has 1 handler(s) for this service…
Service "im_all.sub12092311540.training.oraclevcn.com" has 1 instance(s).
  Instance "DBRAC2", status READY, has 1 handler(s) for this service…
```

从客户端连接 IM1 服务只能连到 RAC 实例 1，连接 IM2 服务只能连到 RAC 实例 2，连接 IM_ALL 则会随机连接到 RAC 实例 1 或实例 2。

```
$ srvctl config scan
SCAN name: dbim-scan.sub12092311540.training.oraclevcn.com, Network: 1
Subnet IPv4: 10.0.0.0/255.255.255.0/ens3, static
Subnet IPv6:
SCAN 1 IPv4 VIP: 10.0.0.116
SCAN VIP is enabled.
SCAN 2 IPv4 VIP: 10.0.0.194
SCAN VIP is enabled.
SCAN 3 IPv4 VIP: 10.0.0.43
SCAN VIP is enabled.

$ export SCANNAME=dbim-scan.sub12092311540.training.oraclevcn.com
$ export SERVICE_IM1=im1.sub12092311540.training.oraclevcn.com
$ export SERVICE_IM2=im2.sub12092311540.training.oraclevcn.com

$ sqlplus sys@$SCANNAME/$SERVICE_IM1 as sysdba
SQL> show parameter instance
instance_mode                        string      READ-WRITE
instance_name                        string      DBRAC1
instance_number                      integer     1
instance_type                        string      RDBMS

$ sqlplus sys@$SCANNAME/$SERVICE_IM2 as sysdba
SQL> show parameter instance
instance_mode                        string      READ-WRITE
instance_name                        string      DBRAC2
instance_number                      integer     2
instance_type                        string      RDBMS
```

接下来需要利用 INMEMORY 的 DISTRIBUTE 子句指定数据库对象对应的服务。为简化测试，所有的表都设置了 INMEMORY 优先级，以便让其自动发布。

```
-- 指定 SUPPLIER 和 PART 表发布到服务 IM1, 也就是 RAC 节点 1
ALTER TABLE SUPPLIER INMEMORY PRIORITY HIGH DISTRIBUTE FOR SERVICE IM1;
ALTER TABLE PART INMEMORY PRIORITY HIGH DISTRIBUTE FOR SERVICE IM1;
```

```
-- 指定 CUSTOMER 和 DATE_DIM 表发布到服务 IM2，也就是 RAC 节点 2
ALTER TABLE CUSTOMER INMEMORY PRIORITY HIGH DISTRIBUTE FOR SERVICE IM2;
ALTER TABLE DATE_DIM INMEMORY PRIORITY HIGH DISTRIBUTE FOR SERVICE IM2;

-- 指定 LINEORDER 表发布到服务 IM_ALL，也就是 RAC 节点 1 和 2
ALTER TABLE LINEORDER INMEMORY PRIORITY HIGH DISTRIBUTE FOR SERVICE IM_ALL;
```

确认所有对象发布完成后，实际的对象分布和之前设定的一样，SUPPLIER 和 PART 表仅分布在 RAC 实例 1 上，CUSTOMER 和 DATE_DIM 表仅分布在 RAC 实例 2 上，LINEORDER 表分布在所有的 RAC 实例上：

```
SQL> SELECT * FROM gv$inmemory_area;
   INST_ID POOL         ALLOC_BYTES  USED_BYTES POPULATE_STATUS      CON_ID
---------- ------------ ----------- ----------- ---------------- ----------
         2 1MB POOL      1089470464   290455552 DONE                      3
         2 64KB POOL      469762048     2752512 DONE                      3
         1 1MB POOL      1089470464   337641472 DONE                      3
         1 64KB POOL      469762048     3145728 DONE                      3

SQL> SELECT segment_name, inst_id, populate_status, inmemory_size, BYTES,
bytes_not_populated FROM gv$im_segments ORDER BY 1,2;

SEGMENT_NAME    INST_ID POPULATE_STAT INMEMORY_SIZE      BYTES BYTES_NOT_POPULATED
--------------- ------- ------------- ------------- ---------- -------------------
CUSTOMER              2 COMPLETED          18087936   30654464                   0
DATE_DIM              2 COMPLETED           1310720     311296                   0
LINEORDER             1 COMPLETED         315949056 1364353024           633004032
LINEORDER             2 COMPLETED         273809408 1364353024           731348992
PART                  1 COMPLETED          22478848  103694336                   0
SUPPLIER              1 COMPLETED           2359296    1966080                   0

6 rows selected.
```

将服务 IM1 从实例 1 迁移到实例 2：

```
$ srvctl relocate service -db $DBNAME -service IM1 -oldinst $INST1 -newinst $INST2

$ srvctl status database -d $DBNAME -v
Instance DBRAC1 is running on node dbim1 with online services IM_ALL. Instance status:
Open.
Instance DBRAC2 is running on node dbim2 with online services IM1,IM2,IM_ALL. Instance
status: Open.
```

由于对象绑定了服务，因此其也会随服务迁移。对象 PART 和 SUPPLIER 被清除出实例 1 的内存列式存储，然后重新发布到实例 2 的内存列式存储中：

```
SQL> SELECT segment_name, inst_id, populate_status, inmemory_size, BYTES,
bytes_not_populated FROM gv$im_segments ORDER BY 1,2;

SEGMENT_NAME    INST_ID POPULATE_STATUS INMEMORY_SIZE      BYTES BYTES_NOT_POPULATED
--------------- ------- --------------- ------------- ---------- -------------------
CUSTOMER              2 COMPLETED            18087936   30654464                   0
DATE_DIM              2 COMPLETED             1310720     311296                   0
```

```
LINEORDER              1 COMPLETED      315949056 1364353024        633004032
LINEORDER              2 COMPLETED      273809408 1364353024        731348992
PART                   2 COMPLETED       22478848  103694336                0
SUPPLIER               2 COMPLETED        2359296    1966080                0

6 rows selected.
```

如果彻底停止服务 IM1，则与其绑定的对象将从内存列式存储中清除：

```
$ srvctl stop service -db $DBNAME -service IM1

SQL> SELECT segment_name, inst_id, populate_status, inmemory_size, BYTES,
bytes_not_populated FROM gv$im_segments ORDER BY 1,2;

SEGMENT_NAME    INST_ID POPULATE_STATUS INMEMORY_SIZE      BYTES BYTES_NOT_POPULATED
--------------- ------- --------------- ------------- ---------- -------------------
CUSTOMER              2 COMPLETED            18087936   30654464                   0
DATE_DIM              2 COMPLETED             1310720     311296                   0
LINEORDER             1 COMPLETED           315949056 1364353024           633004032
LINEORDER             2 COMPLETED           273809408 1364353024           731348992
```

测试结束，执行清理工作。停止并删除所有服务：

```
$ srvctl stop service -db $DBNAME -service "IM1,IM2,IM_ALL"

$ for s in IM1 IM2 IM_ALL; do
    srvctl remove service -db $DBNAME -service "$s"
done
```

通过一个实例来了解通过示例层面的控制实现子集发布的过程。为简化，只使用 LINEORDER 一张表。首先通过 srvctl 命令获取数据库名称和 RAC 集群中两个实例的名称，并为其设置环境变量。然后设置 RAC 服务名称为 IM1。

```
$ export DBNAME=$(srvctl config database)
$ export SERVICENAME=IM1

$ srvctl status database -d $DBNAME -v          # 获取 RAC 集群中的两个实例名
Instance DBRAC1 is running on node dbim1. Instance status: Open.
Instance DBRAC2 is running on node dbim2. Instance status: Open.

$ export INST1=$(srvctl status database -d $DBNAME|awk 'NR==1 {print $2}')
$ export INST2=$(srvctl status database -d $DBNAME|awk 'NR==2 {print $2}')

$ echo $INST1 $INST2
DBRAC1 DBRAC2
```

创建服务 IM1，设置其主实例为实例 1，备实例为实例 2。启动服务，确定服务 IM1 已在实例 1 上启动。

```
$ srvctl add service -db $DBNAME -s $SERVICENAME -preferred $INST1 -available $INST2
-pdb orclpdb1

$ srvctl start service -db $DBNAME -service $SERVICENAME

$ srvctl status service -db $DBNAME
```

```
Service im1 is running on instance(s) DBRAC1
```

在文件 tnsnames.ora 中为服务 IM1 定义网络服务名,然后使用此网络服务名登录,连接将自动导向主实例,即实例 1。

```
# 定义网络服务名 IM1,主机名使用 SCAN IP 地址
$ cat tnsnames.ora
IM1 =
  (DESCRIPTION =
    (ADDRESS =  (PROTOCOL =  TCP)(HOST =
dbim-scan.sub07281611220.training.oraclevcn.com)(PORT = 1521))
    (CONNECT_DATA =
      (SERVER = DEDICATED)
      (SERVICE_NAME = im1.sub07281611220.training.oraclevcn.com)
    )
  )

$ sqlplus ssb@im1
SQL> SHOW PARAMETER instance_number

NAME                                 TYPE        VALUE
------------------------------------ ----------- ------------------------------
instance_number                      integer     1
```

默认情况下,LINEORDER 表将被发布到所有两个实例中。

```
SQL> SELECT inst_id, populate_status, inmemory_size, bytes, bytes_not_populated FROM
gv$im_segments WHERE segment_name = 'LINEORDER';

   INST_ID POPULATE_STAT INMEMORY_SIZE      BYTES BYTES_NOT_POPULATED
---------- ------------- ------------- ---------- -------------------
         1 COMPLETED         314900480 1364353024           633004032
         2 COMPLETED         273809408 1364353024           731348992
```

通过网络服务名 IM1 连接到实例 1,设置实例 1 的 PARALLEL_INSTANCE_GROUP 为服务 IM1,注意 sid 的值需要指定为实例 1 的名称。这表示 LINEORDER 表只会发布到服务 IM1 关联的实例上,即实例 1。

```
$ sqlplus ssb@im1
SQL> SHOW PARAMETER instance_name
NAME                                 TYPE        VALUE
------------------------------------ ----------- ------------------------------
instance_name                        string      DBRAC1

SQL> ALTER SYSTEM SET parallel_instance_group=im1 SCOPE=BOTH SID='DBRAC1';
System altered.
```

重新发布表 LINEORDER,确实其所有数据均发布到了实例 1 上。

```
SQL> ALTER TABLE lineorder NO INMEMORY;
SQL> ALTER TABLE lineorder INMEMORY;
SQL> SELECT /*+ FULL(p) PARALLEL(4) */ COUNT(*) FROM lineorder p;
SQL> SELECT inst_id, populate_status, inmemory_size, bytes, bytes_not_populated FROM
gv$im_segments WHERE segment_name = 'LINEORDER';
```

```
    INST_ID POPULATE_STAT INMEMORY_SIZE        BYTES BYTES_NOT_POPULATED
---------- ------------- ------------- ---------- -------------------
         1 COMPLETED         588644352 1364353024                   0
```

将服务 IM1 从实例 1 切换到实例 2 上,此时实例 2 变为主实例。

```
$ srvctl relocate service -d $DBNAME -s $SERVICENAME -newinst $INST2 -oldinst $INST1
$ srvctl status service -db $DBNAME
Service im1 is running on instance(s) DBRAC2
```

通过网络服务名 IM1 连接,此时将连接到实例 2。

```
$ sqlplus ssb@im1
SQL> SHOW PARAMETER instance_number

NAME                                 TYPE        VALUE
------------------------------------ ----------- ------------------------------
instance_number                      integer     2
```

设置实例 2 的 PARALLEL_INSTANCE_GROUP 为服务 IM1。

```
SQL> SHOW PARAMETER parallel_instance_group

NAME                                 TYPE        VALUE
------------------------------------ ----------- ------------------------------
parallel_instance_group              string

SQL> ALTER SYSTEM SET parallel_instance_group=im1 SCOPE=BOTH SID='DBRAC2';
System altered.

SQL> SHOW PARAMETER parallel_instance_group

NAME                                 TYPE        VALUE
------------------------------------ ----------- ------------------------------
parallel_instance_group              string      IM1
```

重新发布 LINEORDER 表,由于服务 IM1 当前对应的主实例为实例 2,因此 LINEORDER 表被发布到实例 2 上。

```
SQL> ALTER TABLE lineorder NO INMEMORY;
SQL> ALTER TABLE lineorder INMEMORY;
SQL> SELECT /*+ FULL(p) PARALLEL(4) */ COUNT(*) FROM lineorder p;
SQL> SELECT inst_id, populate_status, inmemory_size, bytes, bytes_not_populated FROM
gv$im_segments WHERE segment_name = 'LINEORDER';

    INST_ID POPULATE_STAT INMEMORY_SIZE        BYTES BYTES_NOT_POPULATED
---------- ------------- ------------- ---------- -------------------
         2 COMPLETED         589692928 1364353024                   0
```

停止服务 IM1。此时即使对 LINEORDER 表执行全表扫描,也不会引发发布。

```
$ srvctl stop service -db $DBNAME -service IM1

$ srvctl status service -db $DBNAME
Service im1 is not running.
```

测试完成,最后执行清理,包括删除服务和将初始化参数 PARALLEL_INSTANCE_GROUP

恢复为默认值:

```
$ srvctl remove service -db $DBNAME -service IM1

SQL> ALTER SYSTEM SET parallel_instance_group='' SCOPE=BOTH SID='*';
```

8.2.7 实例失效时的 In-Memory 重新发布

由于 RAC 节点硬件故障或其他原因导致 RAC 实例失效时,如果 RAC 集群中其他实例上的内存列式存储中有足够的内存,失效实例上的 IMCU 会在其他实例发布。当失效 RAC 实例恢复正常,数据会在所有实例上重新发布。为避免频繁发布,在 RAC 实例发生故障和恢复正常后,需要经过约 5 分钟的超时才会进行重新发布。

下面了解 RAC 实例失效时的 In-Memory 发布过程。实验环境为两节点 RAC 集群,并且只使用 LINEORDER 一张表。

```
ALTER TABLE lineorder INMEMORY PRIORITY HIGH;
```

由于设置了发布优先级,LINEORDER 会自动发布。使用脚本 racfail.sql 监控 RAC 集群中的 In-Memory 发布状态:

```
$ cat racfail.sql
SELECT
    inst_id,
    segment_name,
    inmemory_size / 1024 / 1024      im_mb,
    bytes / 1024 / 1024              disk_mb,
    bytes_not_populated,
    populate_status                  AS pop_status
FROM
    gv$im_segments;

SELECT
    inst_id,
    COUNT(inst_id)
FROM
    gv$im_header
WHERE
    is_head_piece = 1
GROUP BY
    inst_id;

SELECT
    inst_id,
    imcu_addr,
    num_rows,
    to_char(timestamp, 'HH24:MI:SS') AS timestamp
FROM
    gv$im_header
WHERE
    is_head_piece = 1
```

```
ORDER BY
    initial_timestamp;
```

其初始发布状态如下，在实例 1 和实例 2 上分别有 13 个和 10 个 IMCU：

```
SQL> @racfail.sql

   INST_ID SEGMENT_NAME      IM_MB    DISK_MB BYTES_NOT_POPULATED POP_STATUS
---------- ------------ ---------- ---------- ------------------- ----------
         1 LINEORDER         314.5 1301.14844           604848128 COMPLETED
         2 LINEORDER       249.9375 1301.14844           759504896 COMPLETED

   INST_ID COUNT(INST_ID)
---------- --------------
         1             13
         2             10

   INST_ID IMCU_ADDR          NUM_ROWS TIMESTAM
---------- ---------------- ---------- --------
         1 0000000127FFFF80     571881 01:46:16
         2 0000000117FFFF00     514068 01:46:16
         1 00000000FFFFFE40     565884 01:46:16
         2 00000000FBFFFE20     523052 01:46:17
         1 00000001299FFF80     514070 01:46:19
         1 00000001019FFE40     504848 01:46:20
         2 00000001197FFF00     550708 01:46:20
         2 00000000FD7FFE20     532276 01:46:21
         1 000000012B1FFF80     550503 01:46:22
         1 00000001031FFE40     550708 01:46:23
         2 000000011B1FFF00     568850 01:46:25
         2 00000000FF0FFE20     532283 01:46:26
         1 0000000124BFFF60     569143 01:46:26
         1 00000000FCBFFE20     532277 01:46:26
         2 0000000114BFFEE0     523069 01:46:29
         2 0000000138000000     504633 01:46:29
         1 00000001265FFF60     569152 01:46:30
         1 00000000FE4FFE20     570788 01:46:32
         2 00000001163FFEE0     505427 01:46:32
         2 0000000139300000     561511 01:46:33
         1 000000011FFFFF40     506292 01:46:34
         1 00000001217FFF40     161067 01:46:34
         1 00000000FB0000B0     515506 01:46:35

23 rows selected.
```

停止实例 2 以模拟 RAC 故障。记录故障时间为 01:50:38。

```
$ srvctl config database
DBRAC_21c

$ export DBNAME=$(srvctl config database)

$ srvctl status database -d $DBNAME -v
Instance DBRAC1 is running on node dbim1. Instance status: Open.
```

```
Instance DBRAC2 is running on node dbim2. Instance status: Open.

$ date; srvctl stop instance -node dbim2 -force
Thu Jul  1 01:50:38 UTC 2021

$ srvctl status database -d $DBNAME -v
Instance DBRAC1 is running on node dbim1. Instance status: Open.
Instance DBRAC2 is not running on node dbim2
```

此时发布状态如下。由于实例 2 失效，之前发布在其内存列式存储中的数据需要从磁盘中读取：

```
SQL> @racfail.sql

   INST_ID SEGMENT_NAME       IM_MB    DISK_MB BYTES_NOT_POPULATED POP_STATUS
---------- ------------ ---------- ---------- ------------------- --------------
         1 LINEORDER         314.5 1301.14844           604848128 COMPLETED

   INST_ID COUNT(INST_ID)
---------- --------------
         1             13
...
```

大约在 4 分钟后，原来在实例 2 上的 IMCU 在实例 1 上重新发布。目前，LINEORDER 表所有的 IMCU 均位于实例 1 的内存列式存储中。

```
SQL> @racfail.sql

   INST_ID SEGMENT_NAME       IM_MB    DISK_MB BYTES_NOT_POPULATED POP_STATUS
---------- ------------ ---------- ---------- ------------------- --------------
         1 LINEORDER       564.375 1301.14844                   0 COMPLETED

   INST_ID COUNT(INST_ID)
---------- --------------
         1             23

   INST_ID IMCU_ADDR          NUM_ROWS TIMESTAM
---------- ---------------- ---------- --------
         1 0000000127FFFF80     571881 01:46:16
         1 00000000FFFFFE40     565884 01:46:16
         1 00000001299FFF80     514070 01:46:19
         1 00000001019FFE40     504848 01:46:20
         1 000000012B1FFF80     550503 01:46:22
         1 00000001031FFE40     550708 01:46:23
         1 0000000124BFFF60     569143 01:46:26
         1 00000000FCBFFE20     532277 01:46:26
         1 00000001265FFF60     569152 01:46:30
         1 00000000FE4FFE20     570788 01:46:32
         1 000000011FFFFF40     506292 01:46:34
         1 00000000FB0000B0     515506 01:46:35
         1 00000001217FFF40     161067 01:50:05
         1 00000001220FFF40     514068 01:54:07
```

```
         1 0000000138800000      523052 01:54:07
         1 000000011BFFFF20     550708 01:54:10
         1 000000013A000000      532276 01:54:10
         1 0000000133FFFFE0      532283 01:54:13
         1 000000011D2FFF20      568850 01:54:13
         1 000000011ECFFF20      504633 01:54:15
         1 00000001351FFFE0      523069 01:54:15
         1 00000001184FFF00      505427 01:54:18
         1 00000001369FFFE0      561511 01:54:18

23 rows selected.
```

启动实例 2，记录恢复时间为 01:59:42：

```
$ date; srvctl start instance -node dbim2
Thu Jul  1 01:59:42 UTC 2021
starting database instances on nodes "dbim2" ...
started resources "ora.dbrac_21c.db" on node "dbim2"
```

经过近 3 分钟后，之前失效时受影响的 IMCU 又重新发布到实例 2 上。LINEORDER 表在内存列式存储中的分布恢复到最初的状态。

```
SQL> @racfail.sql

   INST_ID SEGMENT_NAME      IM_MB    DISK_MB BYTES_NOT_POPULATED POP_STATUS
---------- ------------ ---------- ---------- ------------------- ----------
         1 LINEORDER        314.5  1301.14844           604848128 COMPLETED
         2 LINEORDER      249.9375 1301.14844           759504896 COMPLETED


   INST_ID COUNT(INST_ID)
---------- --------------
         2             10
         1             13


   INST_ID IMCU_ADDR            NUM_ROWS TIMESTAM
---------- -------------------- -------- --------
         1 0000000127FFFF80       571881 01:46:16
         1 00000000FFFFFE40       565884 01:46:16
         1 00000001299FFF80       514070 01:46:19
         1 00000001019FFE40       504848 01:46:20
         1 000000012B1FFF80       550503 01:46:22
         1 00000001031FFE40       550708 01:46:23
         1 0000000124BFFF60       569143 01:46:26
         1 00000000FCBFFE20       532277 01:46:26
         1 00000001265FFF60       569152 01:46:30
         1 00000000FE4FFE20       570788 01:46:32
         1 000000011FFFFF40       506292 01:46:34
         1 00000000FB0000B0       515506 01:46:35
         1 00000001217FFF40       161067 01:50:05
         2 0000000113FFFEE0       514068 02:02:09
         2 0000000133FFFFE0       523052 02:02:09
         2 00000001357FFFE0       532276 02:02:13
         2 00000001157FFEE0       550708 02:02:13
```

```
         2 00000001171FFEE0          532283 02:02:16
         2 00000001370FFFE0          568850 02:02:16
         2 0000000130AFFFC0          504633 02:02:19
         2 0000000110AFFEC0          523069 02:02:19
         2 00000001322FFFC0          505427 02:02:22
         2 00000001122FFEC0          561511 02:02:23

23 rows selected.
```

8.3 Database In-Memory 与 ADG

8.3.1 利用 OCI 搭建 ADG 实验环境

通过 Oracle 公有云（OCI）搭建 ADG 环境需要两个步骤，第一步是创建主数据库系统，第二步是创建备数据库系统并建立与主数据库系统的复制关系。

在以下示例中，将在 Oracle 公有云中建立包含主数据库 primary 和备数据库 standby 的 ADG 复制环境。

首先建立主数据库。登录 OCI 控制台，进入菜单"Oracle 数据库">"裸金属、VM 和 Exadata"，单击"创建数据库系统"按钮，首先输入以下的数据库系统信息。

（1）数据库系统名：adg-primary，此为显示名。

（2）区域：目前 OCI 支持 20 多个区域，可根据可用资源和距离选择适合的区域。

（3）可用性域：AD-1。

（4）配置类型：虚拟机。

（5）配置：VM.Standard2.1。即 1 个 CPU 核，15GB 内存。

（6）节点总数：1。因为是测试环境，选择单实例而非 RAC。

（7）Oracle 数据库软件版本：Enterprise Edition Extreme Performance。这是唯一支持 ADG 的版本，并且包括 Database In-Memory 选件和 RAC 选件。

（8）存储管理软件：Oracle Grid Infrastructure。也可以选择逻辑卷管理器，创建速度更快，但其功能较简单。

（9）网络信息：根据实际配置选择虚拟云网络和客户端子网。

（10）主机名前缀：primary。其实就是不带域名的主机名。

单击"下一步"，然后输入数据库信息。

（1）数据库名称，此处输入 DBADG，系统自动生成的名称格式为 DB 加日期，如 DB0601。

（2）数据库唯一名称后缀，输入 primary，与数据库名称共同组成数据库唯一名称，即 DBADG_primary。

（3）数据库映像，即数据库版本，可以选择从 12.2 到 21c 共 4 个大的版本，本例选择 21c。

（4）PDB 名称：orclpdb1。

（5）管理员身份证明：输入 sys 用户的密码。

在建立主备复制关系前,还需要分别设定主备数据库系统的网络访问控制规则,即允许出网和入网方向对 1521 端口的访问,因为主备数据库需要通过此端口传输数据库日志。为简化设置,可以让主备数据库系统使用同一个区域和同一个虚拟云网络,这样网络访问控制规则只需设置一次。

创建 ADG 主数据库系统的时间约 70 分钟。主数据库系统就绪后,进入数据库系统 adg-primary 配置界面,右击数据库 DBADG 的操作菜单,选择"启用 Data Guard"菜单项,进入创建备数据库系统和建立 ADG 复制关系的流程。为备数据库系统输入以下信息。

(1)显示名称:adg-standby。

(2)区域:选择与主数据库系统相同的区域。OCI 也支持主备数据库系统位于不同区域。

(3)可用性域:建议与主数据库系统不同,如 AD-2 或 AD-3。

(4)配置:VM.Standard2.1,与主数据库系统保持一致。

(5)网络信息:由于之前选择了相同区域,虚拟云网络必须与主数据库系统一致。为简化网络访问配置,子网也选择与主数据库系统一致。

(6)主机名前缀:standby。其实就是不带域名的主机名。

(7)Data Guard 保护模式:最高性能或最高可用性,此处选择最高性能。

(8)Data Guard 传输类型:由于保护模式选择了最高性能,此处只能选择异步。如果保护模式选择最高可用性,则此处只能选择同步。

(9)数据库密码:与主数据库管理员密码相同。

设置完毕后,单击"启用 Data Guard"按钮启动配置流程。整个过程耗时约在 70 分钟左右,最终可以在主数据库端查看 ADG 关联关系,如图 8-8 所示。从底部的 Data Guard 关联部分可以查看对等数据库的详细信息,包括保护模式、传输类型、主备数据库之间延时等。

图 8-8　成功建立 ADG 复制关系的数据库

8.3.2 利用 Vagrant 搭建 ADG 实验环境

8.2.2 节已经介绍了如何通过 Vagrant 搭建 RAC 实验环境，ADG 环境也同样可以使用这种方法方便的搭建。

通过 Vagrant 创建 ADG 环境的简明过程如下。

（1）创建一台宿主主机，ADG 环境需要在宿主主机上创建两个虚拟机，然后在这两个虚拟机上安装数据库软件，并建立主备数据库复制关系。每个虚拟机默认配置为 8GB 内存。如果你的笔记本电脑具有至少 20GB 内存，那么可以直接将其作为宿主主机。如果不满足条件，比较简单的方法是在公有云中创建宿主主机。

（2）在宿主主机上安装虚拟化引擎，Vagrant 支持多种虚拟机引擎，Oracle Vagrant 项目支持 VirtualBox 和 libVirt 两类。

（3）在宿主主机上安装 Vagrant 软件。

（4）下载 Oracle 在 GitHub 上的 Vagrant 项目。

（5）进入 ADG 所在目录，运行 vagrant up。

现在可通过一个示例了解创建 ADG 环境的详细过程。首先创建一台宿主主机，操作系统为 Oracle Linux 7，内存为 30GB，磁盘容量为 256GB。

在开始创建 ADG 环境的 Vagrant Box 之前，需要安装 Vagrant 和 VirtualBox 软件，然后克隆 GitHub 上的 Oracle Vagrant 项目。这 3 个步骤的具体操作与之前创建 RAC 环境的 Vagrant Box 完全相同，此处不再重复，详见 8.2.2 节。

然后从 Oracle 官网[①]下载数据库软件，放置到 ADG 对应的目录下，此软件占用空间约 2.9GB。

```
$ cd ~/vagrant-projects/OracleDG/ORCL_software/

$ ls -lh
total 2.9G
-rw-rw-r--. 1 opc opc 2.9G Feb  9 03:56 LINUX.X64_193000_db_home.zip
-rw-rw-r--. 1 opc opc  154 May 25 09:42 put_Oracle-software_here.txt
```

在建立 ADG Vagrant Box 前，可以预先查看主备 ADG 节点的配置。从配置文件可知，ADG 主备节点的主机名分别为 primary 和 standby，各配置 8GB 内存，2 个虚拟 CPU。

```
$ cd ~/vagrant-projects/OracleDG/config
$ ls
vagrant.yml

$ cat vagrant.yml
# -----------------------------------------------
# vagrant.yml for VirtualBox
# -----------------------------------------------
host1:
  vm_name: primary
  mem_size: 8192
```

[①] https://www.oracle.com/database/technologies/oracle-database-software-downloads.html

```
  cpus: 2
  public_ip: 192.168.56.131
  private_ip: 192.168.200.131
  u01_disk: ./primary_u01.vdi
host2:
  vm_name: standby
  mem_size: 8192
  cpus: 2
  public_ip: 192.168.56.132
  private_ip: 192.168.200.132
  u01_disk: ./standby_u01.vdi
env:
  provider: virtualbox
  # --------------------------------------------
  prefix_name: vgt-ol7-dg
  # --------------------------------------------
  domain: localdomain
  # --------------------------------------------
  non_rotational: 'on'
  oradata_disk_path:
  oradata_disk_num: 2
  oradata_disk_size: 20
  ...
  # --------------------------------------------
  ora_languages: en,en_GB
  # --------------------------------------------
  db_name:          DB193H1
  pdb_name:         PDB1
  cdb:              false
  adg:              false
  # --------------------------------------------
```

在配置文件的末尾，环境变量 adg 设置为 false，这表示默认情况下将创建 DG（Data Guard）环境。和 ADG 作为独立数据库选件不同，DG 是包含在数据库企业版中的功能。ADG 是 DG 的扩展，与 DG 最大的区别是在持续传输重做日志的同时可打开备库进行查询操作。其他 ADG 支持而 DG 不支持的功能包括自动块修复、滚动升级、Far Sync、Sharding、应用连续性、复制级联和 DML 重定向等。在备库端启用内存列式存储也需要 ADG 的支持，因此需要将配置文件中 adg 的值改为 true，以便创建 ADG 的环境。另外，将 cdb 的值由 false 改为 true，pdb_name 的值改为 ORCLPDB1，这样最终的环境将支持容器数据库，并自动创建一个可插拔数据库 ORCLPDB1。

进入 ADG 项目所在目录，开始创建 Vagrant Box。此过程取决于宿主机的性能和网络速度，安装时间大约为 1.5 小时，但整个过程无须人工介入：

```
$ cd ~/vagrant-projects/OracleDG
$ vagrant up
```

如果安装成功，最终会显示类似于如下的信息：

```
host2: DGMGRL>
host2: ------------------------------------------------------------
host2: INFO: 2021-05-25 11:38:04: Setup DB autostart
host2: ------------------------------------------------------------
host2: Created symlink from /etc/systemd/system/multi-user.target.wants/dbora.service
to /usr/lib/systemd/system/dbora.service.
host2: ------------------------------------------------------------
host2: INFO: 2021-05-25 11:38:05: Running user-defined post-setup scripts
host2: ------------------------------------------------------------
```

也可以使用 vagrant 命令确定 ADG 主备节点的运行状态：

```
$ vagrant status

 Oracle DG(Database Dataguard)Vagrant box for KVM/libVirt or VirtualBox
 Copyright(c)1982-2021 Oracle and/or its affiliates
 --------------------------------------------------------------------
 Author: Ruggero Citton <ruggero.citton@oracle.com>
         RAC Pack, Cloud Innovation and Solution Engineering Team

-------------------
Detected virtualbox
-------------------
getting Proxy Configuration from Host…
Current machine states:

host1                     running(virtualbox)
host2                     running(virtualbox)

This environment represents multiple VMs. The VMs are all listed
above with their current state. For more information about a specific
VM, run `vagrant status NAME`.
```

ADG 主机和备机的显示名称分别为 node1 和 node2，使用以下命令登录这两台主机：

```
$ cd ~/vagrant-projects/OracleDG
$ vagrant ssh host1
$ vagrant ssh host2
```

使用 VirtualBox 的命令可以查询到每个虚拟机配置了 4 块磁盘，其中，37GB 和 100GB 磁盘各一块，外加两块 20GB 磁盘。其中，37GB 磁盘为根文件系统，用于安装操作系统；100GB 磁盘为/u01 文件系统，用于安装数据库软件；两块 20GB 磁盘共同构成/u02 文件系统，用于安装数据库数据文件。

```
$ vboxmanage list hdds|egrep "Location|Capacity"
Location:     /home/opc/VirtualBox VMs/vgt-ol7-dg/vgt-ol7-dg-primary/box-disk001.vmdk
Capacity:     37888 MBytes
Location:     /home/opc/vagrant-projects/OracleDG/primary_u01.vdi
Capacity:     102400 MBytes
Location:     /home/opc/vagrant-projects/OracleDG/primary_oradata_disk0.vdi
Capacity:     20480 MBytes
```

```
Location:          /home/opc/vagrant-projects/OracleDG/primary_oradata_disk1.vdi
Capacity:          20480 Mbytes
…
```

实际使用的磁盘空间远比定义的要小。其中操作系统占用 2GB，数据库软件占用近 8GB，数据库数据文件占用约 2.2GB。

```
$ VBoxManage list systemproperties |grep folder
Default machine folder:           /home/opc/VirtualBox VMs

$ cd "/home/opc/VirtualBox VMs"
$ cd vgt-ol7-dg/vgt-ol7-dg-primary
$ ls -lsh *vmdk
2.0G -rw-------. 1 opc opc 2.0G May 25 14:02 box-disk001.vmdk

$ cd ~/vagrant-projects/OracleDG
$ ls -lsh primary*vdi
2.0G -rw-------. 1 opc opc 2.0G May 25 13:55 primary_oradata_disk0.vdi
209M -rw-------. 1 opc opc 210M May 25 13:55 primary_oradata_disk1.vdi
7.9G -rw-------. 1 opc opc 7.9G May 25 13:55 primary_u01.vdi
```

查看虚拟化网络信息，虚拟机共定义了 3 块网卡，其中第 1 块网卡的类型为网络地址转换（NAT），表示从虚拟机可通过宿主机连接互联网。

```
$ VBoxManage showvminfo "vgt-ol7-dg-primary" |grep NIC
NIC 1:                   MAC: 080027531333, Attachment: NAT, Cable connected: on,
Trace: off (file: none), Type: 82540EM, Reported speed: 0 Mbps, Boot priority: 0, Promisc
Policy: deny, Bandwidth group: none
NIC 1 Settings:  MTU: 0, Socket (send: 64, receive: 64), TCP Window (send:64, receive:
64)
NIC 1 Rule(0):   name = ssh, protocol = tcp, host ip = 127.0.0.1, host port = 2222, guest
ip = , guest port = 22
NIC 2:                   MAC: 080027649BA3, Attachment: Host-only Interface 'vboxnet0',
Cable connected: on, Trace: off (file: none), Type: 82540EM, Reported speed: 0 Mbps,
Boot priority: 0, Promisc Policy: deny, Bandwidth group: none
NIC 3:                   MAC: 080027653D12, Attachment: Internal Network 'private',
Cable connected: on, Trace: off (file: none), Type: 82540EM, Reported speed: 0 Mbps,
Boot priority: 0, Promisc Policy: deny, Bandwidth group: none
```

进入虚拟机，在 /etc/resolv.conf 中添加 DNS 服务器，即可访问互联网下载随书示例：

```
sudo bash -c 'echo "nameserver 8.8.8.8" >> /etc/resolv.conf'
git clone https://github.com/XiaoYu-HN/dbimbook
```

8.3.3 ADG 基本概念与 Database In-Memory 参数

Oracle ADG 是实现高可用性，数据保护和灾难恢复的数据库组件。Oracle ADG 维护了一套与主数据库完全相同物理副本，称为备数据库。当主数据库不可用时，可以切换到备数据库继续运行，从而减少业务中断时间。和传统的存储复制相比，ADG 备数据库允许读取操作。这意味着可以将之前运行于主数据库的备份、查询和报表分析等操作转移到备数据库上运行。从 Oracle 数据库 12.2 版本开始，ADG 备数据库开始支持 Database In-Memory，这不仅可以提升备数据库端分析查询的性能，同时也意味着备数据库可以运行更多类型和更复杂的工作负载，从而提升整个灾备系统的投资回报。

ADG 通过传输数据库重做日志实现数据库复制。传输模式分为同步和异步两种。同步是指事务必须等到重做日志成功传输到备数据库后才能提交。尽管 ADG 没有距离的限制，但在同步模式下，更远的距离会引发更长的网络延时。异步是指事务提交不必等待重做日志传输到备数据库，虽然性能优于同步方式，但存在数据丢失的风险。Oracle 12c 版本推出了新的 Far Sync 传输模式，即在靠近主数据库的位置部署 Far Sync 实例，与主数据库形成同步传输关系，与备数据库则形成异步传输关系，这样就可以在远距离灾难备份同时兼顾复制性能和零数据丢失。基于这些传输模式，Oracle ADG 提供 3 种保护模式，即最大性能、最大可用和最大保护模式。最大性能模式采用异步传输模式，对主数据库的性能影响最小，是默认的保护模式；最大可用模式采用同步或 Far Sync 传输模式，如果备数据库不可用，则降级为最大性能模式运行；最大保护模式是最严格的模式，采用同步传输模式，当备数据库不可用时，会将主数据库关闭。

初始化参数 INMEMORY_ADG_ENABLED 可以控制备数据库中是否启用 Database In-Memory，默认为开启。

```
SQL> SHOW PARAMETER inmemory_adg_enabled

NAME                                 TYPE         VALUE
------------------------------------ ------------ ------------------------------
inmemory_adg_enabled                 boolean      TRUE
```

在 ADG 环境下，ENABLE_IMC_WITH_MIRA 是另一个和 Database In-Memory 相关的初始化参数，此参数默认设置为 FALSE。

```
SQL> SHOW PARAMETER enable_imc_with_mira

NAME                                 TYPE         VALUE
------------------------------------ ------------ ------------------------------
enable_imc_with_mira                 boolean      FALSE
```

MIRA（Multi-Instance Redo Apply），也称为多实例重做日志应用，是 Oracle 12.2 版本推出的功能。当备数据库端为 RAC 时，可以通过多实例并行工作加快重做日志应用的速度，MIRA 功能最初和 Database In-Memory 是冲突的，从 Oracle 19c 版本开始，MIRA 可以和 Database In-Memory 同时启用，从而可以同时获得更高的可用性和性能。

8.3.4 ADG 常用管理和监控命令

监控和管理 ADG 可以通过 SQL 查询数据字典，或者使用 DGMGRL 实用程序。

以下是常用的 ADG 相关 SQL 语句：

```
-- 主数据库
SQL> SELECT database_role, open_mode, protection_mode FROM v$database;

DATABASE_ROLE    OPEN_MODE            PROTECTION_MODE
---------------- -------------------- --------------------
PRIMARY          READ WRITE           MAXIMUM PERFORMANCE
```

```
-- 备数据库
SQL> SELECT database_role, open_mode, protection_mode FROM v$database;

DATABASE_ROLE     OPEN_MODE            PROTECTION_MODE
----------------  -------------------  --------------------
PHYSICAL STANDBY  READ ONLY WITH APPLY MAXIMUM PERFORMANCE

-- Data Guard 监控指标,在备数据库执行查询
SQL> SELECT NAME, VALUE FROM v$dataguard_stats;

NAME                          VALUE
----------------------------  -----------------------------------------
transport lag                 +00 00:00:00
apply lag
apply finish time
estimated startup time        22

-- 主备数据库间重做日志传输是否有差距,需在主数据库 CDB 一级执行
SQL> SELECT type, database_mode, status, gap_status, dest_id FROM v$archive_dest_
status WHERE status='VALID';

TYPE             DATABASE_MODE     STATUS    GAP_STATUS              DEST_ID
---------------  ----------------  --------  ----------------------  ----------
LOCAL            OPEN              VALID                             1
PHYSICAL         OPEN_READ-ONLY    VALID     NO GAP                  2

-- 确认主数据库是否可以切换到备数据库,需在主数据库 CDB 一级执行
SQL> SELECT switchover_status FROM v$database;

SWITCHOVER_STATUS
-------------------
TO STANDBY
```

DGMGRL 是 Data Guard 命令行接口,可以进行 Data Guard 复制配置的管理以及数据库切换和回切操作。以下是常用的 DGMGRL 命令:

```
$ dgmgrl sys@DBADG_primary
-- 显示 ADG 配置信息
DGMGRL> show config

Configuration - DBADG_primary_DBADG_standby

  Protection Mode: MaxPerformance
  Members:
  DBADG_primary  - Primary database
    DBADG_standby - Physical standby database

Fast-Start Failover:  Disabled

Configuration Status:
SUCCESS   (status updated 43 seconds ago)
```

```
-- 显示主数据库信息
DGMGRL> show database DBADG_primary

Database - DBADG_primary

  Role:               PRIMARY
  Intended State:     TRANSPORT-ON
  Instance(s):
    DBADG

Database Status:
SUCCESS

-- 显示备数据库信息
DGMGRL> show database DBADG_standby

Database - DBADG_standby

  Role:               PHYSICAL STANDBY
  Intended State:     APPLY-ON
  Transport Lag:      0 seconds(computed 8 seconds ago)
  Apply Lag:          10 seconds(computed 8 seconds ago)
  Average Apply Rate: 4.00 KByte/s
  Real Time Query:    ON
  Instance(s):
    DBADG

Database Status:
SUCCESS

-- 在角色切换前,检查数据库有效性
DGMGRL> validate database DBADG_standby

  Database Role:     Physical standby database
  Primary Database:  DBADG_primary

  Ready for Switchover:  Yes
  Ready for Failover:    Yes(Primary Running)

  Managed by Clusterware:
    DBADG_primary:  YES
    DBADG_standby:  YES

  Standby Apply-Related Information:
    Apply State:    Running
    Apply Lag:      10 seconds(computed 7 seconds ago)
    Apply Delay:    0 minutes

  Parameter Settings:
    Parameter                    DBADG_primary Value         DBADG_standby Value
```

```
DB_BLOCK_CHECKING              FULL              FULL
DB_BLOCK_CHECKSUM              FULL              FULL
DB_LOST_WRITE_PROTECT          TYPICAL           TYPICAL
```

```
-- 主备数据库角色切换，时间近 2 分钟
DGMGRL> switchover to DBADG_standby
2021-06-01T05:23:27.252+00:00
Performing switchover NOW, please wait…

2021-06-01T05:23:30.107+00:00
Operation requires a connection to database "DBADG_standby"
Connecting …
Connected to "DBADG_standby"
Connected as SYSDBA.

2021-06-01T05:23:30.942+00:00
Continuing with the switchover…

2021-06-01T05:24:08.044+00:00
New primary database "DBADG_standby" is opening…

2021-06-01T05:24:08.045+00:00
Oracle Clusterware is restarting database "DBADG_primary" …
Connected to "DBADG_primary"
Connected to "DBADG_primary"
2021-06-01T05:25:13.547+00:00
Switchover succeeded, new primary is "dbadg_standby"

2021-06-01T05:25:13.555+00:00
Switchover processing complete, broker ready.
```

8.3.5　主备数据库发布相同的对象

备数据库可以发布与主数据库相同的 In-Memory 对象，只需在备数据库中启用内存列式存储即可，备数据库中内存列式存储的大小可以与主数据库不同。由于备数据库是只读的，其 INMEMORY 属性只能在主数据库中设置后再复制到备端，但主备数据库可以独立控制 In-Memory 对象的发布。

通过下面的示例了解具体的设置过程。首先确保主备数据库均已启用内存列式存储，然后在主数据库中为对象设置 INMEMORY 属性并发布。

```
SQL> ALTER TABLE lineorder INMEMORY;
Table altered.

SQL> EXEC dbms_inmemory.populate('SSB', 'LINEORDER');
PL/SQL procedure successfully completed.

SQL> SELECT owner, segment_name, populate_status FROM v$im_segments WHERE segment_name
= 'LINEORDER';
```

```
OWNER           SEGMENT_NAME  POPULATE_STATUS
--------------  ------------  ---------------
SSB             LINEORDER     COMPLETED
```

备数据库中对象的发布与主数据库没有关系，可以独立控制。

```
SQL> SELECT database_role,open_mode, protection_mode FROM v$database;

DATABASE_ROLE     OPEN_MODE             PROTECTION_MODE
----------------  --------------------  --------------------
PHYSICAL STANDBY  READ ONLY WITH APPLY  MAXIMUM PERFORMANCE

SQL> SELECT owner, segment_name, populate_status FROM v$im_segments WHERE segment_name = 'LINEORDER';

no rows selected

SQL> EXEC dbms_inmemory.populate('SSB', 'LINEORDER');
PL/SQL procedure successfully completed.

SQL> SELECT owner, segment_name, populate_status FROM v$im_segments WHERE segment_name = 'LINEORDER';
OWNER           SEGMENT_NAME  POPULATE_STATUS
--------------  ------------  ---------------
SSB             LINEORDER     COMPLETED
```

如果在主数据库中将 In-Memory 对象移除出内存列式存储，备数据库中相同的对象也会被移除。

```
-- 主数据库
SQL> ALTER TABLE lineorder NO INMEMORY;
Table altered.

SQL> SELECT owner, segment_name, populate_status FROM v$im_segments WHERE segment_name = 'LINEORDER';

no rows selected

-- 备数据库
SQL> SELECT owner, segment_name, populate_status FROM v$im_segments WHERE segment_name = 'LINEORDER';

no rows selected
```

8.3.6 仅在备数据库发布对象

当主数据库同时运行分析和交易负载，并且负载不断加大时，可以将分析型负载转移到备数据库，而主数据库只运行交易负载。这就是仅在备数据库端发布对象适用的场景。参考 2020 年 IEEE 第 36 届 ICDE 大会论文 Oracle Database In-Memory on Active Data Guard: Real-time

Analytics on a Standby Database[①]的结论，在 ADG 备数据库端启用 In-Memory 后，相对于未启用时，响应时间提速 10～100 倍不等。分析型负载转移到备数据库后，主数据库的 CPU 消耗从 8%降为 0.5%，备数据库的 CPU 消耗从 0.3%增加到 7.9%，成功实现了负载分担和转移。

通过下面的示例来了解具体的设置过程。主数据库中只需将内存列式存储的大小设为 0，而备数据库则将其设为大于等于 100MB 的值。

```
-- 主数据库
SQL> ALTER SYSTEM SET inmemory_size = 0;
System altered.

SQL> SHOW PARAMETER inmemory_size
NAME                                 TYPE        VALUE
------------------------------------ ----------- ------------------------------
inmemory_size                        big integer 0

-- 备数据库
SQL> SHOW PARAMETER inmemory_size
NAME                                 TYPE        VALUE
------------------------------------ ----------- ------------------------------
inmemory_size                        big integer 1504M
```

由于备数据库是只读的，INMEMORY 属性只能在主数据库中设置。虽然主数据库中定义了 INMEMORY 属性，但由于内存列式存储被禁用，因此发布无效：

```
-- 主数据库
SQL> SELECT database_role,open_mode, protection_mode FROM v$database;

DATABASE_ROLE    OPEN_MODE            PROTECTION_MODE
---------------- -------------------- --------------------
PRIMARY          READ WRITE           MAXIMUM PERFORMANCE

SQL> ALTER TABLE lineorder INMEMORY;
Table altered.

SQL> EXEC dbms_inmemory.populate('SSB', 'LINEORDER');
PL/SQL procedure successfully completed.

SQL> SELECT owner, segment_name, populate_status FROM v$im_segments WHERE segment_name = 'LINEORDER';

no rows selected
```

定义的 INMEMORY 属性被复制到备数据库，在备数据库中可以成功发布：

```
-- 备数据库
SQL> SELECT table_name, inmemory, inmemory_service FROM user_tables WHERE table_name = 'LINEORDER';

TABLE_NAME    INMEMORY INMEMORY_SERVICE
```

[①] https://ieeexplore.ieee.org/document/9101796/

```
------------ -------- --------------------
LINEORDER    ENABLED  DEFAULT

SQL> EXEC dbms_inmemory.populate('SSB', 'LINEORDER');
PL/SQL procedure successfully completed.

SQL> SELECT owner, segment_name, populate_status FROM v$im_segments WHERE segment_name
= 'LINEORDER';

OWNER         SEGMENT_NAME   POPULATE_STATUS
------------  -------------  --------------------
SSB           LINEORDER      STARTED
```

由于备数据库为只读，如果希望将对象从内存列式存储中移除，只能在主数据库中操作，否则报错如下：

```
-- 备数据库
SQL> ALTER TABLE lineorder NO INMEMORY;
ALTER TABLE lineorder NO INMEMORY
*
ERROR at line 1:
ORA-16000: database or pluggable database open for read-only access
```

8.3.7 主备数据库发布不同的对象

ADG 主备数据库可以各自发布不同的对象，这适用于主数据库和备数据库分别运行不同的分析场景。或者对于分区表，可以将近期的热点数据分区在主数据库发布，历史数据所在的分区在备数据库发布。和 RAC 类似，通过结合服务可以实现一个对象仅发布在主数据库端、仅发布在备数据库端或在主备数据库端同时发布。

Database In-Memory 与服务的结合可以通过两种方式实现。第一种是使用 srvctl 实用程序，第二种是使用 DBMS_SERVICE PL/SQL 包。srvctl 是推荐的方式，但前提是安装了 Grid Infrastructure。DBMS_SERVICE 已包含在数据库中，但只支持单实例。

先看如何通过 srvctl 实现主备数据库端发布不同的对象。首先在主数据库端定义 3 个服务。其中，primary_only 将只在主数据库端运行，standby_only 将只在备数据库端运行，而 primary_standby 将同时在主备数据库端运行：

```
# 在主数据库端执行
$ DBNAME=$(srvctl config database)
$ srvctl add service -db $DBNAME -service primary_only -role primary -pdb orclpdb1
$ srvctl add service -db $DBNAME -service standby_only -role physical_standby -pdb orclpdb1
$ srvctl add service -db $DBNAME -service primary_standby -pdb orclpdb1
```

然后在主数据库端启动 primary_standby 服务和 primary_only 服务：

```
$ srvctl start service -db $DBNAME -service primary_standby
$ srvctl start service -db $DBNAME -service primary_only

$ srvctl status service -db $DBNAME
```

```
Service primary_only is running on instance(s) DBADG
Service primary_standby is running on instance(s) DBADG
Service standby_only is not running.
```

在备数据库端定义 primary_standby 服务和 primary_only 服务:

```
$ DBNAME=$(srvctl config database)
$ srvctl add service -db $DBNAME -service primary_standby -pdb orclpdb1
$ srvctl add service -db $DBNAME -service standby_only -role physical_standby -pdb orclpdb1
```

接下来需要在备数据库端启动这两个服务。此时如果直接启动,会报类似于如下的错误:

```
$ srvctl add service -db $DBNAME -service primary_standby -pdb orclpdb1
PRCD-1133 : failed to start services primary_standby for database DBADG_standby
PRCR-1095 : Failed to start resources using filter ((NAME == ora.dbadg_standby.primary_standby.svc) AND ((TYPE == ora.service.type) OR (BASE_TYPE == ora.service.type)))
CRS-5017: The resource action "ora.dbadg_standby.primary_standby.svc start" encountered the following error:
ORA-44786: Service operation cannot be completed.
ORA-06512: at "SYS.DBMS_SERVICE", line 76
ORA-06512: at "SYS.DBMS_SERVICE", line 483
ORA-06512: at line 1
. For details refer to "(:CLSN00107:)" in "/u01/app/grid/diag/crs/standby/crs/trace/crsd_oraagent_oracle.trc".
```

由于备库端是只读的,启动服务时还没有获得服务注册信息。解决方法是需要在主数据库端启动服务,服务的注册信息通过重做日志复制到备数据库端后,这些服务即可成功启动。因此先在主数据库端启动这些服务,此时只需要启动 standby_only 服务,因为 primary_standby 服务在之前已经启动过了。

```
# 在主数据库端启动为备数据库端定义的服务,目的是形成注册信息,以便复制到备数据库端
$ srvctl start service -db $DBNAME -service standby_only
# 停止服务,因为主数据库端不需要运行此服务
$ srvctl stop service -db $DBNAME -service standby_only
```

在主数据库中可以看到这些注册信息,稍后在备数据库中也可以看到这些信息,此时这些服务就可以在备数据库端启动了:

```
-- 在 CDB 一级执行
SQL> SELECT service_id, NAME, network_name, pdb FROM service$;

SERVICE_ID NAME                  NETWORK_NAME          PDB
---------- --------------------- --------------------- ------------
         8 ORCLPDB1              ORCLPDB1              ORCLPDB1
         1 primary_standby       primary_standby       ORCLPDB1
         2 standby_only          standby_only          ORCLPDB1
         3 primary_only          primary_only          ORCLPDB1
```

此时,在备数据库端可以成功启动 primary_standby 服务和 primary_only 服务:

```
$ srvctl start service -db $DBNAME -service primary_standby
$ srvctl start service -db $DBNAME -service standby_only

$ srvctl status service -db $DBNAME
```

```
Service primary_standby is running on instance(s) DBADG
Service standby_only is running on instance(s) DBADG
```

至此，主备数据库所需的服务已定义完毕。接下来需要为 In-Memory 对象定义服务属性，以便发布到指定的数据库端，这可以通过 DISTRIBUTE FOR SERVICE 子句实现。为 SSB 示例中的 3 张表定义了服务，余下的两张表使用默认设置。

```
-- 只在主数据库端发布
ALTER TABLE part INMEMORY DISTRIBUTE FOR SERVICE primary_only;

-- 只在备数据库端发布
ALTER TABLE lineorder INMEMORY DISTRIBUTE FOR SERVICE standby_only;

-- 在主数据库端和备数据库端均发布
ALTER TABLE customer INMEMORY DISTRIBUTE FOR SERVICE primary_standby;
```

由于备数据库是只读的，因此以上语句只能在主数据库中定义，否则报错如下：

```
SQL> ALTER TABLE part INMEMORY DISTRIBUTE FOR SERVICE primary_only;
ALTER TABLE part INMEMORY DISTRIBUTE FOR SERVICE primary_only
*
ERROR at line 1:
ORA-16000: database or pluggable database open for read-only access
```

在数据字典中可以验证服务属性已定义成功。由于复制的关系，在备数据库中也可以看到这些信息：

```
SQL> SELECT table_name, inmemory_service, inmemory_service_name FROM user_tables;

TABLE_NAME      INMEMORY_SERVICE        INMEMORY_SERVICE_NAM
------------    --------------------    --------------------
LINEORDER       USER_DEFINED            STANDBY_ONLY
SUPPLIER
PART            USER_DEFINED            PRIMARY_ONLY
CUSTOMER        USER_DEFINED            PRIMARY_STANDBY
DATE_DIM
```

按照之前的设计，在主数据库端和备数据库端均运行以下命令。此时 CUSTOMER 表在主数据库端和备数据库端同时成功发布：

```
SQL> EXEC dbms_inmemory.populate('SSB', 'CUSTOMER');

SQL> SELECT owner, segment_name, populate_status FROM v$im_segments;

OWNER                NAME                              STATUS
------------------   -------------------------------   --------------
SSB                  CUSTOMER                          COMPLETED
```

由于 PART 表对应 primary_only 服务，而此服务只在主数据库端启动，因此在主数据库端可以发布 PART 表，在备数据库端发布无效：

```
-- 主数据库端发布成功
SQL> EXEC dbms_inmemory.populate('SSB', 'PART');

SQL> SELECT owner, segment_name, populate_status FROM v$im_segments WHERE segment_name
```

```
= 'PART';

OWNER        SEGMENT_NAME POPULATE_STATUS
------------ ------------ --------------------
SSB          PART         COMPLETED

-- 备数据库端发布无效
SQL> EXEC dbms_inmemory.populate('SSB', 'PART');

SQL> SELECT owner, segment_name, populate_status FROM v$im_segments WHERE segment_name
= 'PART';

no rows selected
```

同理，由于 LINEORDER 表对应 standby_only 服务，而此服务只在备数据库端启动，因此在备数据库端可以发布 LINEORDER 表，在主数据库端发布无效：

```
-- 主数据库端发布 LINEORDER 无效
SQL> EXEC dbms_inmemory.populate('SSB', 'LINEORDER');

SQL> SELECT owner, segment_name, populate_status FROM v$im_segments WHERE segment_name
= 'LINEORDER';

no rows selected

-- 备数据库端发布 LINEORDER 成功
SQL> EXEC dbms_inmemory.populate('SSB', 'LINEORDER');

SQL> SELECT owner, segment_name, populate_status FROM v$im_segments WHERE segment_name
= 'LINEORDER';

OWNER        SEGMENT_NAME     POPULATE_STATUS
------------ ---------------- --------------------
SSB          LINEORDER        STARTED
```

此时，可以比较在主备数据库端各自发布的对象：

```
-- 主数据库端
SQL> SELECT owner, segment_name, populate_status, inmemory_service_name FROM
v$im_segments;

OWNER        SEGMENT_NAME POPULATE_STATUS      INMEMORY_SERVICE_NAME
------------ ------------ -------------------- --------------------
SSB          PART         COMPLETED            PRIMARY_ONLY
SSB          CUSTOMER     COMPLETED            PRIMARY_STANDBY

-- 备数据库端
SQL> SELECT owner, segment_name, populate_status, inmemory_service_name FROM
v$im_segments;

OWNER        SEGMENT_NAME     POPULATE_STATUS      INMEMORY_SERVICE_NAM
------------ ---------------- -------------------- --------------------
SSB          LINEORDER        COMPLETED            STANDBY_ONLY
```

| SSB | CUSTOMER | COMPLETED | PRIMARY_STANDBY |

在主数据库端停止 primary_only 服务，可以看到此服务关联的对象 PART 被清除出内存列式存储：

```
$ srvctl stop service -db $DBNAME -service primary_only

SQL> SELECT owner, segment_name, populate_status, inmemory_service_name FROM
v$im_segments;

OWNER        SEGMENT_NAME  POPULATE_STATUS      INMEMORY_SERVICE_NAME
------------ ------------  -------------------- ---------------------
SSB          CUSTOMER      COMPLETED            PRIMARY_STANDBY
```

同样，在备数据库端停止 standby_only 服务，可以看到此服务关联的对象 LINEORDER 被清除出内存列式存储：

```
$ srvctl stop service -db $DBNAME -service standby_only

SQL> SELECT owner, segment_name, populate_status, inmemory_service_name FROM
v$im_segments;

OWNER        SEGMENT_NAME     POPULATE_STATUS      INMEMORY_SERVICE_NAM
------------ ---------------- -------------------- --------------------
SSB          CUSTOMER         COMPLETED            PRIMARY_STANDBY
```

在主数据库端停止 primary_standby 服务，可以看到此服务关联的对象 CUSTOMER 被清除出内存列式存储：

```
$ srvctl stop service -db $DBNAME -service primary_standby
SQL> SELECT owner, segment_name, populate_status, inmemory_service_name FROM
v$im_segments;

no rows selected
```

但是，由于备数据库端的 primary_standby 服务仍在运行，因此备数据库中的 CUSTOMER 表不受影响：

```
SQL> SELECT owner, segment_name, populate_status, inmemory_service_name FROM
v$im_segments;

OWNER        SEGMENT_NAME     POPULATE_STATUS      INMEMORY_SERVICE_NAM
------------ ---------------- -------------------- --------------------
SSB          CUSTOMER         COMPLETED            PRIMARY_STANDBY
```

也就是说，通过服务可以实现单独从主数据库或备数据库中移除 In-Memory 对象。

最后，执行清理工作，在主备数据库端删除所有的服务：

```
# 主数据库端
DBNAME=$(srvctl config database)
srvctl remove service -db $DBNAME -service primary_only -force
srvctl remove service -db $DBNAME -service primary_standby -force
srvctl remove service -db $DBNAME -service standby_only -force

# 备数据库端
```

```
DBNAME=$(srvctl config database)
srvctl remove service -db $DBNAME -service standby_only -force
srvctl remove service -db $DBNAME -service primary_standby -force
```

如果主备数据库端都是单实例，也可以使用 DBMS_SERVICE PL/SQL 包来定义服务，这种方式比较简明。来看一个具体的示例。首先，在主数据库端定义服务：

```
SQL> ALTER SESSION SET CONTAINER=orclpdb1;
SQL> EXEC dbms_service.create_service(service_name => 'standby_only', network_name => 'standby_only');
SQL> EXEC dbms_service.create_service(service_name => 'primary_only', network_name => 'primary_only');
SQL> EXEC dbms_service.create_service(service_name => 'primary_standby', network_name => 'primary_standby');
```

在主数据库端启动 primary_only 和 primary_standby 服务：

```
SQL> EXEC dbms_service.start_service(service_name => 'primary_standby');
SQL> EXEC dbms_service.start_service(service_name => 'primary_only');

SQL> SELECT NAME FROM v$active_services;

NAME
-------------------------------------------------------------
primary_standby
orclpdb1
primary_only
```

在备数据库端启动 standby_only 和 primary_standby 服务：

```
SQL> EXEC dbms_service.start_service(service_name => 'primary_standby');
SQL> EXEC dbms_service.start_service(service_name => 'standby_only');

SQL> SELECT NAME FROM v$active_services;

NAME
------------------------------
standby_only
primary_standby
orclpdb1
```

在数据库监听中可以查看到相应的服务：

```
# 主数据库端
[oracle@primary ~]$ lsnrctl status
…
Service "primary_only.sub07281611220.training.oraclevcn.com" has 1 instance(s).
  Instance "DBADG", status READY, has 1 handler(s) for this service…
Service "primary_standby.sub07281611220.training.oraclevcn.com" has 1 instance(s).
  Instance "DBADG", status READY, has 1 handler(s) for this service…
The command completed successfully

# 备数据库端
[oracle@standby ~]$ lsnrctl status
…
Service "primary_standby.sub07281611220.training.oraclevcn.com" has 1 instance(s).
```

```
 Instance "DBADG", status READY, has 2 handler(s) for this service…
Service "standby_only.sub07281611220.training.oraclevcn.com" has 1 instance(s).
 Instance "DBADG", status READY, has 2 handler(s) for this service…
The command completed successfully
```

在主数据库端,通过 DISTRIBUTE FOR SERVICE 子句为 SSB 示例中的 3 张表定义服务属性。

```
-- 只在主数据库端发布
ALTER TABLE part INMEMORY DISTRIBUTE FOR SERVICE primary_only;
-- 只在备数据库端发布
ALTER TABLE lineorder INMEMORY DISTRIBUTE FOR SERVICE standby_only;
-- 在主数据库端和备数据库端均发布
ALTER TABLE customer INMEMORY DISTRIBUTE FOR SERVICE primary_standby;
```

在主数据库和备数据库中均执行以下所有发布命令。尽管其中一些命令不会生效,但都能正常执行:

```
EXEC dbms_inmemory.populate('SSB', 'PART');          # 对备数据库无效
EXEC dbms_inmemory.populate('SSB', 'LINEORDER');     # 对主数据库无效
EXEC dbms_inmemory.populate('SSB', 'CUSTOMER');
```

以下为主备数据库中 In-Memory 对象发布的情况:

```
-- 主数据库
SQL> SELECT owner, segment_name, populate_status FROM v$im_segments;

OWNER         SEGMENT_NAME   POPULATE_STATUS
------------  -------------  --------------------
SSB           PART           COMPLETED
SSB           CUSTOMER       COMPLETED

-- 备数据库
SQL> SELECT owner, segment_name, populate_status FROM v$im_segments;

OWNER         SEGMENT_NAME   POPULATE_STATUS
------------  -------------  --------------------
SSB           LINEORDER      COMPLETED
SSB           CUSTOMER       COMPLETED
```

在主数据库中停止 primary_only 和 primary_standby 服务,在备数据库中停止 standby_only 服务:

```
-- 主数据库
EXEC dbms_service.stop_service(service_name => 'primary_only');
EXEC dbms_service.stop_service(service_name => 'primary_standby');

-- 备数据库
EXEC dbms_service.stop_service(service_name => 'standby_only');
```

以下为主备数据库中 In-Memory 对象发布的情况。由于备数据库中保留了 primary_standby 服务,因此 CUSTOMER 表仍保持发布状态:

```
-- 主数据库
SQL> SELECT owner, segment_name, populate_status FROM v$im_segments;
no rows selected
```

```
-- 备数据库
SQL> SELECT owner, segment_name, populate_status FROM v$im_segments;

OWNER           SEGMENT_NAME    POPULATE_STATUS
--------------- --------------- --------------------
SSB             CUSTOMER        COMPLETED
```

最后执行清理工作，停止和删除所有的服务。由于备数据库是只读的，因此删除服务操作只能在主数据库中执行。在删除服务前，必须在主备数据库中停止所有的服务。

```
-- 主数据库
EXEC dbms_service.delete_service(service_name => 'standby_only');
EXEC dbms_service.delete_service(service_name => 'primary_only');
EXEC dbms_service.delete_service(service_name => 'primary_standby');
```

到目前为止，使用 DBMS_SERVICE 的效果和使用 srvctl 都是一致的。但 DBMS_SERVICE 功能比较简单，而且存在使用限制。因此，在高可用环境下，应使用 srvctl 来进行服务相关操作。

8.3.8　Database In-Memory 与 ADG 主备切换

Oracle 数据库在 11.2 版本提出了基于角色的数据库服务概念，即无论 ADG 主备数据库如何切换，可以定义服务始终指向既定的角色，如主数据库或只读的备数据库。由于 In-Memory 对象的发布可以跟随数据库服务，因此当应用使用数据库服务进行连接时，可以保证其在内存列式存储中始终可以访问到所需的数据库对象。

通过下面的示例了解具体的配置过程。假设 ADG 环境由一主一备两个单实例数据库组成。首先在主备数据库端同时定义主数据库服务 primary_only 和备数据库服务 standby_only：

```
# 主数据库端和备数据库端均执行
DBNAME=$(srvctl config database)

srvctl add service -db $DBNAME -service primary_only -role PRIMARY -failovertype SESSION
-failovermethod BASIC -failoverdelay 10 -failoverretry 10

srvctl add service -db $DBNAME -service standby_only -role PHYSICAL_STANDBY
-failovertype SESSION -failovermethod BASIC -failoverdelay 10 -failoverretry 10
```

依次执行以下脚本，最终在主数据库端只运行 primary_only 服务，在备数据库端只运行 standby_only 服务。其中在主数据库端临时启停了 standby_only 服务，目的是为了让服务信息复制到备端，否则备数据库端启动 standby_only 服务时会报错。

```
# 主数据库端
srvctl start service -db $DBNAME -service primary_only

srvctl start service -db $DBNAME -service standby_only

# 备数据库端
srvctl start service -db $DBNAME -service standby_only

# 主数据库端
```

```
srvctl stop service -db $DBNAME -service standby_only
```

目前为止，主备数据库端的服务运行情况如下。记录这两个服务的完整服务名称，在下一个步骤将使用其定义网络服务名：

```
# 主数据库
$ srvctl status service -db $DBNAME
Service primary_only is running on instance(s) DBADG
Service standby_only is not running.

$ lsnrctl status | grep primary_only
Service "primary_only.sub07281611220.training.oraclevcn.com" has 1 instance(s).

# 备数据库
$ srvctl status service -db $DBNAME
Service primary_only is not running.
Service standby_only is running on instance(s) DBADG

$ lsnrctl status | grep standby_only
Service "standby_only.sub07281611220.training.oraclevcn.com" has 1 instance(s).
```

在客户端配置网络服务名 PRIM_DB 和 STBY_DB，它们分别关联刚才定义的 primary_only 和 standby_only 服务。如果没有多余的客户端，可以在主备数据库服务器上同时定义。在文件 tnsnames.ora 中的网络服务名定义如下，其中 primary 和 standby 分别为主备数据库服务器的主机名。

```
PRIM_DB =
  (DESCRIPTION =
    (ADDRESS_LIST =
      (FAILOVER = ON)
      (LOAD_BALANCE = OFF)
      (ADDRESS = (PROTOCOL = TCP)(HOST = primary)(PORT = 1521))
      (ADDRESS = (PROTOCOL = TCP)(HOST = standby)(PORT = 1521))
    )
    (CONNECT_DATA =
      (SERVICE_NAME = primary_only.sub07281611220.training.oraclevcn.com)
    )
  )

STBY_DB =
  (DESCRIPTION =
    (ADDRESS_LIST =
      (FAILOVER = ON)
      (LOAD_BALANCE = OFF)
      (ADDRESS = (PROTOCOL = TCP)(HOST = standby)(PORT = 1521))
      (ADDRESS = (PROTOCOL = TCP)(HOST = primary)(PORT = 1521))
    )
    (CONNECT_DATA =
      (SERVICE_NAME = standby_only.sub07281611220.training.oraclevcn.com)
    )
  )
```

此时，使用 PRIM_DB 将始终连接到可读写的主数据库，使用 STBY_DB 将始终连接到只读的备数据库。使用两个终端连接，但不要退出，因为后续要验证服务切换不会影响这些连接。

```
# 终端1，连接主数据库服务
$ sqlplus sys@prim_db as sysdba

SQL> SET SQLPROMPT "PRIM_DB_SQL> "

PRIM_DB_SQL> SHOW PARAMETER uniq

NAME                                 TYPE        VALUE
------------------------------------ ----------- ------------------------------
db_unique_name                       string      DBADG_primary

PRIM_DB_SQL> SELECT database_role, open_mode FROM v$database;

DATABASE_ROLE    OPEN_MODE
---------------- --------------------
PRIMARY          READ WRITE

# 终端2，连接备数据库服务
$ sqlplus sys@stby_db as sysdba

SQL> SET SQLPROMPT "STBY_DB_SQL> "

STBY_DB_SQL> SHOW PARAMETER uniq

NAME                                 TYPE        VALUE
------------------------------------ ----------- ------------------------------
db_unique_name                       string      DBADG_standby

STBY_DB_SQL> SELECT database_role, open_mode FROM v$database;
DATABASE_ROLE    OPEN_MODE
---------------- --------------------
PHYSICAL STANDBY READ ONLY WITH APPLY
```

利用 DGMGRL 执行 ADG 主备切换：

```
$ dgmgrl sys@DBADG_primary
DGMGRL> show config

Configuration - DBADG_primary_DBADG_standby

  Protection Mode: MaxPerformance
  Members:
  DBADG_primary - Primary database
    DBADG_standby - Physical standby database

Fast-Start Failover:  Disabled

Configuration Status:
SUCCESS   (status updated 17 seconds ago)
```

```
DGMGRL> switchover to DBADG_standby
2021-06-01T07:51:54.045+00:00
Performing switchover NOW, please wait…

2021-06-01T07:51:56.726+00:00
Operation requires a connection to database "DBADG_standby"
Connecting …
Connected to "DBADG_standby"
Connected as SYSDBA.

2021-06-01T07:51:57.601+00:00
Continuing with the switchover…

2021-06-01T07:52:38.358+00:00
New primary database "DBADG_standby" is opening…

2021-06-01T07:52:38.358+00:00
Oracle Clusterware is restarting database "DBADG_primary"…
Connected to "DBADG_primary"
Connected to "DBADG_primary"
2021-06-01T07:53:30.528+00:00
Switchover succeeded, new primary is "DBADG_standby"

2021-06-01T07:53:30.532+00:00
Switchover processing complete, broker ready.
```

在之前连接的两个终端继续执行命令，尽管后端的数据库实例发生了变化，但连接始终连向其服务所对应的角色。

```
-- 终端 1, 始终连接到主数据库服务
PRIM_DB_SQL> SHOW PARAMETER uniq

NAME                                 TYPE        VALUE
------------------------------------ ----------- ------------------------------
db_unique_name                       string      DBADG_standby

PRIM_DB_SQL> SELECT database_role, open_mode FROM v$database;

DATABASE_ROLE    OPEN_MODE
---------------- --------------------
PRIMARY          READ WRITE

-- 终端 2, 始终连接到备数据库服务
STBY_DB_SQL> SHOW PARAMETER uniq
NAME                                 TYPE        VALUE
------------------------------------ ----------- ------------------------------
db_unique_name                       string      DBADG_primary

STBY_DB_SQL> SELECT database_role, open_mode FROM v$database;

DATABASE_ROLE    OPEN_MODE
```

```
-------------------   --------------------
PHYSICAL STANDBY      READ ONLY WITH APPLY
```

至此,已经验证基于角色的服务运行正常。在此基础上,可以进一步定义 In-Memory 对象并绑定服务属性,这样对象将始终跟随服务,发布到指定角色的内存列式存储中。例如,希望 PART 表始终在主数据库端发布,而 LINEORDER 表始终在备数据库端发布,可以使用以下命令进行定义:

```
-- 只在主数据库发布
ALTER TABLE part INMEMORY PRIORITY HIGH DISTRIBUTE FOR SERVICE primary_only;

-- 只在备数据库发布
ALTER TABLE lineorder INMEMORY PRIORITY HIGH DISTRIBUTE FOR SERVICE standby_only;
```

之后,无论 ADG 主备数据库角色如何切换,PART 表将始终在主数据库发布,LINEORDER 表将始终在备数据库发布。

实验结束,最后在主备数据库端均执行以下命令进行清理:

```
DBNAME=$(srvctl config database)
srvctl remove service -db $DBNAME -service primary_only -force
srvctl remove service -db $DBNAME -service standby_only -force
```

8.4　In-Memory FastStart

内存列式存储是 Oracle 数据库中唯一支持列格式的地方,因此当实例重启时,Oracle 必须将传统的行格式数据转换为列格式,然后实施压缩后存放,这是一个 I/O 和 CPU 密集型的操作。Oracle 12.2 版本推出 In-Memory FastStart 功能,通过将已发布到内存列式存储中的数据定期转储到磁盘,在重新发布时可以省去格式转换和压缩的步骤,减少资源消耗,加速发布过程,从而使业务应用可以尽快享用分析提速的好处。

In-Memory FastStart 需要指定一个表空间来存放内存列式存储中的数据,称为 FastStart 表空间。每一个可插拔数据库只能有一个 FastStart 表空间,表空间的大小建议为内存列式存储大小的两倍。对于 RAC 集群而言,所有的实例共享一个 FastStart 表空间。

以下通过一个示例来了解 In-Memory FastStart 的配置过程。首先建立 FastStart 表空间,由于内存列式存储的大小约为 1.5GB,将 FastStart 表空间的大小设为 3GB。

```
SQL> ALTER SESSION SET CONTAINER = orclpdb1;

SQL> CREATE TABLESPACE fs_tbs DATAFILE 'fs_tbs.dbf' SIZE 1500M REUSE AUTOEXTEND ON NEXT 500K MAXSIZE 3G;
```

然后启用 In-Memory FastStart,注意表空间名要大写。

```
SQL> EXEC dbms_inmemory_admin.faststart_enable('FS_TBS');
```

此时可以查看 FastStart 表空间的状态。最初分配了 1500MB,只使用了 1MB:

```
SQL> SELECT tablespace_name, status, ((allocated_size/1024) / 1024) AS alloc_mb,
((used_size/1024) / 1024) AS used_mb  FROM v$inmemory_faststart_area;
```

```
TABLESPACE_NAME        STATUS        ALLOC_MB      USED_MB
-------------------    ----------    ----------    ----------
FS_TBS                 ENABLE              1500             1
```

为所有 SSB 示例表启用 INMEMORY，然后发布这些对象：

```
$ cd chap04
$ ./dbim_alter_table_on.sh
$ ./dbim_pop_by_query.sh
```

再次查询 FastStart 表空间的状态，可以看到已使用了超过 600MB 空间：

```
SQL> SELECT tablespace_name, status, ((allocated_size/1024) / 1024) AS alloc_mb,
((used_size/1024) / 1024) AS used_mb  FROM v$inmemory_faststart_area;

TABLESPACE_NAME        STATUS        ALLOC_MB      USED_MB
-------------------    ----------    ----------    ----------
FS_TBS                 ENABLE              1500        673.25
```

此时 In-Memory FastStart 已经生效。重启可插拔数据库 ORCLPDB1，发布所有 5 张 SSB 示例表的时间为 4.21 秒：

```
SQL> ALTER SESSION SET CONTAINER=orclpdb1;
Session altered.

SQL> SHUTDOWN IMMEDIATE
Pluggable Database closed.

SQL> STARTUP
Pluggable Database opened.

-- 发布所有 SSB 示例表
SQL> !./dbim_pop_by_query.sh

SQL> SELECT SUM(time_to_populate) FROM v$im_header;

SUM(TIME_TO_POPULATE)
---------------------
                 4210
```

从 V$SYSSTAT 表中可以查看和 In-Memory FastStart 相关的执行统计信息：

```
SQL> SELECT name, value FROM v$sysstat WHERE name LIKE 'IM faststart%';

NAME                                                          VALUE
------------------------------------------------------------ ----------
IM faststart read data accumulated time(ms)                      40
IM faststart read CUs problems                                    0
IM faststart read headers accumulated time(ms)                    1
IM faststart read accumulated time(ms)                            0
IM faststart read verify accumulated time(ms)                     0
IM faststart read CUs requested                                  29
IM faststart read CUs                                             5
IM faststart read CUs not accessible                              0
IM faststart read bytes                                    39227023
IM faststart read CUs incompatible                                0
```

```
IM faststart read CUs invalid                                    24
IM faststart write bytes                                  573338963
IM faststart write CUs                                           24
IM faststart write CUs requested                                 24
IM faststart write CUs problems                                   0
IM faststart write accumulated time(ms)                        1156
...
```

运行以下命令，禁用 In-Memory FastStart 和清理 FastStart 表空间：

```
SQL> EXEC dbms_inmemory_admin.faststart_disable;
PL/SQL procedure successfully completed.

SQL> DROP TABLESPACE fs_tbs INCLUDING CONTENTS AND DATAFILES;
Tablespace dropped.
```

此时，可以再次测试 In-Memory 对象的发布速度。可以看到，未启用 In-Memory FastStart 时的发布速度为 33.58 秒，慢了将近 8 倍：

```
$ ./dbim_alter_table_off.sh
$ ./dbim_alter_table_on.sh
$ ./dbim_pop_by_query.sh
$ ./dbim_is_pop_ok.sh

-- 确认 5 个对象发布完成后，执行以下查询
SQL> SELECT SUM(time_to_populate) FROM v$im_header;

SUM(TIME_TO_POPULATE)
---------------------
                33586
```

在大部分情形，In-Memory FastStart 都可以提高发布的速度。但需要注意，由于每次重新发布时，受影响的 IMCU 都需要转储到 FastStart 表空间，当 IMCU 的 DML 操作频繁时，Oracle 会选择直接从数据文件中读取数据。因此对于改动频繁的表，需要综合评估 In-Memory FastStart 带来的好处和开销。另外，选用 I/O 性能较高的磁盘来建立 FastStart 表空间，可以减少 FastStart 操作的开销，进一步提升对象的发布速度。

第9章

Database In-Memory 与可管理性

9.1 In-Memory 自动数据优化

信息生命周期管理既是指数据从创建、获取、归档直至删除的数据管理过程，也是指确保数据业务价值与经济高效 IT 基础架构保持一致的策略、流程、实践和工具。Oracle 数据库 12.1.0.1 版发布了两个和信息生命周期管理相关的特性，分别是热图（heat map）和自动数据优化（automatic data optimization），这两个特性均属于高级压缩选件。自动数据优化可以创建信息生命周期管理策略，并基于热图所搜集的数据访问统计信息自动执行这些策略。

9.1.1 自动数据优化基本概念

热图在行级和段级自动跟踪表和分区的使用信息。热图跟踪行级数据修改时间并汇总到块级，在段级跟踪修改时间、全表扫描时间和索引查询时间。通过热图可以详细了解数据的访问方式、模式和频度。热图自动排除了执行系统任务时的内部访问信息，包括统计信息收集、DDL或表重定义等。

自动数据优化可以创建数据压缩和数据移动策略，以便实施存储分层和压缩分层。例如将不常访问的数据从昂贵的高性能存储层迁移到成本与性能较低的存储层，又如对于较少访问的数据可以进一步实施更高级别的压缩。在自动数据优化策略中定义了自动数据优化操作的触发条件，包括对象创建的时间、多长时间没有访问和多长时间没有修改，这些信息均来自于热图。另外，数据移动策略也可以基于表空间的空间使用率来实施数据迁移。Oracle 数据库在维护时段自动在后台评估和执行策略，也可以手动或通过脚本来评估和执行自动数据优化策略。

以下为一些常见的自动数据优化策略定义示例：

```
-- orders 表 30 天没有修改则实施行级压缩
ALTER TABLE orders ILM ADD POLICY ROW STORE COMPRESS ADVANCED ROW AFTER 30 DAYS OF NO
MODIFICATION;
```

```
-- orders 表 60 天没有修改则实施混合列压缩，需要底层存储支持，如 Exadata
ALTER TABLE orders ILM ADD POLICY COLUMN STORE COMPRESS FOR QUERY HIGH SEGMENT AFTER
60 DAYS OF NO MODIFICATION;

-- orders 表空间满时迁移到表空间 lowcost_tablespace
-- 表空间满的定义在视图 DBA_ILMPARAMETERS 中
ALTER TABLE orders ILM ADD POLICY TIER TO lowcost_tablespace;
```

通过初始化参数 HEAT_MAP 可以开启或关闭热图和自动数据优化特性，默认为关闭状态。

```
SQL> SHOW PARAMETER heat_map

NAME                                 TYPE        VALUE
------------------------------------ ----------- -----
heat_map                             string      OFF
```

视图 DBA_ILMPARAMETERS 定义了自动数据优化相关参数：

```
SQL> SELECT * FROM dba_ilmparameters;

NAME                 VALUE
-------------------- ----------
ENABLED                       1
RETENTION TIME               30
JOB LIMIT                     2
EXECUTION MODE                2
EXECUTION INTERVAL           15
TBS PERCENT USED             85
TBS PERCENT FREE             25
POLICY TIME                   0

8 rows selected.
```

视图 DBA_ILMPARAMETERS 中各参数定义如表 9-1 所示。

表 9-1 自动数据优化参数说明

参 数	描 述
ENABLED	控制自动数据优化后台评估与执行，默认值 1 表示启用
RETENTION TIME	已完成自动数据优化任务的保留时间，默认为 30 天
JOB LIMIT	控制最大自动数据优化任务的数量，默认值为 2。最大并发自动数据优化任务数为(JOB LIMIT)×(实例数)×(每实例 CPU 数)
EXECUTION MODE	在线或离线执行控制自动优化任务，默认为在线
EXECUTION INTERVAL	执行自动数据优化后台评估的频率，默认为 15 分钟
TBS PERCENT USED	认为表空间满的百分比阈值，默认为 85
TBS PERCENT FREE	认为表空间空闲的百分比阈值，默认为 25
POLICY TIME	自动数据优化使用的时间单位，默认值 0 表示天，1 表示秒。通常在测试时将时间单位设为秒

通过下面的示例来了解自动数据优化的使用过程。测试场景为：假设有高性能和低成本两个存储层，当高性能层空间满时，自动数据优化会自动将其上的数据库对象迁移到低成本层。

首先开启热图特性，此特性可在系统级或会话级开启，也支持在 PDB 一级开启：

```
ALTER SYSTEM SET heat_map = ON SCOPE = BOTH;
```
然后建立两个表空间 highperf_tbs 和 lowcost_tbs，分别表示高性能和低成本存储层，它们的大小均设置为固定值 10MB。
```
-- 建立固定大小为 10MB 的两个表空间
CREATE TABLESPACE highperf_tbs DATAFILE 'highperf.dbf' SIZE 10m REUSE AUTOEXTEND OFF
EXTENT MANAGEMENT LOCAL;

CREATE TABLESPACE lowcost_tbs DATAFILE 'lowcost.dbf' SIZE 10m REUSE AUTOEXTEND OFF
EXTENT MANAGEMENT LOCAL;
```
基于 SSB 示例中的 SUPPLIER 表，在高性能表空间上创建表 TEST，大约 2MB。
```
SQL> CREATE TABLE TEST TABLESPACE highperf_tbs AS SELECT * FROM supplier;

SQL> SELECT tablespace_name, segment_name FROM dba_segments WHERE segment_name = 'TEST';

TABLESPACE_NAME                SEGMENT_NAME
------------------------------ ----------------
HIGHPERF_TBS                   TEST
```
向测试表 TEST 中不断插入数据，以模拟表空间满：
```
INSERT INTO TEST SELECT * FROM supplier;
INSERT INTO TEST SELECT * FROM supplier;
INSERT INTO TEST SELECT * FROM supplier;
COMMIT;
```
此时高性能表空间的剩余空间为 10%，空间占有率为 90%，已经超过了默认的阈值 85%。
```
-- 高性能表空间的大小
SQL> SELECT bytes FROM dba_data_files WHERE tablespace_name = 'HIGHPERF_TBS';

     BYTES
----------
  10485760

-- 高性能表空间的剩余空间
SQL> SELECT SUM(bytes) FROM dba_free_space WHERE tablespace_name = 'HIGHPERF_TBS';

SUM(BYTES)
----------
   1048576

-- 高性能表空间还剩余 10%空间
SQL> SELECT 1048576*100.0/10485760 AS "%Free" FROM dual;

     %Free
----------
        10
```
此时从系统视图中可以查询到热图信息，其中记录了对象读、写和全表扫描的时间：
```
SQL> SELECT object_name,segment_write_time , segment_read_time, full_scan
FROM user_heat_map_segment WHERE object_name = 'TEST';
```

```
OBJECT_NAME   SEGMENT_W SEGMENT_R FULL_SCAN
------------  --------- --------- ---------
TEST                    01-JUL-21 01-JUL-21
```

为测试表 TEST 创建数据移动策略,将其移动到低成本存储层。在系统视图中可以看到策略信息,其中,策略名 P101 是自动生成的:

```
SQL> ALTER TABLE TEST ILM ADD POLICY TIER TO lowcost_tbs;

SQL> SELECT policy_name, action_type, tier_to, tier_tablespace
FROM user_ilmdatamovementpolicies;

POLICY_NAME   ACTION_TYPE  TIER_TO     TIER_TABLESP
------------  -----------  ----------  ------------
P101          STORAGE      TABLESPACE  LOWCOST_TBS
```

正常情况下,只有在数据库维护窗口才会检查和触发自动数据优化任务。作为测试,选择手工触发自动数据优化任务的执行。

```
DECLARE
    v_task_id NUMBER;
BEGIN
    dbms_ilm.execute_ilm (ilm_scope => dbms_ilm.scope_schema,
           execution_mode => dbms_ilm.ilm_execution_offline,
           task_id   => v_task_id);
END;
/
```

在系统视图中可以确认任务已成功执行:

```
SQL> SELECT task_id, STATE, start_time, completion_time FROM user_ilmtasks;

  TASK_ID STATE     START_TIME                      COMPLETION_TIME
---------- --------- ------------------------------- -------------------------------
      101 COMPLETED 01-JUL-21 10.16.19.177170000 PM 01-JUL-21 10.16.20.551194000 PM
```

此时测试表 TEST 已经迁移到低成本表空间,并且高性能表空间的剩余空间已恢复:

```
SQL> SELECT tablespace_name, segment_name FROM dba_segments WHERE segment_name = 'TEST';

TABLESPACE_NAME                SEGMENT_NAME
------------------------------ ----------------
LOWCOST_TBS                    TEST

SQL> SELECT SUM(bytes) AS free_space FROM dba_free_space WHERE tablespace_name = 'HIGHPERF_TBS';

FREE_SPACE
----------
   9437184
```

测试完成后,执行清理任务,删除测试表和表空间:

```
DROP TABLE test PURGE;
```

```
DROP TABLESPACE highperf_tbs INCLUDING CONTENTS AND DATAFILES;
DROP TABLESPACE lowcost_tbs INCLUDING CONTENTS AND DATAFILES;
```

9.1.2 In-Memory 自动数据优化

Oracle 从 12.1.0.1 版本开始支持自动数据优化，在 12.2.0.1 版本提供了多项增强功能。这其中包括支持多租户架构，支持混合列压缩作为新的存储层，以及 In-Memory 自动数据优化，也就是可以将内存列式存储作为新的自动数据优化策略。由于内存是稀缺和昂贵的资源，In-Memory 自动数据优化可以保证将最适合的对象发布到内存中，并减少人工干预和 DBA 的管理负担。

基于热图统计信息，In-Memory 自动数据优化可以将数据库对象自动发布到内存列式存储，或从内存列式存储中移除。In-Memory 自动数据优化只依赖于 Database In-Memory 选件，和高级压缩选件无关。

In-Memory 自动数据优化策略支持三种操作，即将"热"对象发布到内存列式存储，将"冷"对象移除出内存列式存储，以及对"冷"对象实施更高的内存压缩级。以下我们通过示例来了解这些操作的配置过程。

首先需要开启热图，因为自动数据优化依赖热图统计信息：

```
SQL> ALTER SYSTEM SET heat_map = ON;
System altered.

SQL> SHOW PARAMETER heat_map

NAME                                 TYPE        VALUE
------------------------------------ ----------- ------
heat_map                             string      ON
```

默认情况下，自动数据优化评估使用的时间单位为天。为便于测试，将其修改为秒。同时将后台评估的周期改为 1，也就是每 1 秒执行一次评估。

```
-- 自动数据优化策略使用的时间单位设为秒
SQL> EXEC dbms_ilm_admin.customize_ilm(dbms_ilm_admin.policy_time,
 dbms_ilm_admin.ilm_policy_in_seconds);
PL/SQL procedure successfully completed.

-- 自动数据优化后台评估周期改为 1
SQL> EXEC dbms_ilm_admin.customize_ilm(dbms_ilm_admin.execution_interval, 1);
PL/SQL procedure successfully completed.

-- 确认修改
SQL> SELECT * FROM dba_ilmparameters;

NAME                      VALUE
------------------------- ----------
ENABLED                            1
RETENTION TIME                    30
JOB LIMIT                          2
```

```
EXECUTION MODE              2
EXECUTION INTERVAL          1
TBS PERCENT USED            85
TBS PERCENT FREE            25
POLICY TIME                 1

8 rows selected.
```

使用 LINEORDER 作为测试表，确保其 INMEMORY 属性为未开启状态：

```
SQL> SELECT inmemory, inmemory_compression FROM user_tables
WHERE table_name = 'LINEORDER';

INMEMORY INMEMORY_COMPRESS
-------- -----------------
DISABLED
```

第一个示例需要实现的自动数据优化策略为：当 LINEORDER 表创建 10 天后，自动为其开启 INMEMORY 属性。对应的 DDL 语句如下：

```
ALTER TABLE lineorder ILM ADD POLICY SET INMEMORY AFTER 10 DAYS OF CREATION;
```

此策略实际上并不依赖于热图信息，因为对象创建时间在系统视图中就可以查到：

```
SQL> SELECT created FROM user_objects WHERE object_name = 'LINEORDER';

CREATED
---------
06-MAY-21
```

设置完成后，从系统视图中可以查看策略的定义和状态。其中 P21 是系统自动生成的策略名：

```
-- 策略概览，状态为启用
SQL> SELECT * FROM dba_ilmpolicies;

POLICY_NAME  POLICY_TYPE     TABLESPACE                      ENA DEL
-----------  --------------  ------------------------------  --- ---
P21          DATA MOVEMENT                                   YES NO

-- 策略关联的对象
SQL> SELECT policy_name, object_name, object_type, inherited_from, enabled, deleted
FROM user_ilmobjects;

POLICY_NAME  OBJECT_NAME  OBJECT_TYPE   INHERITED_FROM          ENA DEL
-----------  -----------  ------------  ----------------------  --- ---
P21          LINEORDER    TABLE         POLICY NOT INHERITED    YES NO

-- 策略详细定义
SQL> SELECT policy_name, action_type, condition_type, condition_days, action_clause
FROM user_ilmdatamovementpolicies;

POLICY_NAME  ACTION_TYPE  CONDITION_TYPE  CONDITION_DAYS ACTION_CLAUS
-----------  -----------  --------------  -------------- ------------
P21          ANNOTATE     CREATION TIME               10 INMEMORY
```

手工评估和执行策略。由于之前已将评估单位从天改为秒，因此策略很快可以被执行。注意指定的策略名必须与之前输出保持一致。

```
SET SERVEROUTPUT ON
DECLARE
  v_task_id NUMBER;
BEGIN
  dbms_ilm.execute_ilm(
    OWNER => 'SSB',
    object_name => 'LINEORDER',
    task_id    => v_task_id,
    policy_name => 'P21'
    );
    dbms_output.put_line('Task id is: ' || v_task_id);
END;
/
```

查询系统视图，可以确认自动数据优化任务执行成功，并且 LINEORDER 表成功开启了 INMEMORY 属性。

```
SQL> SELECT task_id, task_owner, state, start_time FROM dba_ilmtasks;

   TASK_ID TASK_OWNER STATE      START_TIME
---------- ---------- ---------- --------------------------------
         1 SSB        COMPLETED  20-JUN-21 10.35.20.152678000 AM

SQL> SELECT task_id, job_state, completion_time FROM dba_ilmresults;

   TASK_ID JOB_STATE                             COMPLETION_TIME
---------- ------------------------------------- --------------------------------
         1 COMPLETED SUCCESSFULLY                20-JUN-21 10.35.32.150374000 AM

SQL> SELECT inmemory, inmemory_compression FROM user_tables WHERE table_name = 'LINEORDER';

INMEMORY INMEMORY_COMPRESS
-------- -----------------
ENABLED  FOR QUERY LOW
```

注意，策略执行完成后，其状态变为禁用。也就是说，自动数据优化策略是一次性的，需要重新定义策略以执行新的操作。

```
SQL> SELECT policy_name, object_name, object_type, inherited_from, enabled, deleted FROM user_ilmobjects;

POLICY_NAME  OBJECT_NAME  OBJECT_TYPE  INHERITED_FROM         ENA  DEL
-----------  -----------  -----------  ---------------------  ---  ---
P21          LINEORDER    TABLE        POLICY NOT INHERITED   NO   NO
```

在上一个示例的基础上，为 LINEORDER 表添加第二个自动数据优化策略：当对象 20 天没有修改时将压缩级调高。对应的 DDL 语句如下：

```
ALTER TABLE lineorder ILM ADD POLICY MODIFY INMEMORY MEMCOMPRESS FOR CAPACITY HIGH AFTER 20 DAYS OF NO MODIFICATION;
```

查看自动数据优化策略状态和详细定义，P41 为系统自动生成的自动数据优化策略名：

```
SQL> SELECT * FROM dba_ilmpolicies;

POLICY_NAME   POLICY_TYPE     TABLESPACE                      ENA DEL
-----------   -------------   ------------------------------  --- ---
P41           DATA MOVEMENT                                   YES NO

SQL> SELECT policy_name, object_name, object_type, inherited_from, enabled, deleted
FROM user_ilmobjects;

POLICY_NAME   OBJECT_NAME   OBJECT_TYPE       INHERITED_FROM          ENA DEL
-----------   -----------   ---------------   ----------------------  --- ---
P41           LINEORDER     TABLE             POLICY NOT INHERITED    YES NO

SQL> SELECT action_type, compression_level, condition_type, condition_days FROM
user_ilmdatamovementpolicies where policy_name = 'P41';

ACTION_TYPE  COMPRESSION_LEVEL            CONDITION_TYPE          CONDITION_DAYS
-----------  ---------------------------  ----------------------  --------------
COMPRESSION  MEMCOMPRESS FOR CAPACITY HIGH LAST MODIFICATION TIME              20
```

和第一个示例不同，判断对象是否修改需要依赖热图信息。因此需要针对表执行一些操作以产生热图统计信息：

```
-- 生成优化器统计信息
EXEC DBMS_STATS.GATHER_TABLE_STATS('SSB','LINEORDER')

-- 针对表执行操作，如全表扫描
SELECT /*+ FULL(p) */ COUNT(*) FROM lineorder p;

-- 将热图统计信息从内存冲刷到磁盘
EXEC dbms_ilm.flush_all_segments
```

确保可以查询到 LINEORDER 表的热图统计信息，然后手工评估和执行策略。注意指定的策略名 P41 需要与之前输出的策略名保持一致。同时记录输出的任务 ID（本例为 41），后续可以查询到此任务执行的状态。

```
SQL> SELECT segment_read_time, segment_write_time, full_scan, lookup_scan FROM
DBA_HEAT_MAP_SEGMENT WHERE object_name = 'LINEORDER';

SEGMENT_R  SEGMENT_W  FULL_SCAN  LOOKUP_SC
---------  ---------  ---------  ---------
20-JUN-21  20-JUN-21

SQL>
DECLARE
    v_task_id NUMBER;
BEGIN
    dbms_ilm.execute_ilm(owner => 'SSB', object_name => 'LINEORDER', task_id =>
v_task_id, policy_name => 'P41');
    DBMS_OUTPUT.PUT_LINE('Task id is: ' || v_task_id);
END;
```

/
Task id is: **41**

PL/SQL procedure successfully completed.

查询系统视图，可以确认自动数据优化任务执行成功，并且 LINEORDER 表的内存压缩级别已由默认值调整为更高的 FOR CAPACITY HIGH。

```
-- 任务被执行
SQL> SELECT task_id, policy_name, object_name, selected_for_execution FROM
user_ilmevaluationdetails ORDER BY task_id;

   TASK_ID POLICY_NAME OBJECT_NAME SELECTED_FOR_EXECUTION
---------- ----------- ----------- -----------------------------------
        41 P41         LINEORDER   SELECTED FOR EXECUTION

-- 任务完成
SQL> SELECT task_id, task_owner, state, start_time FROM dba_ilmtasks;

   TASK_ID TASK_OWNER STATE      START_TIME
---------- ---------- ---------  ----------------------------------
        41 SSB        COMPLETED  20-JUN-21 01.18.30.026415000 PM

-- 任务执行成功
SQL> SELECT task_id, job_state, completion_time FROM dba_ilmresults;

   TASK_ID JOB_STATE                         COMPLETION_TIME
---------- --------------------------------  ----------------------------------
        41 COMPLETED SUCCESSFULLY            20-JUN-21 01.18.37.232157000 PM

-- 任务执行后，策略状态变为禁用
SQL> SELECT policy_name, object_name, object_type, inherited_from, enabled, deleted
FROM user_ilmobjects;

POLICY_NAME OBJECT_NAME OBJECT_TYPE    INHERITED_FROM        ENA DEL
----------- ----------- -------------- --------------------- --- ---
P41         LINEORDER   TABLE          POLICY NOT INHERITED  NO  NO

-- 确认压缩级被调高
SQL> SELECT inmemory, inmemory_compression FROM user_tables WHERE table_name =
'LINEORDER';

INMEMORY INMEMORY_COMPRESS
-------- -----------------
ENABLED  FOR CAPACITY HIGH
```

第三个示例的自动数据优化策略被设置为：当 LINEORDER 表 30 天没有被访问时，将其从内存列式存储中移除。对应的 DDL 语句如下：

```
ALTER TABLE lineorder ILM ADD POLICY NO INMEMORY AFTER 30 DAYS OF NO ACCESS;
```

查看自动数据优化策略状态和详细定义，P61 为系统自动生成的自动数据优化策略名：

```
SQL> SELECT * FROM dba_ilmpolicies;

POLICY_NAME      POLICY_TYPE        TABLESPACE                          ENA DEL
-----------      -------------      ------------------------------      --- ---
P61              DATA MOVEMENT                                          YES NO

SQL> SELECT policy_name, object_name, object_type, inherited_from, enabled, deleted
FROM user_ilmobjects;

POLICY_NAME      OBJECT_NAME    OBJECT_TYPE        INHERITED_FROM          ENA DEL
-----------      -----------    -----------        --------------------    --- ---
P61              LINEORDER      TABLE              POLICY NOT INHERITED    YES NO

SQL> SELECT action_type, condition_type, condition_days, action_clause FROM
user_ilmdatamovementpolicies where policy_name = 'P61';

ACTION_TYPE   CONDITION_TYPE              CONDITION_DAYS ACTION_CLAUSE
-----------   -------------------------   -------------- ----------------
EVICT         LAST MODIFICATION TIME                  20 NO INMEMORY
```

在之前评估和执行自动数据优化策略时，使用的是 DBMS_ILM PL/SQL 包中的 EXECUTE_ILM 过程。以下使用另一种方式，即使用 PREVIEW_ILM 和 EXECUTE_ILM_TASK 两个过程的组合，来评估当前用户下所有的自动数据优化策略：

```
DECLARE
    v_task_id NUMBER;
BEGIN
    dbms_ilm.preview_ilm(task_id => v_task_id, ilm_scope => dbms_ilm.SCOPE_SCHEMA);
    dbms_output.put_line('Task id is: ' || v_task_id);
END;
/
```

在系统视图中，可以查询到系统自动生成的自动数据优化策略名 P61，以及新的任务 ID 61。由于 LINEORDER 表之前已定义了两个策略，因此任务 ID 61 实际对应 3 个任务，但由于 INMEMORY 属性已经开启，压缩级别已经调高，因此系统只会选择执行最后一个任务。

```
SQL> SELECT task_id, task_owner, state, start_time, complete_time FROM dba_ilmtasks;

  TASK_ID TASK_OWNER STATE      START_TIME              COMPLETION_TIME
---------- ---------- ---------- ----------------------- --------------------
       61 SSB        INACTIVE

SQL> SELECT task_id, policy_name, object_name, selected_for_execution from
user_ilmevaluationdetails order by task_id;

  TASK_ID POLICY_NAME  OBJECT_NAME  SELECTED_FOR_EXECUTION
---------- ------------ ------------ ----------------------------------------
...
       61 P21          LINEORDER    IM ATTRIBUTE ALREADY SET
       61 P61          LINEORDER    SELECTED FOR EXECUTION
       61 P41          LINEORDER    TARGET COMPRESSION NOT HIGHER THAN CURRENT
```

```
6 rows selected.
```

指定任务 ID 61 并执行此任务，从系统视图中可以确认 INMEMORY 属性被关闭：

```
SQL>    EXEC    dbms_ilm.execute_ilm_task(61,    dbms_ilm.ILM_EXECUTION_ONLINE,
dbms_ilm.SCHEDULE_IMMEDIATE);

PL/SQL procedure successfully completed.

SQL> SELECT inmemory, inmemory_compression FROM user_tables WHERE table_name =
'LINEORDER';

INMEMORY INMEMORY_COMPRESS
-------- -----------------
DISABLED
```

在以上 3 个示例中，自动优化策略使用了创建时间、访问时间和修改时间作为策略评估条件。实际上，用户还可以使用函数自定义评估条件。例如以下示例函数判断当前是否是 6 月，如果是则返回 TRUE。

```
CREATE OR REPLACE FUNCTION F_POLICYTEST(objn IN NUMBER)RETURN BOOLEAN AS
l_month NUMBER;
BEGIN
  select extract(month from SYSDATE)into l_month from dual;
  if l_month = 6 then
    RETURN TRUE;
  else
    RETURN FALSE;
  end if;

END F_POLICYTEST;
/
```

然后基于此函数定义策略，即当前月份为 6 月时为表开启 INMEMORY 属性：

```
SQL> ALTER TABLE lineorder ILM ADD POLICY SET INMEMORY  ON f_policytest;
ALTER TABLE lineorder ILM ADD POLICY SET INMEMORY  ON f_policytest
*
ERROR at line 1:
ORA-38341: policy conflicts with policy 21
```

执行失败是由于一个自动数据优化操作不能关联多个相同策略，因此需要删除之前的策略：

```
SQL> ALTER TABLE lineorder ILM DISABLE POLICY P21;
SQL> ALTER TABLE lineorder ILM DELETE POLICY P21;
SQL> ALTER TABLE lineorder ILM ADD POLICY SET INMEMORY ON f_policytest;
```

在系统视图中，可以查询到系统自动生成的策略名 P81，以及评估条件的类型为自定义：

```
SQL> SELECT policy_name, object_name, object_type, inherited_from, enabled, deleted
FROM user_ilmobjects;

POLICY_NAME  OBJECT_NAME  OBJECT_TYPE       INHERITED_FROM        ENA DEL
-----------  -----------  ----------------  --------------------  --- ---
P81          LINEORDER    TABLE             POLICY NOT INHERITED  YES NO
```

```
SQL> SELECT action_type, condition_type, condition_days, action_clause FROM
user_ilmdatamovementpolicies where policy_name = 'P81';

ACTION_TYPE  CONDITION_TYPE              CONDITION_DAYS ACTION_CLAUSE
-----------  --------------------------  -------------- ---------------
ANNOTATE     USER DEFINED                             0 INMEMORY
```

手工评估和执行自动数据优化策略:

```
DECLARE
    v_task_id NUMBER;
BEGIN
    dbms_ilm.execute_ilm(owner => 'SSB', object_name => 'LINEORDER', task_id =>
v_task_id, policy_name => 'P81');
    DBMS_OUTPUT.PUT_LINE('Task id is: ' || v_task_id);
END;
/

Task id is: 81

PL/SQL procedure successfully completed.
```

最终可以确认 LINEORDER 表被设置了 INMEMORY 属性:

```
SQL> SELECT inmemory, inmemory_compression FROM user_tables WHERE table_name =
'LINEORDER';

INMEMORY  INMEMORY_COMPRESS
--------  -----------------
ENABLED   FOR QUERY LOW
```

完成所有示例后,使用以下 SQL 清理实验环境:

```
ALTER TABLE lineorder NO INMEMORY;
ALTER TABLE lineorder ILM DISABLE_ALL;
ALTER TABLE lineorder ILM DELETE_ALL;
ALTER SYSTEM SET heat_map = OFF SCOPE=BOTH;

EXEC dbms_ilm_admin.customize_ilm(
    dbms_ilm_admin.POLICY_TIME, dbms_ilm_admin.ILM_POLICY_IN_DAYS
);

EXEC dbms_ilm_admin.customize_ilm(
    dbms_ilm_admin.EXECUTION_INTERVAL, 15
);
```

9.2 Automatic In-Memory

AIM(Automatic In-Memory)是 Oracle 18c Database In-Memory 中新增的特性。AIM 可以对内存列式存储中的对象实现自动化的管理,例如自动移除"冷"对象或自动发布"热"对象。在 Oracle 21c 版本,AIM 特性得到进一步增强,可以实现完全自动、无须任何人工干预的管理。

为区分，将 18c 版本的 AIM 特性称为自动 In-Memory 管理，将 21c 版本的 AIM 特性称为自治 In-Memory 管理。

Automatic In-Memory 原本是 Exadata 专属的特性。2021 年 6 月，Oracle 宣布在非 Exadata 环境也可以支持 Automatic In-Memory，目前支持的数据库版本为 19.9 及以上企业版。

9.2.1 自动 In-Memory 管理

Oracle 从 18c 版本开始支持自动 In-Memory 管理。自动 In-Memory 管理利用热图、列统计及其他统计信息来管理内存列式存储中的对象。当面临内存压力时，内存列式存储自动移除"冷"对象，腾出内存空间以发布频繁访问的"热"对象。

在 Database In-Memory 中有一个工作数据集（Working Data Set）的概念，指数据库查询频繁访问的段对象集合，如表和分区。显然，工作数据集的内容会随着时间和应用负载的变化而变化。在 Oracle 18c 版本以前，只能通过手工发布和移除，或者是自动数据优化策略来管理内存列式存储中的对象，这些方式需要对应用负载有透彻的理解，并且伴随大量的人为操作。自动 In-Memory 管理利用实时更新的统计信息，可尽量保证工作数据集中的所有对象发布到内存列式存储中，同时避免了人工干预和猜测。

自动 In-Memory 管理是自动数据优化特性的延伸和补充，这两个功能可以共同作用于同一数据库对象，但自动数据优化的优先级更高。

自动 In-Memory 管理只针对设置了 INMEMORY 属性,并且发布优先级为 NONE 的段对象，如表和分区。启用自动 In-Memory 管理需要将 INMEMORY_AUTOMATIC_LEVEL 初始化参数设置为 LOW 或 MEDIUM，其默认值为 OFF。其中 LOW 表示存在内存压力时，内存列式存储将移除"冷"对象。MEDIUM 在 LOW 的基础上增加额外的优化，即优先发布之前因内存不足未成功发布的对象。虽然自动 In-Memory 管理使用了热图的信息，但并不要求设置 HEAT_MAP 参数。

```
SQL> SHOW PARAMETER inmemory_automatic_level

NAME                                 TYPE        VALUE
------------------------------------ ----------- -----
inmemory_automatic_level             string      OFF
```

自动 In-Memory 管理启用后的基本工作流程如下。

（1）由于 In-Memory Area 内存空间不足导致对象发布失败，如不完整发布。

（2）自动 In-Memory 管理根据内部统计信息确定需要移除的"冷"对象。

（3）如果在"冷"对象上已经定义了要求其保持发布状态的自动数据优化策略，则忽略此对象。否则执行下一步。

（4）由空间管理工作进程 W*nnn* 移除符合条件的"冷"对象，释放内存空间。

参考的统计信息窗口范围可通过 DBMS_INMEMORY_ADMIN PL/SQL 包设定，默认为 31 天。

```
SQL>
```

```
SET SEVEROUTPUT ON
VARIABLE b_interval NUMBER
BEGIN
    dbms_inmemory_admin.aim_get_parameter(dbms_inmemory_admin.aim_statwindow_days,
 :b_interval);
END;
/

PRINT b_interval

B_INTERVAL
----------
        31
```

通过下面的示例了解自动 In-Memory 管理的工作过程与原理。首先需要设置初始化参数 INMEMORY_AUTOMATIC_LEVEL，然后将 In-Memory Area 区域的设置调小，以便后续模拟内存不足的情形。

```
ALTER SYSTEM SET inmemory_automatic_level = MEDIUM;

ALTER SYSTEM SET inmemory_size = 1000 M;
```

为 SSB 示例中的 5 张表开启 INMEMORY 属性，然后发布。从输出可知，LINEORDER 表占用内存超过 500MB，内存列式存储剩余不到 400MB 空间。

```
$ ./dbim_alter_table_on.sh
$ ./dbim_pop_by_procedure.sh
$ ./dbim_get_popstatus.sh

              SEGMENT          IN MEM      ON DISK  COMPRESSION  BYTES NOT   POPULATED
OWNER         NAME             SIZE(KB)    SIZE(KB)       RATIO  POPULATED   STATUS
------        ----------      ---------   ---------  -----------  ---------  ----------
SSB           SUPPLIER            2,304       1,920          .83          0  COMPLETED
SSB           CUSTOMER           17,664      29,936         1.69          0  COMPLETED
SSB           LINEORDER         575,872   1,332,376         2.31          0  COMPLETED
SSB           PART               20,928     101,264         4.84          0  COMPLETED
SSB           DATE_DIM            1,280         304          .24          0  COMPLETED


TOTAL_DISK_SIZE(KB)  TOTAL_IM_SIZE(KB) OVERALL_COMPRESSION_RATIO
-------------------  ----------------- -------------------------
            1465800             618048                      2.37
```

基于 LINEORDER 表复制一张新表 LINEORDER2，然后通过全表扫描发布。

```
CREATE TABLE lineorder2 NOLOGGING AS SELECT * FROM lineorder;

ALTER TABLE lineorder2 INMEMORY;

SELECT /*+ FULL(p) NO_PARALLEL(p) */ COUNT(*) FROM lineorder2 p;
```

显然，剩余内存空间不足以容纳 LINEORDER2 表。从输出中可以确认 LINEORDER2 表为不完整发布：

```
$ ./dbim_get_popstatus.sh

         SEGMENT       IN MEM        ON DISK   COMPRESSION   BYTES NOT   POPULATED
OWNER    NAME          SIZE(KB)      SIZE(KB)  RATIO         POPULATED   STATUS
-----    --------      --------      --------  -----------   ---------   ---------
SSB      SUPPLIER      2,304         1,920     .83                   0   COMPLETED
SSB      CUSTOMER      17,664        29,936    1.69                  0   COMPLETED
SSB      LINEORDER     575,872       1,332,376 2.31                  0   COMPLETED
SSB      PART          20,928        101,264   4.84                  0   COMPLETED
SSB      DATE_DIM      1,280         304       .24                   0   COMPLETED
SSB      LINEORDER2    95,040        1,332,696 14.02        1141047296   COMPLETED

6 rows selected.


TOTAL_DISK_SIZE(KB)  TOTAL_IM_SIZE(KB)  OVERALL_COMPRESSION_RATIO
-------------------  -----------------  -------------------------
            2798496             713088                       3.92
```

为精确控制此测试，将统计信息窗口从默认的 31 天改为 1 天：

```
SQL> connect sys@orclpdb1 as sysdba
Enter password:
Connected.

SQL> EXEC DBMS_INMEMORY_ADMIN.AIM_SET_PARAMETER(
     DBMS_INMEMORY_ADMIN.AIM_STATWINDOW_DAYS, 1 );

PL/SQL procedure successfully completed.
```

对表 LINEORDER2 查询 10 次，以使其变"热"。通过 SQLcl 可以很方便地实现重复执行：

```
SQL> SELECT /*+ FULL(p) NO_PARALLEL(p) */ COUNT(*) FROM lineorder2 p;

  COUNT(*)
----------
  11997996

SQL> repeat 10 1
```

从系统视图中，可以查询到 LINEORDER2 表有 11 次全表扫描，相对于其他表而言访问更频繁。

```
SQL> SELECT object_name, full_scan, n_full_scan FROM V$IMHMSEG;
OBJECT_NAME                    FUL N_FULL_SCAN
------------------------------ --- -----------
SUPPLIER                       NO            0
LINEORDER                      NO            0
DATE_DIM                       NO            0
PART                           NO            0
LINEORDER2                     YES          11
CUSTOMER                       NO            0
```

```
6 rows selected.
```

通过系统视图查询 AIM 任务详情，发现"冷"对象 DATE_DIM 被移除出内存列式存储，LINEORDER 表被部分发布。这两个操作使得内存空间被释放，从而成功发布"热"对象 LINEORDER2。

```
SQL> SELECT * FROM DBA_INMEMORY_AIMTASKS ORDER BY creation_time DESC;
  TASK_ID CREATION_TIME                   STATE
---------- ------------------------------- -------
    29763 02-JUL-21 08.28.54.000000000 AM RUNNING
    29762 02-JUL-21 08.26.54.000000000 AM DONE
    29761 02-JUL-21 08.24.53.000000000 AM DONE
    29760 02-JUL-21 08.22.53.000000000 AM DONE
...

SQL> SELECT object_owner, object_name, action, state FROM DBA_INMEMORY_AIMTASKDETAILS
WHERE task_id = 29763;
...
SSB           SUPPLIER           POPULATE           DONE
SSB           CUSTOMER           POPULATE           DONE
SSB           PART               POPULATE           DONE
SSB           LINEORDER          PARTIAL POPULATE   DONE
SSB           DATE_DIM           EVICT              DONE
SSB           LINEORDER2         POPULATE           DONE

27 rows selected.
```

由于 LINEORDER 表不完整发布并非最佳状态，因此可以手动或利用自动数据优化策略调高内存压缩级，使其完整发布：

```
ALTER TABLE lineorder INMEMORY MEMCOMPRESS FOR CAPACITY HIGH;
```

测试完成后，可执行以下脚本清理环境：

```
DROP TABLE lineorder2 PURGE;

ALTER SYSTEM SET inmemory_automatic_level = OFF;

EXEC dbms_inmemory_admin.aim_set_parameter(dbms_inmemory_admin.aim_statwindow_days, 31);

ALTER SYSTEM SET inmemory_size = 1500 M;
```

9.2.2 自治 In-Memory 管理

在 18c 自动 In-Memory 管理基础上，Oracle 21c 版本的自治 In-Memory 管理更进一步，数据库对象即使未设置 INMEMORY 属性也可以通过 Automatic In-Memory 进行管理。

Oracle 21c 版本的自治 In-Memory 管理的大部分功能与 18c 版本时相同，最主要的改变是初始化参数 INMEMORY_AUTOMATIC_LEVEL 增加了 HIGH 级别。在此级别下，除属于 Oracle 保留用户的表和外部表，所有没有设置过 INMEMORY 属性的对象均纳入自治 In-Memory 管理的范畴。数据库根据内部统计参数自动移除"冷"对象和发布"热"对象。此外,某些不常访问的表或列可能会被实施更高级别的压缩,从而为"热"对象释放更多内存空间。

通过以下查询可以列出 Oracle 保留用户:

```
SQL> SELECT username FROM dba_users WHERE oracle_maintained = 'Y';

USERNAME
--------
SYS
SYSTEM
XS$NULL
LBACSYS
...
```

如果希望某些对象不受自治 In-Memory 管理的控制,可以使用以下两种方法。

(1)显式设置 NO INMEMORY 属性的对象将不会被发布到内存列式存储中。

(2)设置了内存发布优先级(非 NONE)的对象将不会从内存列式存储中移除。

21c 版本的自治 In-Memory 管理还有一个小的改变,即参考统计信息窗口的默认值由 18c 版本的 31 天变为 1 天。

```
SQL>
SET SEVEROUTPUT ON
VARIABLE b_interval NUMBER

BEGIN
    dbms_inmemory_admin.aim_get_parameter(dbms_inmemory_admin.aim_statwindow_days,
:b_interval);
END;
/

PRINT b_interval

B_INTERVAL
----------
         1
```

通过下面的示例了解自治 In-Memory 管理的工作原理和过程。首先需要将初始化参数 INMEMORY_AUTOMATIC_LEVEL 设置为 HIGH:

```
SQL> ALTER SYSTEM SET inmemory_automatic_level = HIGH;
System altered.
```

使用 SSB 示例中的 5 张表及一张新建的表 PART2 作为测试对象,PART2 基于 PART 表创建:

```
CREATE TABLE part2 AS SELECT * FROM part;
```

查询这些表的 INMEMORY 属性,发现只有新建的 PART2 表纳入自治 In-Memory 管理:

```
SQL> SELECT table_name, inmemory, inmemory_priority, inmemory_compression FROM user_tables;

TABLE_NAME    INMEMORY  INMEMORY  INMEMORY_COMPRESSION
----------    --------  --------  --------------------
SUPPLIER      DISABLED            NO INMEMORY
```

```
PART          DISABLED                  NO INMEMORY
CUSTOMER      DISABLED                  NO INMEMORY
DATE_DIM      DISABLED                  NO INMEMORY
PART2         ENABLED      NONE         AUTO
LINEORDER     DISABLED                  NO INMEMORY

6 rows selected.
```

这是由于其他的表之前都设置了 NO INMEMORY 属性，我们需要使用以下 SQL 来清除这些属性：

```
ALTER TABLE lineorder INMEMORY MEMCOMPRESS AUTO;
ALTER TABLE part INMEMORY MEMCOMPRESS AUTO;
ALTER TABLE customer INMEMORY MEMCOMPRESS AUTO;
ALTER TABLE supplier INMEMORY MEMCOMPRESS AUTO;
ALTER TABLE date_dim INMEMORY MEMCOMPRESS AUTO;
```

再次查询，现在所有的表均启用了自治 In-Memory 管理：

```
SQL> SELECT table_name, inmemory, inmemory_priority, inmemory_compression FROM user_tables;

TABLE_NAME    INMEMORY   INMEMORY_PRIORITY  INMEMORY_COMPRESS
-----------   --------   -----------------  -----------------
SUPPLIER      ENABLED    NONE               AUTO
PART          ENABLED    NONE               AUTO
CUSTOMER      ENABLED    NONE               AUTO
DATE_DIM      ENABLED    NONE               AUTO
PART2         ENABLED    NONE               AUTO
LINEORDER     ENABLED    NONE               AUTO

6 rows selected.
```

目前内存列式存储中还未发布任何表，需要执行一些工作负载来触发发布。运行随书示例中的 runall.sh 脚本，此脚本将运行所有 SSB 示例 SQL。这些 SQL 会访问 SSB 示例中的 5 张表，但不包括 PART2 表：

```
$ cd ~/dbimbook/chap02/SSB/sql
$ ./runall.sh
```

从系统视图中可以查询到对象的访问类型和访问次数：

```
SQL> SELECT object_name, full_scan, n_full_scan, track_time FROM V$IMHMSEG;

OBJECT_NAME                    FUL  N_FULL_SCAN  TRACK_TIM
-----------------------------  ---  -----------  ---------
SUPPLIER                       YES           10  02-JUL-21
DATE_DIM                       YES           13  02-JUL-21
LINEORDER                      YES           13  02-JUL-21
PART                           YES            6  02-JUL-21
CUSTOMER                       YES            7  02-JUL-21

5 rows selected.
```

根据对象访问的热度，系统自动发布了以下对象：

```
$ ./dbim_get_popstatus.sh

              SEGMENT           IN MEM      ON DISK  COMPRESSION  BYTES NOT  POPULATED
OWNER         NAME              SIZE(KB)    SIZE(KB)       RATIO  POPULATED  STATUS
-----         --------          --------    --------   ---------  ---------  ---------
SSB           SUPPLIER             1,280       1,920         1.5          0  COMPLETED
SSB           CUSTOMER             9,472      29,936        3.16          0  COMPLETED
SSB           LINEORDER          577,920   1,332,696        2.31          0  COMPLETED
SSB           DATE_DIM             1,280         304         .24          0  COMPLETED


TOTAL_DISK_SIZE(KB)  TOTAL_IM_SIZE(KB)  OVERALL_COMPRESSION_RATIO
-------------------  -----------------  -------------------------
            1364856             589952                       2.31
```

和 18c 版本的自动数据管理类似，从系统视图 DBA_INMEMORY_AIMTASKS 和 DBA_INMEMORY_AIMTASKDETAILS 中可以查询所有 AIM 任务和执行详情。

测试完成后，执行以下脚本清理环境：

```
SQL> DROP TABLE part2 PURGE;

SQL> ALTER SYSTEM SET inmemory_automatic_level = OFF;
```

此时 SSB 示例表的 INMEMORY 属性如下：

```
SQL> SELECT table_name, inmemory, inmemory_priority, inmemory_compression FROM user_tables;

TABLE_NAME    INMEMORY  INMEMORY_PRIORITY  INMEMORY_COMPRESSION
-----------   --------  -----------------  --------------------
SUPPLIER      DISABLED
PART          DISABLED
CUSTOMER      DISABLED
DATE_DIM      DISABLED
LINEORDER     DISABLED
```

9.3 Database In-Memory 与分区

Oracle 分区技术于 1997 年在 Oracle 8.0 版本中首次引入，最初支持范围分区、分区裁剪和分区交换功能。Oracle 8i 版本增加了对哈希分区、范围哈希混合分区和智能分区联结的支持。Oracle 9i 版本新增支持列表分区和范围列表复合分区。之后 10g、11g 和 12c 版本功能不断丰富和完善，18c 版本和 19c 版本新增支持分区外部表和混合分区表。分区是 Oracle 数据库最基础、最重要的数据库选件之一，同时也是与 Database In-Memory 配合最为紧密的选件。这两个选件兼具改善数据库性能和可管理性的功能，并且均具有应用透明性。以下介绍的分区裁剪、智能分区联结和分区交换功能，均非专为 Database In-Memory 而设计，但都可以与 Database In-Memory 配合使用，并取得良好的效果。

9.3.1 分区发布

分区表可以在表或分区一级设置 INMEMORY 属性。如果在表一级设置，新加入的分区继承表一级设置的属性。在以下示例中，SH 用户下的 SALES 表为分区表，来自 Oracle 标准示例。

```
-- 分区表在表一级的 INMEMORY 属性设置
SQL> CONNECT sh@orclpdb1
SQL> ALTER TABLE sales INMEMORY;

SQL> SELECT table_name, def_inmemory FROM user_part_tables WHERE table_name = 'SALES';

TABLE_NAME      DEF_INME
------------    --------
SALES           ENABLED


SQL> SELECT table_name, partition_name, INMEMORY FROM user_tab_partitions WHERE table_name = 'SALES';

TABLE_NAME      PARTITION_NAME          INMEMORY
------------    --------------------    --------
SALES           SALES_1996              ENABLED
SALES           SALES_1995              ENABLED
SALES           SALES_Q1_2002           ENABLED
SALES           SALES_Q2_2002           ENABLED
...
28 rows selected.

-- 分区表在分区一级的 INMEMORY 属性设置
SQL> ALTER TABLE sales NO INMEMORY;
SQL> ALTER TABLE sales MODIFY PARTITION sales_1996 INMEMORY;

SQL> SELECT table_name, def_inmemory FROM user_part_tables WHERE table_name = 'SALES';

TABLE_NAME      DEF_INME
------------    --------
SALES           DISABLED

SQL> SELECT table_name, partition_name, INMEMORY FROM user_tab_partitions WHERE table_name = 'SALES';

TABLE_NAME      PARTITION_NAME          INMEMORY
------------    --------------------    --------
SALES           SALES_1996              ENABLED
SALES           SALES_1995              DISABLED
...
28 rows selected.
```

分区表的发布与普通表一样，既可以通过全表扫描引发发布，也可以通过 PL/SQL 触发发布。

```
SQL>
ALTER TABLE sales NO INMEMORY;
ALTER TABLE sales INMEMORY;
```

```
-- 全表扫描引发发布
SELECT /*+ FULL(p) NO_PARALLEL(p) */ COUNT(*) FROM sales partition(SALES_Q4_1999) p;

-- 通过 PL/SQL 过程指定分区发布
EXEC DBMS_INMEMORY.POPULATE('SH', 'SALES', 'SALES_Q4_2000');

SQL> SELECT partition_name, segment_type, populate_status FROM v$im_segments;

PARTITION_NAME        SEGMENT_TYPE         POPULATE_STAT
------------------    ------------------   --------------
SALES_Q4_1999         TABLE PARTITION      COMPLETED
SALES_Q4_2000         TABLE PARTITION      COMPLETED
```

9.3.2 分区裁剪

分区裁剪是数据仓库的基本性能特性。优化器在构建分区访问列表时，通过分析 SQL 语句中的 WHERE 条件可以排除不必要的分区，从而使 Oracle 数据库仅对与 SQL 语句相关的分区执行操作。分区裁剪减少了从磁盘检索的数据量，可提高查询性能并优化资源利用率。

根据实际的 SQL 语句，Oracle 数据库可能使用静态或动态裁剪。静态裁剪发生在编译时，预先可以决定需要排除的分区。动态裁剪发生在运行时，必须结合 WHERE 条件中的绑定变量、子查询、星型转换或嵌套查询才能确定需要访问和排除的分区。

Database In-Memory 可以很好地与分区裁剪功能结合。以 SH 用户下的 COSTS 表为例，将表发布到内存列式存储，然后执行查询。根据 WHERE 条件，Oracle 数据库选择了静态分区裁剪方法，最终只使用了第 13 ~ 24 个分区：

```
SQL> ALTER TABLE COSTS INMEMORY;
SQL> EXEC DBMS_INMEMORY.POPULATE('SH', 'COSTS');
SQL> EXPLAIN PLAN FOR SELECT * FROM costs
WHERE time_id >= TO_DATE('2000-01-01 00:00:00', 'SYYYY-MM-DD HH24:MI:SS')
AND time_id <= TO_DATE('2002-12-31 00:00:00', 'SYYYY-MM-DD HH24:MI:SS')
AND prod_id = 16;

SQL> SELECT plan_table_output
FROM TABLE(dbms_xplan.display('plan_table',NULL,'all -cost -bytes -rows'));

---------------------------------------------------------------------------
| Id  | Operation                    | Name  | Time     | Pstart| Pstop |
---------------------------------------------------------------------------
|   0 | SELECT STATEMENT             |       | 00:00:01 |       |       |
|   1 |  PARTITION RANGE ITERATOR    |       | 00:00:01 |    13 |    24 |
|*  2 |   TABLE ACCESS INMEMORY FULL | COSTS | 00:00:01 |    13 |    24 |
---------------------------------------------------------------------------
```

查看系统视图可知，COSTS 表具有 28 个分区。第 13 ~ 24 个分区正好与 WHERE 条件对应，即 2000 年到 2002 年间 12 个季度所对应的分区。通过分区裁剪，排除了 16 个分区，节省了大量磁盘 I/O。

```
SQL> SELECT partition_name, high_value FROM user_tab_partitions WHERE table_name =
'COSTS';

PARTITION_NAME          GET_HIGH_
--------------------    ---------
COSTS_1995              01-JAN-96
COSTS_1996              01-JAN-97
COSTS_H1_1997           01-JUL-97
COSTS_H2_1997           01-JAN-98
COSTS_Q1_1998           01-APR-98
COSTS_Q2_1998           01-JUL-98
COSTS_Q3_1998           01-OCT-98
COSTS_Q4_1998           01-JAN-99
COSTS_Q1_1999           01-APR-99
COSTS_Q2_1999           01-JUL-99
COSTS_Q3_1999           01-OCT-99
COSTS_Q4_1999           01-JAN-00
COSTS_Q1_2000           01-APR-00
COSTS_Q2_2000           01-JUL-00
COSTS_Q3_2000           01-OCT-00
COSTS_Q4_2000           01-JAN-01
COSTS_Q1_2001           01-APR-01
COSTS_Q2_2001           01-JUL-01
COSTS_Q3_2001           01-OCT-01
COSTS_Q4_2001           01-JAN-02
COSTS_Q1_2002           01-APR-02
COSTS_Q2_2002           01-JUL-02
COSTS_Q3_2002           01-OCT-02
COSTS_Q4_2002           01-JAN-03
COSTS_Q1_2003           01-APR-03
COSTS_Q2_2003           01-JUL-03
COSTS_Q3_2003           01-OCT-03
COSTS_Q4_2003           01-JAN-04

28 rows selected.
```

9.3.3 智能分区联结

智能分区联结也称为 Partition-Wise Join，是指采用相同分区键和分区策略的两张表之间的联结操作。当联结被并行执行时，智能分区联结通过最小化并行执行服务器之间交换的数据来加速查询，并减少了内存和 CPU 的需求。在 Oracle RAC 环境下，智能分区联结可以减少或消除私有互连上的数据流量，这是大规模联结操作实现良好可扩展性的关键。

以下通过示例了解 Database In-Memory 如何与智能分区联结配合。首先创建分区表 t1 和 t2，它们采用相同的分区键，并且具有相同的分区数量和分区范围。另外，这两张表在创建时均开启了 INMEMORY 属性。

```
CREATE TABLE t1(x, y)INMEMORY
PARTITION BY RANGE(x)
```

```
(
PARTITION p1 VALUES LESS THAN(2500),
PARTITION p2 VALUES LESS THAN(5000),
PARTITION p3 VALUES LESS THAN(7500),
PARTITION p4 VALUES LESS THAN(10000)
)
AS
SELECT ROWNUM, round(dbms_random.VALUE(1,10)) FROM dual CONNECT BY LEVEL < 10000;

CREATE TABLE t2(x, y) INMEMORY
PARTITION BY RANGE(x)
(
PARTITION p1 VALUES LESS THAN(2500),
PARTITION p2 VALUES LESS THAN(5000),
PARTITION p3 VALUES LESS THAN(7500),
PARTITION p4 VALUES LESS THAN(10000)
)
AS
SELECT ROWNUM, round(dbms_random.VALUE(1,10)) FROM dual CONNECT BY LEVEL < 10000;
```

执行以下 SQL 语句，即表 t1 和表 t2 之间的联结操作：

```
SELECT SUM(t1.y + t2.y) FROM t1,t2 WHERE t1.x = t2.x;
```

以下为串行执行时的执行计划，可以看到表联结操作被拆分为 4 个小的哈希联结，并循环执行。这表明使用了智能分区联结：

```
-----------------------------------------------------------------------
| Id  | Operation                    | Name | Time     | Pstart| Pstop |
-----------------------------------------------------------------------
|  0  | SELECT STATEMENT             |      | 00:00:01 |       |       |
|  1  |  SORT AGGREGATE              |      |          |       |       |
|  2  |   PARTITION RANGE ALL        |      | 00:00:01 |   1   |   4   |
|* 3  |    HASH JOIN                 |      | 00:00:01 |       |       |
|  4  |     TABLE ACCESS INMEMORY FULL| T1  | 00:00:01 |   1   |   4   |
|  5  |     TABLE ACCESS INMEMORY FULL| T2  | 00:00:01 |   1   |   4   |
-----------------------------------------------------------------------
```

以下为并行执行时的执行计划，同样也使用了智能分区联结。其中 P->S 表示并行进程将数据发送给查询协调器。PCWP（Parallel Combine With Parent）和 PCWC（Parallel Combine With Child）成对出现，由同一个并行进程执行。在本例中，并行执行的 CPU 成本为 4，远低于串行执行时的 54。

```
-----------------------------------------------------------------------
| Id  | Operation                | Name    | Pstart| Pstop |IN-OUT|
-----------------------------------------------------------------------
|  0  | SELECT STATEMENT         |         |       |       |      |
|  1  |  SORT AGGREGATE          |         |       |       |      |
|  2  |   PX COORDINATOR         |         |       |       |      |
|  3  |    PX SEND QC(RANDOM)    | :TQ10000|       |       | P->S |
|  4  |     SORT AGGREGATE       |         |       |       | PCWP |
|  5  |      PX PARTITION RANGE ALL|       |   1   |   4   | PCWC |
|* 6  |       HASH JOIN          |         |       |       | PCWP |
-----------------------------------------------------------------------
```

```
|   7 |         TABLE ACCESS INMEMORY FULL| T1           |       1 |       4 | PCWP |
|   8 |         TABLE ACCESS INMEMORY FULL| T2           |       1 |       4 | PCWP |
----------------------------------------------------------------------------------------
```

9.3.4 分区交换

通过交换数据段，分区交换可以将分区或子分区转换为非分区表，或者将非分区表转换为分区表的分区或子分区，这实际上是通过交换元数据来实现的。分区交换是一种将数据快速移入和移出分区表的有效手段。例如在数据仓库系统中将新数据快速加载到分区表，或者在数据仓库或 OLTP 系统中，将历史数据清除出分区表。这些数据并没有真正被删除，后续可以进行归档。

分区交换早在 Oracle 8 版本最初推出分区技术时就已经支持，和 Database In-Memory 技术可以很好地结合。以下将演示如何利用分区交换将新数据加载到 In-Memory 分区表。

首先基于 SUPPLIER 创建范围分区表 P_SUPPLIER 并插入数据，然后设置 INMEMORY 属性并发布。最终表 P_SUPPLIER 拥有 3 个分区，其中分区 P1 暂时没有数据，后续会将数据加载到此分区。

```sql
SQL> CREATE TABLE p_supplier PARTITION BY RANGE (pkey) INTERVAL(1)
   (PARTITION p1 VALUES LESS THAN(1)) AS SELECT 1 AS pkey, supplier.* FROM supplier;

SQL> INSERT INTO p_supplier SELECT 2, supplier.* FROM supplier;

SQL> ALTER TABLE p_supplier INMEMORY;

SQL> EXEC dbms_stats.gather_table_stats('','P_SUPPLIER')

SQL> SELECT table_name, partition_name, inmemory, num_rows FROM user_tab_partitions
WHERE table_name = 'P_SUPPLIER';

TABLE_NAME           PARTITION_NAME       INMEMORY   NUM_ROWS
-------------------- -------------------- ---------- ----------
P_SUPPLIER           P1                   DISABLED   0
P_SUPPLIER           SYS_P666             DISABLED   16000
P_SUPPLIER           SYS_P667             DISABLED   16000
```

为分区表开启 INMEMORY 属性并发布，分区 P1 由于没有数据，因此不会发布到内存列式存储中：

```sql
SQL> ALTER TABLE p_supplier INMEMORY;

SQL> SELECT /*+ FULL(s) NO_PARALLEL(s) */ COUNT(*) FROM p_supplier s;

  COUNT(*)
----------
     32000

SQL> SELECT segment_name, partition_name, inmemory_size, bytes, populate_status FROM
v$im_segments;
```

```
SEGMENT_NAME       PARTITION_NAME         INMEMORY_SIZE      BYTES POPULATE_STAT
----------------   --------------------   -------------   -------- -------------
P_SUPPLIER         SYS_P666                     2359296    2015232 COMPLETED
P_SUPPLIER         SYS_P667                     2359296    8241152 COMPLETED
```

创建临时表 TMP_SUPPLIER，并插入数据，开启 INMEMORY 属性并发布。此表后续会与分区 P1 进行交换：

```
SQL> CREATE TABLE tmp_supplier FOR EXCHANGE WITH TABLE p_supplier;

SQL> INSERT INTO tmp_supplier SELECT 0 AS key_no, supplier.* FROM supplier;

SQL> ALTER TABLE tmp_supplier INMEMORY;

SQL> SELECT /*+ FULL(s) NO_PARALLEL(s) */ COUNT(*) FROM tmp_supplier s;

  COUNT(*)
----------
     16000

SQL> SELECT segment_name, partition_name, inmemory_size, BYTES, populate_status FROM
v$im_segments;

SEGMENT_NAME       PARTITION_NAME         INMEMORY_SIZE      BYTES POPULATE_STAT
----------------   --------------------   -------------   -------- -------------
P_SUPPLIER         SYS_P666                     2359296    2015232 COMPLETED
TMP_SUPPLIER                                    2359296    1998848 COMPLETED
P_SUPPLIER         SYS_P667                     2359296    8241152 COMPLETED
```

通过 EXCHANGE PARTITION 命令启动分区交换。交换完成后，分区 P1 中的行数由 0 变为 16000，而表 TMP_SUPPLIER 中的行数变为 0。在分区交换前，关闭了分区 P1 的 INMEMORY 属性，此操作不是必需的，只是为了证明除数据外，分区交换也会交换段对象的属性。例如在分区交换后，分区 P1 具有 INMEMORY 属性并已发布，而表 TMP_SUPPLIER 的 INMEMORY 属性已关闭。

```
SQL> SELECT COUNT(*) FROM p_supplier PARTITION(p1);

  COUNT(*)
----------
         0

SQL> ALTER TABLE p_supplier MODIFY PARTITION p1 NO INMEMORY;

SQL> ALTER TABLE p_supplier EXCHANGE PARTITION p1 WITH TABLE tmp_supplier;

SQL> SELECT COUNT(*) FROM p_supplier PARTITION(p1);

  COUNT(*)
----------
     16000
```

```
SQL> SELECT COUNT(*) FROM tmp_supplier;

  COUNT(*)
----------
         0

SQL> SELECT segment_name, partition_name, inmemory_size, BYTES, populate_status FROM
v$im_segments;

SEGMENT_NAME    PARTITION_NAME         INMEMORY_SIZE      BYTES POPULATE_STAT
--------------- ---------------------- ------------- ---------- -------------
P_SUPPLIER      SYS_P666                     2359296    2015232 COMPLETED
P_SUPPLIER      P1                           2359296    1998848 COMPLETED
P_SUPPLIER      SYS_P667                     2359296    8241152 COMPLETED

SQL> SELECT INMEMORY FROM user_tables WHERE table_name = 'TMP_SUPPLIER';

INMEMORY
--------
DISABLED
```

第10章

Database In-Memory 与大数据

10.1 Database In-Memory 与外部表

10.1.1 外部表基本概念

外部表是 Oracle 9i 版本引入的功能，使得用户可以用访问内部表相同的方式来访问外部数据源。外部表最初是作为 SQL Loader 实用程序的补充，随着支持外部数据源类型不断丰富，包括二进制的 Data Pump 文件、HIVE、HDFS 和对象存储等，外部表成为拓展数据库外沿的重要手段，使得 Oracle 在以关系型数据为核心的基础上延伸到丰富多元的大数据领域。

创建外部表有两种方式，第一种是基于已有的外部文件，此时需要在建表语句中指定参数去对应外部文件的格式；第二种方式是基于数据库中已经存在的表，具体实现上有两种方法。

先来看第一种方式。先通过 SQLcl 中的 unload 命令生成示例外部表文件，数据来自于 Oracle 标准示例 SH 用户下的 SALES 表。其中 set loadformmat 可以设置丰富的文件格式和参数定义，默认文件格式为 CSV。

```
$ sql ssb@orclpdb1
SQL> SET LOADFORMAT COLUMNNAMES OFF
SQL> UNLOAD sh.sales dir /home/oracle/dbimbook/chap10

format csv

column_names off
delimiter ,
enclosure_left "
enclosure_right "
encoding UTF8
row_terminator default

** UNLOAD Start ** at 2021.07.12-10.31.25
Export Separate Files to /home/oracle/dbimbook/chap10
```

```
DATA TABLE SALES
File Name: /home/oracle/dbimbook/chap10/SALES_DATA_TABLE.csv
Number of Rows Exported: 918843
** UNLOAD End ** at 2021.07.12-10.31.36
```

输出文件为 SALES_DATA_TABLE.csv，位于目录/home/oracle/dbimbook/chap10。在数据库中创建对应的目录对象，然后根据外部文件格式和文件名得到以下的外部表建表语句：

```
CREATE DIRECTORY default_dir AS '/home/oracle/dbimbook/chap10';

CREATE TABLE sales_ext(
    "PROD_ID"          NUMBER,
    "CUST_ID"          NUMBER,
    "TIME_ID"          DATE,
    "CHANNEL_ID"       NUMBER,
    "PROMO_ID"         NUMBER,
    "QUANTITY_SOLD"    NUMBER(10, 2),
    "AMOUNT_SOLD"      NUMBER(10, 2)
)
ORGANIZATION EXTERNAL(
    TYPE ORACLE_LOADER
    DEFAULT DIRECTORY default_dir
    ACCESS PARAMETERS()
    LOCATION('SALES_DATA_TABLE.csv')
);
```

其中 ORGANIZATION EXTERNAL 表明这是一个外部表，TYPE 和 ACCESS PARAMETERS 指定了外部表访问驱动类型和驱动中的参数，如字段分隔符、字段包围符、是否忽略表头等。LOCATION 和 DEFAULT DIRECTORY 指定了外部文件名和所在的目录。需要注意，即使外部文件不存在或文件名不正确，上述表定义也可以成功。因为外部表只有在被访问时，才会真正使用访问驱动去读取外部文件。

第二种方式可以使用 Oracle 10g 版本新增的外部表访问驱动：ORACLE_DATAPUMP。这种方法结合 CTAS，可以在一个 DDL 语句中同时生成外部表以及外部表文件。其更强大之处在于，可以将复杂的多表联结 SQL 语句的查询结果存为外部表，后续可以将此结果导入到其他数据库。

```
CREATE TABLE sales_ext
ORGANIZATION EXTERNAL
(
  TYPE ORACLE_DATAPUMP
  DEFAULT DIRECTORY default_dir
  LOCATION('sales_ext.dmp')
)
AS SELECT * FROM sh.sales;
```

Oracle 数据库在 12.2.0.1 版本对外部表特性做了大幅改进。首先是新增了 ORACLE_HDFS 和 ORACLE_HIVE 两个外部表访问驱动，从而可以访问存储在 Hadoop 上的文件，包括 HDFS 和 Apache Hive。另一个重要的改进是支持对外部表进行分区，这不仅可以提升外部表查询的性能，同时也可以简化外部表数据的维护。

Oracle 19c 版本新增了外部表访问驱动 ORACLE_BIGDATA,从而可以访问存储在对象存储上的数据,包括 Oracle 对象存储、Amazon S3 和 Azure Blob Storage。从 Oracle 19c 版本开始,也支持将内部分区和外部分区集成到一个分区表中,这项功能称为混合分区表。混合分区表进一步深化了 Oracle 数据生命周期管理,可以将大量访问较少的数据存在外部文件或对象存储,而将频繁存取的数据存于数据库内部。

Oracle Database In-Memory 从 18c 版本开始支持外部表,也就是 In-Memory 外部表功能。最初支持 ORACLE_LOADER 和 ORACLE_DATAPUMP 两个外部表访问驱动,并且只能通过 PL/SQL 过程实现发布。In-Memory 外部表在 Oracle 19c 版本新增了对 ORACLE_HIVE 和 ORACLE_BIGDATA 驱动的支持,并且支持 RAC、ADG 配置,支持并行查询和全表扫描引发发布。从 Oracle 21c 版本开始,In-Memory 外部表支持分区和混合分区,在之前的版本中,只有内部分区才能设置 INMEMORY 属性。In-Memory 外部表最初仅在 Exadata 上支持,从 Oracle 21c 版本开始,在非 Exadata 环境也可以支持此特性。

10.1.2　In-Memory 普通外部表

在 Oracle 18c 版本之前,作为一种临时解决方案,外部表可以通过与物化视图结合发布到内存列式存储。下面通过 Oracle 标准示例中 SH 用户下的外部表 SALES_TRANSACTIONS_EXT 来演示其实现过程。首先基于外部表创建物化视图,其 DDL 语句如下:

```
CREATE MATERIALIZED VIEW mv_sales_transactions_ext
    BUILD IMMEDIATE
    REFRESH
        COMPLETE
        ON DEMAND
AS
    SELECT
        *
    FROM
        sh.sales_transactions_ext;
```

为物化视图开启 INMEMORY 属性,然后通过全表扫描发布:

```
SQL> ALTER MATERIALIZED VIEW mv_sales_transactions_ext INMEMORY;

SQL> SELECT /*+ full(p) */ COUNT(*) FROM mv_sales_transactions_ext p;
```

其发布情况和执行计划如下,可以看到物化视图的段类型实际上就是 TABLE。执行计划中出现了 MAT_VIEW ACCESS INMEMORY FULL 关键字,说明外部表间接通过物化视图发布到了内存列式存储中:

```
SQL> SELECT segment_name, segment_type, inmemory_size, bytes, populate_status FROM
v$im_segments WHERE segment_name = 'MV_SALES_TRANSACTIONS_EXT';

SEGMENT_NAME                    SEGMENT_TYPE  INMEMORY_SIZE      BYTES POPULATE_STAT
------------------------------  ------------  -------------  --------- --------------
MV_SALES_TRANSACTIONS_EXT       TABLE              10944512   44793856 COMPLETED
```

```
SQL> @../showplan

PLAN_TABLE_OUTPUT
----------------------------------------
SQL_ID  9d29a2n31uzs4, child number 0
----------------------------------------
SELECT /*+ full(p) */ COUNT(*) FROM mv_sales_transactions_ext p

Plan hash value: 364821825

--------------------------------------------------------------------------------
| Id  | Operation                     | Name                      | Rows  |
--------------------------------------------------------------------------------
|   0 | SELECT STATEMENT              |                           |       |
|   1 |  SORT AGGREGATE               |                           |     1 |
|   2 |   MAT_VIEW ACCESS INMEMORY FULL| MV_SALES_TRANSACTIONS_EXT |  916K |
--------------------------------------------------------------------------------

14 rows selected.
```

In-Memory 外部表功能要求初始化参数 QUERY_REWRITE_INTEGRITY 设置为 STALE_TOLERATED，默认值为 ENFORCED。此参数最初用于物化视图查询改写，也就是将原本查询表的 SQL 语句改写为查询物化视图。In-Memory 外部表和物化视图类似，当发布到内存列式存储中后，从数据库内部是不允许修改的，但不能防止外部数据文件被修改。因此有可能出现外部数据文件和内存列式存储中内容不一致的情况，将 QUERY_REWRITE_INTEGRITY 设置为 STALE_TOLERATED 即表示容忍这种不一致的情况发生。不过，当外部文件确实发生改变后，应重新发布外部表以保持数据一致性。

以下为设置初始化参数 QUERY_REWRITE_INTEGRITY 的过程：

```
SQL> SHOW PARAMETER query_rewrite_integrity

NAME                                 TYPE        VALUE
------------------------------------ ----------- ------------------------------
query_rewrite_integrity              string      enforced

SQL> ALTER SESSION SET query_rewrite_integrity = stale_tolerated;

Session altered.
```

以 SH 用户下的 SALES_TRANSACTIONS_EXT 外部表为例，为其开启 INMEMORY 属性。在系统视图可以查询到对应的信息，其中列 TYPE_NAME 表示外部表访问驱动，本例为 ORACLE_LOADER：

```
SQL> ALTER TABLE sales_transactions_ext INMEMORY MEMCOMPRESS FOR CAPACITY HIGH;
Table altered.

SQL> SELECT owner, table_name, type_name, default_directory_name, inmemory FROM
all_external_tables;

OWNER    TABLE_NAME              TYPE_NAME        DEFAULT_DIRECTORY_NAME   INMEMORY
```

```
------ ------------------------ ---------------- ------------------------ --------
SH     SALES_TRANSACTIONS_EXT   ORACLE_LOADER    DATA_FILE_DIR            ENABLED
```

在 Oracle 18c 版本时，In-Memory 外部表的发布只能通过调用 PL/SQL 过程触发。在 Oracle 19c 版本，通过全表扫描也可以引发发布，但前提是 QUERY_REWRITE_INTEGRITY 初始化参数必须设置为 STALE_TOLERATED。

```
SQL> EXEC dbms_inmemory.populate('SH','SALES_TRANSACTIONS_EXT');

/* 需要 Oracle 19c 才能支持
SQL> SELECT /*+ FULL(p) */ COUNT(*) FROM sales_transactions_ext p;
*/

SQL> @./dbim_get_popstatus.sh

              SEGMENT               IN MEM        ON DISK   COMPRESSION    BYTES NOT  POPULATED
OWNER         NAME                  SIZE(KB)      SIZE(KB)        RATIO    POPULATED  STATUS
------        ------------          --------      --------   -----------   ---------  -------------
SH            SALES_TRANSA           9,344              0             0            0  COMPLETED
              CTIONS_EXT


TOTAL_DISK_SIZE(KB)  TOTAL_IM_SIZE(KB)  OVERALL_COMPRESSION_RATIO
-------------------  -----------------  -------------------------
                  0               9344                          0
```

从系统视图中可以确认表已成功发布，注意其中的 IS_EXTERNAL 属性为 TRUE。

```
SQL> SELECT segment_type, inmemory_size, inmemory_compression, bytes, populate_status,
is_external FROM v$im_segments where segment_name = 'SALES_TRANSACTIONS_EXT';

SEGMENT_TYPE     INMEMORY_SIZE  INMEMORY_COMPRESS          BYTES  POPULATE_STAT  IS_EXTERNAL
-------------    -------------  -----------------     ----------  -------------  -----------
TABLE                  9568256  FOR QUERY LOW                  0  COMPLETED      TRUE
```

通过对比执行计划可以看出，只有当初始化参数 QUERY_REWRITE_INTEGRITY 设置为 STALE_TOLERATED 时，系统才会将外部表发布到内存列式存储中。

```
SQL> SHOW PARAMETER query_rewrite_integrity

NAME                                 TYPE         VALUE
------------------------------------ ------------ ------------------------------
query_rewrite_integrity              string       enforced

SQL> SELECT /*+ FULL(p) */ count(*) FROM sales_transactions_ext  p;

  COUNT(*)
----------
    916039


SQL> @../showplan

PLAN_TABLE_OUTPUT
-----------------------------------------
```

```
SQL_ID  gxpagcabazwmz, child number 0
-------------------------------------
SELECT /*+ FULL(p) */ count(*) FROM sales_transactions_ext  p

Plan hash value: 1947249939

---------------------------------------------------------------------
| Id  | Operation                   | Name                  | Rows  |
---------------------------------------------------------------------
|   0 | SELECT STATEMENT            |                       |       |
|   1 |  SORT AGGREGATE             |                       |     1 |
|   2 |   EXTERNAL TABLE ACCESS FULL| SALES_TRANSACTIONS_EXT |  916K|
---------------------------------------------------------------------

14 rows selected.

SQL> ALTER SESSION SET query_rewrite_integrity = stale_tolerated;

Session altered.

SQL> SELECT /*+ FULL(p) */ count(*) FROM sales_transactions_ext p;

  COUNT(*)
----------
    916039

SQL> @../showplan

PLAN_TABLE_OUTPUT
-------------------------------------
SQL_ID  gxpagcabazwmz, child number 1
-------------------------------------
SELECT /*+ FULL(p) */ count(*) FROM sales_transactions_ext p

Plan hash value: 1947249939

------------------------------------------------------------------------------
| Id  | Operation                            | Name                  | Rows  |
------------------------------------------------------------------------------
|   0 | SELECT STATEMENT                     |                       |       |
|   1 |  SORT AGGREGATE                      |                       |     1 |
|   2 |   EXTERNAL TABLE ACCESS INMEMORY FULL| SALES_TRANSACTIONS_EXT |  916K|
------------------------------------------------------------------------------

14 rows selected.
```

在 SALES_TRANSACTIONS_EXT 表已经发布到内存列式存储的前提下，在接下来的示例中修改外部表文件，可以了解初始化参数 QUERY_REWRITE_INTEGRITY 对执行计划和查询结果的影响。

首先获取外部文件的文件名及所在目录：

```
SQL> SELECT * FROM user_external_locations;

TABLE_NAME                  LOCATION        DIRECTORY_OWNER   DIRECTORY_NAME
--------------------------  --------------  ----------------  -----------------
SALES_TRANSACTIONS_EXT      sale1v3.dat     SYS               DATA_FILE_DIR

SQL> SELECT directory_path FROM all_directories WHERE directory_name = 'DATA_FILE_DIR';

DIRECTORY_PATH
------------------------------------------------------------------------------
/home/oracle/db-sample-schemas/sales_history/
```

将外部数据文件备份，以便后续恢复。然后将外部数据文件放大一倍：

```
# 备份文件
$ cd /home/oracle/db-sample-schemas/sales_history
$ cp sale1v3.dat sale1v3.dat.orig
$ wc -l sale1v3.dat
916039 sale1v3.dat

# 利用备份文件将外部表文件放大一倍
$ cat sale1v3.dat.orig >> sale1v3.dat
$ wc -l sale1v3.dat
1832078 sale1v3.dat
```

此时出现了一个有趣的现象。当 QUERY_REWRITE_INTEGRITY 设置为 ENFORCED 时，查询结果为 1 832 078 行，与外部文件最新内容保持一致。而当其设置为 STALE_TOLERATED 时，查询结果为 916 039，反映的是数据文件放大前的状态，也就是之前陈旧的信息。通过查询执行计划，可以发现不同的 QUERY_REWRITE_INTEGRITY 设置会影响查询优化器在行式存储和内存列式存储中做出选择。

```
SQL> SHOW PARAMETER query_rewrite_integrity

NAME                                 TYPE         VALUE
-----------------------------------  -----------  --------
query_rewrite_integrity              string       enforced

SQL> SELECT COUNT(*) FROM sales_transactions_ext;

  COUNT(*)
----------
   1832078

SQL> @showplan

PLAN_TABLE_OUTPUT
-----------------------------------
SQL_ID  93x43r9m87tnp, child number 0
-----------------------------------
```

```
SELECT COUNT(*) FROM sales_transactions_ext

Plan hash value: 1947249939

---------------------------------------------------------------------
| Id | Operation                   | Name                  | Rows |
---------------------------------------------------------------------
|  0 | SELECT STATEMENT            |                       |      |
|  1 |  SORT AGGREGATE             |                       |   1  |
|  2 |   EXTERNAL TABLE ACCESS FULL| SALES_TRANSACTIONS_EXT| 916K |
---------------------------------------------------------------------

14 rows selected.

SQL> ALTER SESSION SET query_rewrite_integrity = stale_tolerated;

Session altered.

SQL> SELECT COUNT(*) FROM sales_transactions_ext;

  COUNT(*)
----------
    916039

SQL> @showplan

PLAN_TABLE_OUTPUT
-------------------------------------
SQL_ID  93x43r9m87tnp, child number 1
-------------------------------------
SELECT COUNT(*) FROM sales_transactions_ext

Plan hash value: 1947249939

------------------------------------------------------------------------------
| Id | Operation                           | Name                   | Rows |
------------------------------------------------------------------------------
|  0 | SELECT STATEMENT                    |                        |      |
|  1 |  SORT AGGREGATE                     |                        |   1  |
|  2 |   EXTERNAL TABLE ACCESS INMEMORY FULL| SALES_TRANSACTIONS_EXT| 916K |
------------------------------------------------------------------------------

14 rows selected.
```
所以当外部文件发生变化时，应重新发布外部表以使数据保持一致，否则会出现不一致的查询结果：
```
SQL> EXEC dbms_inmemory.populate('SH','SALES_TRANSACTIONS_EXT');

PL/SQL procedure successfully completed.

SQL> SELECT COUNT(*) FROM sales_transactions_ext;
```

```
  COUNT(*)
----------
   1832078

SQL> @showplan

...
---------------------------------------------------------------------------
| Id  | Operation                         | Name                  | Rows  |
---------------------------------------------------------------------------
|   0 | SELECT STATEMENT                  |                       |       |
|   1 |  SORT AGGREGATE                   |                       |     1 |
|   2 |   EXTERNAL TABLE ACCESS INMEMORY FULL| SALES_TRANSACTIONS_EXT |  916K |
---------------------------------------------------------------------------
...
```

实验完成后，使用备份还原外部数据文件：

```
$ cp sale1v3.dat.orig sale1v3.dat
```

ORACLE_BIGDATA 外部表访问驱动依赖于 DBMS_CLOUD PL/SQL 包，在 Oracle 自治数据库中此包已安装和配置。对于传统的 Oracle 数据库，从 19c 版本中的 19.9 和 21c 版本的 21.3 开始，DBMS_CLOUD 已包含在数据库中，但需要额外的安装和配置，过程可参考 Oracle 支持文档 2748362.1[①]。

假设 Oracle 数据库已成功安装和配置 DBMS_CLOUD PL/SQL 包，可以有两种方式来建立使用 ORACLE_BIGDATA 驱动的外部表。使用传统方式的外部表定义语句如下：

```
CREATE TABLE customer_ext_bd_cloud(
    c_custkey       NUMBER,
    c_name          VARCHAR(25),
    c_address       VARCHAR(25),
    c_city          CHAR(10),
    c_nation        CHAR(15),
    c_region        CHAR(12),
    c_phone         CHAR(15),
    c_mktsegment    CHAR(10)
)
ORGANIZATION EXTERNAL
  (TYPE oracle_bigdata
   DEFAULT DIRECTORY default_dir
   ACCESS PARAMETERS
   (
   COM.ORACLE.bigdata.fileformat = textfile
   COM.ORACLE.bigdata.CSV.SKIP.header=0
   COM.ORACLE.bigdata.CSV.rowformat.quotecharacter='"'
   )
   LOCATION ('https://objectstorage.us-ashburn-1.oraclecloud.com/n/ocichina001/b/
```

---

① https://support.oracle.com/epmos/faces/DocContentDisplay?id=2748362.1

```
b01/o/customer.csv')
  ) REJECT LIMIT UNLIMITED;
```
也可以使用新型的 DBMS_CLOUD PL/SQL 包进行定义,例如:
```
BEGIN
    DBMS_CLOUD.CREATE_EXTERNAL_TABLE(
    TABLE_NAME => 'customer_ext_bd_cloud',
    FILE_URI_LIST => 'https://objectstorage.us-ashburn-1.oraclecloud.com/n/ocichina001/b/b01/o/customer.csv',
    FORMAT => JSON_OBJECT('type' VALUE 'csv', 'skipheaders' VALUE '0'),
    COLUMN_LIST => 'c_custkey       NUMBER,
    c_name          VARCHAR(25),
    c_address       VARCHAR(25),
    c_city          CHAR(10),
    c_nation        CHAR(15),
    c_region        CHAR(12),
    c_phone         CHAR(15),
    c_mktsegment    CHAR(10)'
    );
END;
/
```
在以上两种外部表定义中,LOCATION 和 FILE_URI_LIST 指定的对象存储位置不能是任意形式的 URI,必须满足一定的格式要求。目前支持以下 3 种格式之一,在以上外部表定义中使用的是第一种格式:
```
-- Oracle Cloud Infrastructure 原生 URI 格式
https://objectstorage.云区域.oraclecloud.com/n/命名空间/b/存储桶/o/文件名
-- Swift URI 格式
https://swiftobjectstorage.云区域.oraclecloud.com/v1/命名空间/存储桶/文件名
-- 预认证请求(PAR) URI 格式:
https://objectstorage.云区域.oraclecloud.com/p/加密串/n/命名空间/b/存储桶/o/文件名 e
```
Oracle 也支持使用 ORACLE_BIGDATA 访问驱动来引用存放在本地的文件,这可以用来测试驱动是否配置正确。例如:
```
CREATE TABLE customer_ext_bd_local(
    c_custkey       NUMBER,
    c_name          VARCHAR(25),
    c_address       VARCHAR(25),
    c_city          CHAR(10),
    c_nation        CHAR(15),
    c_region        CHAR(12),
    c_phone         CHAR(15),
    c_mktsegment    CHAR(10)
)
ORGANIZATION EXTERNAL
  (TYPE oracle_bigdata
   DEFAULT DIRECTORY default_dir
   ACCESS PARAMETERS
   (
```

```
    COM.ORACLE.bigdata.fileformat = textfile
    COM.ORACLE.bigdata.CSV.SKIP.header=0
    COM.ORACLE.bigdata.CSV.rowformat.quotecharacter='"'
  )
  LOCATION('customer.csv')
) REJECT LIMIT UNLIMITED;
```

为这两张表启用 INMEMORY 属性并发布到内存列式存储。注意，外部表可以额外设置内存压缩级，但不支持设置发布优先级：

```
ALTER TABLE customer_ext_bd_cloud INMEMORY;
ALTER TABLE customer_ext_bd_local INMEMORY;
EXEC dbms_inmemory.populate('SSB', 'CUSTOMER_EXT_BD_CLOUD');
EXEC dbms_inmemory.populate('SSB', 'CUSTOMER_EXT_BD_LOCAL');
```

查询系统视图，确认已成功发布。由于是外部表，可以看到本地磁盘空间占用为 0。

```
SQL> SELECT owner, table_name, type_name, default_directory_name, inmemory FROM
all_external_tables;

OWNER      TABLE_NAME              TYPE_NAME       DEFAULT_DIRECTORY_NAME   INMEMORY
---------- ----------------------- --------------- ------------------------ ----------
SSB        CUSTOMER_EXT_BD_LOCAL   ORACLE_BIGDATA  DEFAULT_DIR              ENABLED
SSB        CUSTOMER_EXT_BD_CLOUD   ORACLE_BIGDATA  DEFAULT_DIR              ENABLED

SQL> SELECT segment_name, segment_type, inmemory_size, bytes, populate_status FROM
v$im_segments;

SEGMENT_NAME            SEGMENT_TYPE    INMEMORY_SIZE    BYTES  POPULATE_STAT
----------------------- --------------- --------------- ------- -------------
CUSTOMER_EXT_BD_LOCAL   TABLE                  1179648       0  COMPLETED
CUSTOMER_EXT_BD_CLOUD   TABLE                  1179648       0  COMPLETED
```

由于外部表位于云端的对象存储，发布到本地内存列式存储后，访问模式由远程变为本地，访问速度也会相应提升。另外，和之前的测试类似，只有当初始化参数 QUERY_REWRITE_INTEGRITY 设置为 STALE_TOLERATED 时，外部表才可以被发布到内存列式存储中：

```
SQL> SHOW PARAMETER query_rewrite_integrity

NAME                                  TYPE         VALUE
------------------------------------- ------------ -----------------------
query_rewrite_integrity               string       enforced

SQL> SELECT /*+ full(p) */ COUNT(*) FROM customer_ext_bd_cloud p;

  COUNT(*)
----------
       100
Elapsed: 00:00:02.26

SQL> @../showplan

--------------------------------------------------------------------------------
```

```
| Id  | Operation                       | Name                 | Rows |
-----------------------------------------------------------------------
|   0 | SELECT STATEMENT                |                      |      |
|   1 |  SORT AGGREGATE                 |                      |    1 |
|*  2 |   EXTERNAL TABLE ACCESS STORAGE FULL| CUSTOMER_EXT_BD_CLOUD |  164 |
-----------------------------------------------------------------------

Predicate Information (identified by operation id):
---------------------------------------------------

   2 - filter(SYS_OP_VECTOR_GROUP_BY(NULL,COUNT(*)))

SQL> ALTER SESSION SET query_rewrite_integrity = stale_tolerated;

SQL> SELECT /*+ full(p) */ COUNT(*) FROM customer_ext_bd_cloud p;

  COUNT(*)
----------
       100
Elapsed: 00:00:01.09

SQL> @../showplan


| Id  | Operation                       | Name                 | Rows |
-----------------------------------------------------------------------
|   0 | SELECT STATEMENT                |                      |      |
|   1 |  SORT AGGREGATE                 |                      |    1 |
|   2 |   EXTERNAL TABLE ACCESS INMEMORY FULL| CUSTOMER_EXT_BD_CLOUD |  164 |
-----------------------------------------------------------------------
```

10.1.3 In-Memory 分区外部表

Oracle 数据库从 12.2 版本开始支持分区外部表，即外部表可以由多个分区组成。从 Oracle 21c 版本开始，分区外部表支持发布到内存列式存储，此功能也称为 In-Memory 分区外部表。

通过一个示例来说明如何创建分区外部表，以及如何使用 In-Memory 分区外部表特性。在随书示例目录下，已经包含两个 CSV 文件，分别包含了 2020 年和 2021 年的数据：

```
$ cd ~/dbimbook/chap10
$ ls -l *.csv
-rw-r--r--. 1 oracle oinstall 1292 Jul 11 18:41 2020.csv
-rw-r--r--. 1 oracle oinstall 1400 Jul 11 18:48 2021.csv
```

其中，2020.csv 文件通过在 SQLcl 中运行以下脚本生成：

```
CREATE TABLE t1(id int, t date);

BEGIN
    FOR i IN 1..100 LOOP
        INSERT INTO t1
            SELECT
```

```
                i,
                to_date(trunc(dbms_random.value(to_char(DATE  '2020-01-01',  'J'),
to_char(DATE '2020-12-31', 'J'))), 'J')
            FROM
                dual;

    END LOOP;

    COMMIT;
END;
/

set loadformat columnnames off
unload t1 dir .

!mv T1_DATA_TABLE.csv 2020.csv
```

与之类似，文件 2021.csv 通过在 SQLcl 中运行以下脚本生成。和上一个脚本的区别在于日期和序号范围不同：

```
CREATE TABLE t1(id int, t date);

BEGIN
    FOR i IN 101..200 LOOP
        INSERT INTO t1
            SELECT
                i,
                to_date(trunc(dbms_random.value(to_char(DATE  '2021-01-01',  'J'),
to_char(DATE '2021-12-31', 'J'))), 'J')
            FROM
                dual;

    END LOOP;

    COMMIT;
END;
/

set loadformat columnnames off
unload t1 dir .

!mv T1_DATA_TABLE.csv 2021.csv
```

将 2 个文件移动到 /tmp 目录：

```
$ mv 2020.csv 2021.csv /tmp
```

以下为创建分区外部表的 DDL 语句，注意已经为表设置了 INMEMORY 属性：

```
CREATE DIRECTORY default_dir AS '/tmp';

CREATE TABLE test_ext_part(
    id   NUMBER,
    t    DATE
```

```
)
ORGANIZATION EXTERNAL(
    TYPE ORACLE_LOADER
    DEFAULT DIRECTORY default_dir
)
PARTITION BY RANGE(t)
(
    PARTITION p2020 VALUES LESS THAN (TO_DATE('01-Jan-2021','dd-MON-yyyy'))LOCATION
('2020.csv'),
    PARTITION p2021 VALUES LESS THAN (TO_DATE('01-Jan-2022','dd-MON-yyyy'))LOCATION
('2021.csv')
) INMEMORY;
```

以上分区外部表包括 p2020 和 p2021 共 2 个分区。外部分区不允许修改，因此执行以下 DML 语句会报错：

```
SQL> INSERT INTO test_ext_part VALUES(
    10000,
    TO_DATE('01-Feb-2021', 'dd-MON-yyyy')
);

INSERT INTO test_ext_part VALUES(
            *
ERROR at line 1:
ORA-30657: operation not supported on external organized table
```

从系统视图中可以查询到外部表和分区的信息，输出中的 EXTERNAL 为 YES 表明这是一个外部表：

```
SQL> SELECT external, hybrid FROM user_tables WHERE table_name = 'TEST_EXT_PART';

EXTERNAL   HYBRID
---------- ------
YES        NO

SQL>    SELECT    table_name,    type_name,    default_directory_name    FROM
all_xternal_part_tables;

TABLE_NAME              TYPE_NAME              DEFAULT_DIRECTORY_NAME
------------------      ------------------     ----------------------
TEST_EXT_PART           ORACLE_LOADER          DEFAULT_DIR

SQL>   SELECT   table_name,   partition_name,   partition_position,   high_value   FROM
user_tab_partitions;

TABLE_NAME         PARTITION_NAME PARTITION_POSITION HIGH_VALUE
------------------ -------------- ------------------ ------------------------
TEST_EXT_PART      P2020                           1 TO_DATE(' 2021-01-01 00…
TEST_EXT_PART      P2021                           2 TO_DATE(' 2022-01-01 00…
```

通过全表扫描将表发布到内存列式存储，在执行计划中可以看到 EXTERNAL TABLE ACCESS INMEMORY FULL 关键字：

```
SQL> ALTER SESSION SET query_rewrite_integrity = stale_tolerated;

Session altered.

SQL> SELECT /*+ full(p) */ COUNT(*) FROM test_ext_part p;

  COUNT(*)
----------
       200

SQL> @../showplan

---------------------------------------------------------------------------------
| Id | Operation                         | Name          | Pstart| Pstop |
---------------------------------------------------------------------------------
|  0 | SELECT STATEMENT                  |               |       |       |
|  1 |  SORT AGGREGATE                   |               |       |       |
|  2 |   PARTITION RANGE ALL             |               |   1   |   2   |
|  3 |    EXTERNAL TABLE ACCESS INMEMORY FULL| TEST_EXT_PART |   1   |   2   |
---------------------------------------------------------------------------------
```

在系统视图中可以查询到发布成功的信息:

```
SQL> SELECT segment_name, partition_name, populate_status, is_external FROM
v$im_segments;

SEGMENT_NAME           PARTITION_NAME      POPULATE_STAT  IS_EX
--------------------   -----------------   -------------  -----
TEST_EXT_PART          P2020               COMPLETED      TRUE
TEST_EXT_PART          P2021               COMPLETED      TRUE
```

10.1.4 In-Memory 混合分区表

Oracle 数据库从 19c 版本开始支持混合分区表,也就是表的分区同时包含来自数据库内部和外部的分区。从 Oracle 21c 版本开始,混合分区表支持发布到内存列式存储中,此特性也称为 In-Memory 混合分区表。

通过一个示例说明如何创建混合分区表,以及如何使用 In-Memory 混合分区表特性。

仍沿用 10.1.3 节中的数据库目录和外部数据文件,创建混合分区表的 DDL 语句如下:

```
CREATE TABLE test_hypt_part(
    id  NUMBER,
    t   DATE
)
EXTERNAL PARTITION ATTRIBUTES(
TYPE ORACLE_LOADER
DEFAULT DIRECTORY default_dir
)
PARTITION BY RANGE(t)
(
PARTITION p2020 VALUES LESS THAN(TO_DATE('01-Jan-2021','dd-MON-yyyy'))EXTERNAL
```

```
LOCATION('2020.csv'),
PARTITION p2021 VALUES LESS THAN(TO_DATE('01-Jan-2022','dd-MON-yyyy'))EXTERNAL
LOCATION('2021.csv'),
PARTITION pmax VALUES LESS THAN(MAXVALUE)
) INMEMORY;
```

以上创建的混合分区表共有 3 个分区,其中 pmax 为内部分区,p2020 和 p2021 为外部分区。虽然外部分区不允许修改,但以下插入的数据只涉及到内部分区,因此可以执行成功:

```
INSERT INTO test_hypt_part VALUES(
    10000,
    TO_DATE('01-Jan-2028', 'dd-MON-yyyy')
);

UPDATE test_hypt_part
SET
    t = TO_DATE('01-Jan-2029', 'dd-MON-yyyy')
WHERE
    id = 10000;
```

从系统视图中可以查询到外部表和分区的信息,输出中的 HYBRID 为 YES 表明这是一个混合分区表:

```
SQL> SELECT external, hybrid FROM user_tables WHERE table_name = 'TEST_HYPT_PART';

EXTERNAL   HYBRID
---------- ----------
NO         YES

SQL> SELECT table_name, type_name, default_directory_name FROM all_xternal_part_
tables;

TABLE_NAME          TYPE_NAME              DEFAULT_DIRECTORY_NAME
---------------     --------------------   ------------------------------
TEST_HYPT_PART      ORACLE_LOADER          DEFAULT_DIR

TABLE_NAME        PARTITION_NAME PARTITION_POSITION HIGH_VALUE
---------------   -------------- ------------------ ------------------------
TEST_HYPT_PART    P2020                           1 TO_DATE(' 2021-01-01 00…
TEST_HYPT_PART    P2021                           2 TO_DATE(' 2022-01-01 00…
TEST_HYPT_PART    PMAX                            3 MAXVALUE
```

通过全表扫描将表发布到内存列式存储,在执行计划中可以看到 HYBRID PART INMEMORY FULL 关键字:

```
SQL> ALTER SESSION SET query_rewrite_integrity = stale_tolerated;

Session altered.

SQL> SELECT /*+ full(p) */ COUNT(*) FROM test_hypt_part p;

  COUNT(*)
----------
       201
```

```
SQL> @../showplan

---------------------------------------------------------------------------------
| Id  | Operation                             | Name           | Pstart| Pstop |
---------------------------------------------------------------------------------
|  0  | SELECT STATEMENT                      |                |       |       |
|  1  |  SORT AGGREGATE                       |                |       |       |
|  2  |   PARTITION RANGE ALL                 |                |   1   |   3   |
|  3  |    TABLE ACCESS HYBRID PART INMEMORY FULL| TEST_HYPT_PART |   1   |   3   |
|  4  |    TABLE ACCESS INMEMORY FULL         | TEST_HYPT_PART |   1   |   3   |
---------------------------------------------------------------------------------
```

在系统视图中可以查询到发布成功的信息：

```
SELECT segment_name, partition_name, populate_status, is_external FROM v$im_segments;

SEGMENT_NAME              PARTITION_NAME          POPULATE_STAT  IS_EX
------------------------  ----------------------  -------------  -----
TEST_EXT_PART             P2020                   COMPLETED      TRUE
TEST_EXT_PART             P2021                   COMPLETED      TRUE
```

10.2 内存优化行存储

物联网是指由网络联结到一起，并相互传递信息的系统。这其中的"物"可以指人或设备。物联网和我们的生活关系密切，从可穿戴设备、智能电表到电子不停车收费、物流和车队管理等，都属于物联网应用的范畴。物联网和大数据也有着非常紧密的联系。特别是对于大型物联网络，联结的设备众多，数据产生频率很高，生成的数据量巨大。因此要求底层的数据处理系统具有快速的数据查询和数据摄入能力。

Oracle 在 18c 版本推出了内存优化行存储（Memoptimized Rowstore）功能，可以提供快速的数据查询。在随后的 19c 版本中，又补充了快速数据摄入的功能，可以加速物联网应用中海量数据摄入的速度和减轻后端数据库的压力。内存优化行存储最初仅在 Exadata 上支持，从 Oracle 21c 版本开始，在非 Exadata 环境也可以支持此特性。严格来说，内存优化行存储并不属于 Database In-Memory 的范畴，因为它们分别使用了不同的行格式和列格式。但由于它们同属于 Oracle 内存计算技术，并且都由 Database In-Memory 团队研发。从内存计算的角度和完整性考虑，将在本节对其进行简要介绍。

10.2.1 行存储快速查询

Oracle 数据库在 18c 版本推出了内存优化行存储，此时仅支持数据快速查询功能。在 19c 版本支持数据快速摄入特性后，为表示区别，此项功能更名为内存优化行存储快速查询，简称为行存储快速查询。

启用行存储快速查询的表只支持基于主键的查询，并且表不能被压缩。其最大的特点是需要将查询的数据加载到特定的内存区域，并且用内存中的哈希索引替代传统二叉树索引来实现查询提速。

这个特殊的内存区域称为内存优化池，是 SGA 的一部分，用来存储需要查询的数据和哈希索引。内存优化池通过初始化参数 MEMOPTIMIZE_POOL_SIZE 设置，此参数只能在 CDB 中设置，设置后需要重启数据库。

```
ALTER SYSTEM SET memoptimize_pool_size = 100M SCOPE=SPFILE;
SHUTDOWN IMMEDIATE
STARTUP
```

设置完成后，可以在系统视图 V$SGA_DYNAMIC_COMPONENTS 中查询确认其设置：

```
SQL> SELECT component, current_size FROM v$sga_dynamic_components WHERE component LIKE
'memopti%';

COMPONENT                                                        CURRENT_SIZE
---------------------------------------------------------------- ------------
memoptimize buffer cache                                             104857600

SQL> SHOW PARAMETER memoptimize

NAME                                 TYPE        VALUE
------------------------------------ ----------- ------------------------------
memoptimize_pool_size                big integer 100M
```

先建立一张测试表 IOT_READ，此表必须具有主键。在 11g 版本以后，表在初始创建时并不会在数据文件中生成实际的物理段。但是行存储快速查询功能要求表的物理段必须存在，因此在建表语句中加入 SEGMENT CREATION IMMEDIATE 子句以生成物理段。

```
CREATE TABLE iot_read(
    id      NUMBER PRIMARY KEY,
    value   CHAR(8)
)
SEGMENT CREATION IMMEDIATE
MEMOPTIMIZE FOR READ;
```

接下来生成 100 万条测试数据，然后搜集优化器统计信息：

```
BEGIN
    FOR i IN 1..1000000 LOOP
        INSERT INTO iot_read VALUES(
            i,
            dbms_random.string('L', 4)
        );

        IF MOD(i, 5000) = 0 THEN
            COMMIT;
        END IF;
    END LOOP;

    COMMIT;
END;
/
```

```
EXEC dbms_stats.gather_table_stats(ownname=>NULL, tabname=>'IOT_READ');
```

执行基于主键的查询。执行计划中的 READ OPTIM 关键字表明优化器使用了内存优化行存储。但在统计信息中的 consistent gets 却表明数据是从传统的内存缓存中获取的。memopt r lookups 和 memopt r misses 的值大于 0，表明在内存优化行存储中进行了查询，但并没有查询到相应的数据。

```
SQL> SELECT value FROM iot_read WHERE id = 10000;

VALUE
--------
yqzs

Explain Plan
---------------------------------------------------------------------------
| Id  | Operation                                   | Name        | Rows | Cost (%CPU)|
---------------------------------------------------------------------------
|   0 | SELECT STATEMENT                            |             |    1 |    3   (0)|
|   1 |  TABLE ACCESS BY INDEX ROWID READ OPTIM     | IOT_READ    |    1 |    3   (0)|
|*  2 |   INDEX UNIQUE SCAN READ OPTIM              | SYS_C007966 |    1 |    2   (0)|
---------------------------------------------------------------------------

Predicate Information(identified by operation id):
---------------------------------------------------

   2 - access("ID"=10000)

Statistics
----------------------------------------------------------
...
          4  consistent gets
          4  consistent gets examination
          4  consistent gets examination(fastpath)
          4  consistent gets from cache
...
          1  memopt r lookups
          1  memopt r misses
...
```

通过以下的 SQL 查询也可以得到相同的执行统计信息，说明在内存优化行存储中的查询失败：

```
SQL>
SELECT
    n.name,
    s.value
FROM
    v$sysstat   s,
    v$statname  n
WHERE
        n.statistic# = s.statistic#
```

```
        AND n.name LIKE 'memopt r%' AND s.value != 0;

NAME                                                           VALUE
-------------------------------------------------------------- ----------
memopt r lookups                                                   1
memopt r misses                                                    1
```

这是由于查询的对象尚未发布到内存优化池中，需要使用以下的 PL/SQL 过程进行发布：

```
SQL> EXEC dbms_memoptimize.populate('SSB', 'IOT_READ');

PL/SQL procedure successfully completed.
```

再次查询，这一次在执行统计信息中没有再出现 consistent gets，而且 memopt r hits 的值大于 0，表明查询命中，也就是行存储快速查询成功：

```
SQL> SELECT value FROM iot_read WHERE id = 10000;

VALUE
----------
yqzs

Explain Plan
--------------------------------------------------------------------------------
| Id  | Operation                          | Name       | Rows  | Cost (%CPU)|
--------------------------------------------------------------------------------
|   0 | SELECT STATEMENT                   |            |    1  |     3   (0)|
|   1 |  TABLE ACCESS BY INDEX ROWID READ OPTIM| IOT_READ |    1  |     3   (0)|
|*  2 |   INDEX UNIQUE SCAN READ OPTIM     | SYS_C007966|    1  |     2   (0)|
--------------------------------------------------------------------------------

Predicate Information (identified by operation id):
---------------------------------------------------

   2 - access("ID"=10000)

Statistics
----------------------------------------------------------
...
          1  memopt r hits
          1  memopt r lookups
         55  non-idle wait count
          2  opened cursors cumulative
          1  opened cursors current
          2  parse count (total)
          1  process last non-idle time
          1  sorts (memory)
       2010  sorts (rows)
         45  user calls
```

之前在视图 V$SGA_DYNAMIC_COMPONENTS 中，只能查询内存优化池的总大小。为估算内存池大小，我们需要知道数据库对象在其中占用的空间。由于内存优化池是 SGA 的一部分，因此可以通过 V$BH 视图进行查询：

```
SQL> SELECT object_id FROM user_objects WHERE object_name = 'IOT_READ';

OBJECT_ID
----------
     75284

SQL> SELECT status, COUNT(block#) FROM v$bh WHERE objd = 75284 GROUP BY ROLLUP(status);

STATUS     COUNT(BLOCK#)
---------- -------------
free                2638
xcur                2688
                    5326
```

最后，可以使用以下 DDL 语句禁用内存优化行存储：

```
ALTER TABLE iot_read NO MEMOPTIMIZE FOR READ;
```

通过将数据置于内存优化行存储，数据查询避免了缓存争用和 consistent gets 开销，并且内存中哈希索引可快速定位数据，从而可以提升数据查询的性能。

10.2.2 行存储快速摄入

在 Oracle 18c 版本行存储快速查询的基础上，内存优化行存储在 19c 版本中新增了行存储快速摄入功能。此特性非常适合于物联网应用中海量数据采集的场景，插入的数据首先被缓存到 SGA 的 Large Pool 中，后续再异步、批量地写入数据库。

需要特别指出，为了最大化数据摄入吞吐量，行存储快速摄入采用了与传统数据库事务处理完全不同的机制。在传统数据库事务处理中，一旦事务提交，数据将持久化并且不会丢失。而行存储快速摄入绕过了 Oracle 事务处理机制，因此 COMMIT 和 ROLLBACK 操作对其不起作用。

必须由应用来判断数据是否最终写到数据库，应用可以调用 Oracle 提供的 API 来进行确认。有两种情况可能导致数据不能成功写入数据库，首先是数据库或主机故障导致缓存在 Large Pool 中的数据丢失，其次是数据从 Large Pool 批量写入数据库时，可能未通过数据库约束检查而导致写入失败。作为最佳实践，应用应保留一份本地数据副本，在确认数据成功写入数据库后再删除本地副本，这样在失败时可以保留后续插入这些数据的可能性。

下面通过一个示例来了解如何使用行存储快速摄入特性，以及其带来的性能提升。

首先来看一下传统方式下的数据摄入速度。执行以下脚本，先建立表 IOT_INGEST，然后连续插入 100 万条数据，耗时约 16 秒。注意此示例采用了批量提交事务的方式，即每插入 5000 条数据后提交一次。批量提交是事务提交的最佳实践，常见的错误是在每次插入数据后提交事务，这会给数据库带来更多开销。如果采用逐一提交方式，插入耗时会增加到 44 秒。

```
SQL>
CREATE TABLE iot_ingest(
    id      NUMBER PRIMARY KEY,
```

```
    value   CHAR(8)
);

SET TIMING ON
BEGIN
    FOR i IN 1..1000000 LOOP
        INSERT INTO iot_ingest VALUES(
            i,
            'ABCDEFGH'
        );

        IF MOD(i, 5000) = 0 THEN
            COMMIT;
        END IF;
    END LOOP;

    COMMIT;
END;
/

Elapsed: 00:00:15.91
```

接下来测试行存储快速摄入功能。重新建立 IOT_INGEST 表，此表需要有主键，同时需要指定 MEMOPTIMIZE FOR WRITE 子句。在 INSERT 语句中需要加入 MEMOPTIMIZE_WRITE 提示，以确保其使用行存储快速摄入功能。整个过程执行时间为 5.48 秒，比之前批量提交方式提速近 3 倍。

```
SQL>
DROP TABLE iot_ingest PURGE;

CREATE TABLE iot_ingest(
    id      NUMBER PRIMARY KEY,
    value   CHAR(8)
)
SEGMENT CREATION IMMEDIATE
MEMOPTIMIZE FOR WRITE;

SET TIMING ON
BEGIN
    FOR i IN 1..1000000 LOOP
        INSERT /*+ MEMOPTIMIZE_WRITE */ INTO iot_ingest VALUES(
            i,
            'ABCDEFGH'
        );

    END LOOP;

END;
/

Elapsed: 00:00:05.48
```

位于 Large Pool 缓存区中的数据后续会自动写入数据库，也可以使用 PL/SQL 过程手动将缓冲区中的数据立即冲刷到数据库。

```
EXEC DBMS_MEMOPTIMIZE_ADMIN.WRITES_FLUSH;
```

通过 DBMS_MEMOPTIMIZE PL/SQL 包和 V$MEMOPTIMIZE_WRITE_AREA 系统视图，可以确认数据是否已经写入数据库。例如在以下输出中，NUM_WRITES 表示位于缓存区，等待后续写入数据库的 INSERT 数量，其值为 0 表示所有数据均已写入数据库。另外，GET_WRITE_HWM_SEQID 和 GET_APPLY_HWM_SEQID 过程的输出分别表示已经写入缓存区的序列号和已经写入数据库的序列号，如果后者大于前者，也表示所有数据均已写入数据库。

```
SQL> SELECT * FROM V$MEMOPTIMIZE_WRITE_AREA;

      TOTAL_SIZE         USED_SPACE         FREE_SPACE         NUM_WRITES
---------------    ---------------    ---------------    ---------------
     333,447,168             25,440        333,421,728                  0

SQL> SELECT DBMS_MEMOPTIMIZE.GET_WRITE_HWM_SEQID FROM dual;

GET_WRITE_HWM_SEQID
-------------------
     67,511,234,948

SQL> SELECT DBMS_MEMOPTIMIZE.GET_APPLY_HWM_SEQID FROM dual;

GET_APPLY_HWM_SEQID
-------------------
     67,511,238,833
```

需要注意，行存储快速摄入特性对插入的目标表有一定的要求。例如不支持加密和压缩，不支持触发器等，具体的限制可参见 Oracle 官方文档[①]。

对于物联网数据的快速摄入场景，Oracle 数据库提供了丰富的优化手段，行存储快速摄入特性只是其中一种。数据库优化也不仅仅是数据库单方面的事情，开发层面也有许多需要注意的地方。如前面提到的使用批量提交而非逐行提交，以及 Oracle 提供的数组绑定（Array Binding）特性。数组绑定特性可以将多个 SQL 语句一次性发往数据库执行，从而减少客户端与服务器间的往返次数，极大提升数据摄入的速度和吞吐量。以下为使用 Python 实现的数组绑定特性示例：

```
$ cat arraybinding.py
import cx_Oracle

conn = cx_Oracle.connect("ssb", "Welcome1", "localhost/orclpdb1")

sqlstmt = 'insert into iot_ingest values(:id, :value)';
```

[①] https://docs.oracle.com/en/database/oracle/oracle-database/19/tgdba/tuning-system-global-area.html#GUID-CFADC9EA-2E2F-4EBB-BA2C-3663291DCC25

```
cursor = conn.cursor()

cursor.prepare(sqlstmt)

data_array = []

for i in range(1, 1000000 + 1):
        data_array.append((i, 'ABCDEFGH'))
        if i%5000 == 0:
                cursor.executemany(sqlstmt, data_array)
                conn.commit()
                data_array = []

cursor.executemany(sqlstmt, data_array)
conn.commit()

cursor.close()
conn.close()
```

使用与之前相同的测试表和摄入负载，使用数组绑定时的插入速度比逐行提交快 150 倍，比使用批量提交快 63 倍。

其他优化手段还包括分区、直接路径加载和并行执行等，具体可参考 Oracle 官方白皮书：使用 Oracle Database 19c 实现高容量物联网工作负载的最佳实践[1]。

10.3 Exadata In-Memory 列格式支持

Oracle Exadata 是专门用于运行 Oracle 数据库的一体化工程整合平台。第一代诞生于 2008 年，迄今为止已发展到第十代，最新的型号为 X9M。Exadata 非常适合于运行大型的数据仓库和联机事务处理系统，以及作为数据库整合平台来对数据库统一管理和控制，最终实现数据库环境的标准化和简化。

Exadata 由计算节点和存储节点组成，每一类节点都可以单独在线横向扩展。除了在计算节点安装数据库外，Exadata 在存储节点上安装了 Exadata 系统软件，这也是与其他数据库一体化平台最为本质的区别。Exadata 系统软件中的典型功能包括智能扫描、智能闪存缓存和混合列压缩。由于 Exadata 系统软件可以识别存储节点中储存的数据库数据，因此原先必须在计算层进行的部分操作可以下沉到存储层处理，从而可以将尽可能少的数据返回计算层，避免 I/O 成为数据库性能的瓶颈。在第 6 章介绍了操作下推的概念，Exadata 是较早将这种优化手段付诸实施的系统。这种软硬一体且相互配合的架构也是 Exadata 被称为工程整合系统的原因。

和 Oracle 数据库一样，Exadata 系统软件也有自己的版本。Exadata 系统软件 12.1.2.1.0 版本支持将混合列压缩的数据转换为闪存缓存中的纯列格式，这项功能也称为 Exadata 列式缓存。

---

[1] https://www.oracle.com/technetwork/database/in-memory/overview/twp-bp-for-iot-with-12c-042017-3679918.html

这是 Exadata 支持的第 1 代纯列格式，可以节省磁盘 I/O 和提升查询性能。但第 1 代纯列格式并不是 Database In-Memory 所使用的纯列格式，Exadata 对 In-Memory 格式的支持要等到 Exadata 系统软件 12.2.1.1.0 版本。这是 Exadata 支持的第 2 代纯列格式，也称为 In-Memory 列式缓存。由于 In-Memory 列格式所应用的闪存位于存储节点，存储节点在 Exadata 的术语中也称为 CELL，因此这项功能有时也被称为 CELLMEMORY。最初，CELLMEMORY 只支持混合列压缩表，并且需要启用内存列式存储，尽管其并没有真正使用。到 Exadata 系统软件 18.1.0 版本，In-Memory 列式缓存支持的对象不再限于混合列压缩表，普通的非压缩表和 OLTP 压缩表均可以支持。到了 Oracle 数据库 19.8 版本，In-Memory 列式缓存功能不再依赖于 INMEMORY_SIZE 初始化参数，只需将 INMEMORY_FORCE 初始化参数设置为 CELLMEMORY_LEVEL 即可启用。

In-Memory 列式缓存使用与 Database In-Memory 完全相同的纯列格式，因此可以享受到 In-Memory 列格式所有的好处，包括 SIMD 向量计算、字典编码、In-Memory 联结、In-Memory 聚合等。从磁盘格式到 In-Memory 列式缓存需要两个步骤。首先，符合扫描条件的数据将被转换为第 1 代纯列格式，也就是 Exadata 列式缓存。其次，如果启用了 In-Memory 列式缓存功能，后台进程会将闪存上的第 1 代纯列格式数据转换为第 2 代纯列格式。

In-Memory 列式缓存将 In-Memory 列格式从内存延伸到 Exadata 上的闪存，极大地扩展了 Database In-Memory 应用的领域。以 Exadata X8M-2 高容量 1/4 标准配置为例，所有计算节点上的内存合计为 768GB，而存储节点上的闪存容量合计为 76.8TB，约为内存总容量的 100 倍。尽管闪存的性能不如内存，但仍能带来相当可观的性能提升。

10.3.1 In-Memory 列式缓存基本操作

启用 In-Memory 列式缓存可以使用以下两种方法之一。

（1）将初始化参数 INMEMORY_SIZE 设置为大于 0 的值。

（2）如果数据库版本为 19.8.0.0.200714 或以上，设置初始化参数 INMEMORY_FORCE 为 cellmemory_level。如果此时 INMEMORY_SIZE 为 0，表示仅启用 CELLMEMORY。

启用 In-Memory 列式缓存后，还需要为对象设置 CELLMEMORY 属性。In-Memory 列式缓存支持 FOR QUERY 和 FOR CAPACITY 两种压缩级，默认为 FOR CAPACITY。FOR CAPACITY 的压缩比更高，而 FOR QUERY 的查询性能更好，压缩解压的开销更小。在此基础上可以指定 HIGH 或者 LOW 关键字，但并不会发生实际作用。以下为一些常见用法的示例：

```
-- 为表启用 In-Memory 列式缓存, 使用默认压缩级 MEMCOMPRESS FOR CAPACITY
ALTER TABLE lineorder CELLMEMORY;

-- 为表启用 In-Memory 列式缓存, 压缩级为 MEMCOMPRESS FOR QUERY
ALTER TABLE lineorder CELLMEMORY MEMCOMPRESS FOR QUERY;

-- 为表启用 In-Memory 列式缓存, 压缩级为 MEMCOMPRESS FOR QUERY, LOW 和 HIGH 关键字无效
ALTER TABLE lineorder CELLMEMORY MEMCOMPRESS FOR QUERY LOW;
ALTER TABLE lineorder CELLMEMORY MEMCOMPRESS FOR QUERY HIGH;
```

可以为一个对象同时指定 CELLMEMORY 和 INMEMORY 关键字，这并不表示一个对象可

以同时位于内存和闪存中。因为对于一个段对象而言，在任一时刻只能发布到一种 In-Memory 列格式存储中，要么是内存列式存储，要么是 In-Memory 列式缓存，并且内存列式存储优先。由于 Database In-Memory 按优先级发布对象，因此可以对低优先级对象同时指定两个关键字。当内存足够时，会发布到内存列式存储；而当内存不够时，对象至少还可以发布到闪存中以提升性能。并且对于分区表，可以对"热分区"指定 INMEMORY 属性，而对"冷"分区指定 CELLMEMORY 属性。以下为同时指定两种属性的示例。注意，CELLMEMORY 并不支持发布优先级设置：

```
ALTER TABLE lineorder INMEMORY CELLMEMORY;
ALTER TABLE lineorder INMEMORY PRIORITY LOW CELLMEMORY;
ALTER TABLE lineorder INMEMORY CELLMEMORY MEMCOMPRESS FOR QUERY;
```

当对象为临时使用或查询次数较少时，或者系统已经比较繁忙，不想增加系统负担时，可以使用以下命令禁用 CELLMEMORY。

```
-- 仅禁止 CELLMEMORY
ALTER TABLE lineorder NO CELLMEMORY;
-- 同时禁止 INMEMORY
ALTER TABLE lineorder NO INMEMORY NO CELLMEMORY;
```

CELLMEMORY 和 INMEMORY 在使用上有一些差异，下面通过一个示例来说明。在 PDB 一级执行以下脚本，禁用内存列式存储，仅启用 In-Memory 列式缓存。

```
-- 设置 cellmemory_level 需数据库版本大于 19.8.0.0.200714
ALTER SYSTEM SET inmemory_force = cellmemory_level SCOPE = SPFILE;
SHUTDOWN IMMEDIATELY
STARTUP
ALTER SYSTEM SET inmemory_size = 0;
ALTER TABLE lineorder CELLMEMORY NO INMEMORY;
```

查询 CELLMEMORY 属性，可知默认压缩级为 FOR CAPACITY：

```
SQL> SELECT cellmemory FROM user_tables WHERE table_name = 'LINEORDER';

CELLMEMORY
----------------------
MEMCOMPRESS FOR CAPACITY
```

由于没有 INMEMORY 属性，Database In-Memory 相关的 PL/SQL 过程并不能用来发布对象，因此只能通过全表扫描来发布。并且 V$IM 系列视图只针对发布到内存列式存储中的对象，对于 In-Memory 列式缓存无效：

```
SQL> EXEC DBMS_INMEMORY.POPULATE('SSB', 'LINEORDER');
BEGIN DBMS_INMEMORY.POPULATE('SSB', 'LINEORDER'); END;

*
ERROR at line 1:
ORA-64399: In-Memory population or repopulation cannot be run for this segment.
ORA-06512: at "SYS.DBMS_INMEMORY", line 22
ORA-06512: at "SYS.DBMS_INMEMORY", line 265
ORA-06512: at line 1
```

```
SQL> SELECT /*+ FULL(p) */ COUNT(*) FROM lineorder p;

  COUNT(*)
----------
  11997996

-- V$IM 视图中查不到发布对象的信息
SQL> SELECT * FROM v$im_segments;

no rows selected
```

此时的执行计划如下：

```
SQL> @../showplan

PLAN_TABLE_OUTPUT
---------------------------------------
SQL_ID  fdqum4n00u6hz, child number 0
---------------------------------------
SELECT /*+ FULL(p)  */ COUNT(*) FROM lineorder p

Plan hash value: 2267213921

---------------------------------------------------------------------------
| Id | Operation                  | Name      | Rows | Cost (%CPU)| Time     |
---------------------------------------------------------------------------
|  0 | SELECT STATEMENT           |           |      | 45429 (100)|          |
|  1 |  SORT AGGREGATE            |           |    1 |            |          |
|  2 |   TABLE ACCESS STORAGE FULL| LINEORDER |  11M | 45429   (1)| 00:00:02 |
---------------------------------------------------------------------------

Note
-----
   - dynamic statistics used: dynamic sampling (level=2)


18 rows selected.
```

执行计划中的 TABLE ACCESS STORAGE FULL 说明 Exadata 的智能扫描（smart scan）功能生效，表示部分操作被下推到 Exadata 的存储层（CELL 节点）执行。但是这并不能确定 In-Memory 列式缓存生效，我们需要进一步查看执行统计信息：

```
SQL> SET ECHO ON
SQL> @cellstat
SQL>
SELECT
    name,
    value
FROM
    v$sesstat   a,
    v$statname  b
```

```
WHERE
    (a.statistic# = b.statistic#)
    AND(a.sid) = userenv('sid')
    AND(name LIKE '%cell%')
    AND value != 0
ORDER BY
    name;

NAME                                                            VALUE
--------------------------------------------------------------  ----------
HCC scan cell CUs processed for uncompressed                    1304
HCC scan cell CUs sent uncompressed                             1304
cell IO uncompressed bytes                                      73760978
cell RDMA reads                                                 54
cell blocks helped by minscn optimization                       166547
cell blocks processed by cache layer                            166547
cell blocks processed by data layer                             370
cell blocks processed by txn layer                              166547
cell blocks returned by data layer                              1674
cell flash cache read hits                                      2619
cell flash cache read hits for smart IO                         2619
cell num smartio automem buffer allocation attempts             1
cell physical IO bytes eligible for predicate offload           1364353024
cell physical IO bytes eligible for smart IOs                   515522560
cell physical IO bytes processed for IM capacity                85458944
cell physical IO bytes saved by columnar cache                  427032576
cell physical IO interconnect bytes                             3296560
cell physical IO interconnect bytes returned by smart scan      2796848
cell pmem cache read hits                                       7
cell scans                                                      1
cellmemory IM scan CUs processed for capacity                   1304

21 rows selected.
```

在输出中，cell physical IO bytes saved by columnar cache 表示使用列式缓存所避免的物理 I/O，而 cellmemory IM scan CUs processed for capacity 则表示转换为压缩级为 FOR CAPACITY 的 In-Memory 列格式的 1MB CU 数量，这表明 In-Memory 列式缓存已生效。如果压缩级为 FOR QUERY，此参数名称则变为 cellmemory IM scan CUs processed for query。

10.3.2　RAC 环境下的 In-Memory 列式缓存

在 RAC 环境下，每一个实例都拥有自己的内存列式存储，发布的对象会分布到各个实例的内存列式存储中。而 In-Memory 列式缓存位于 Exadata 存储节点，是全共享架构，为各计算节点所共享。因此 RAC 节点的失效对其没有影响。如果存储节点失效，In-Memory 列式缓存可以在其他存储节点上的闪存中建立，对于客户端而言是透明的。

例如在一个两节点 RAC 集群中，最初在实例 1 中执行查询，从执行统计信息中可确认 In-Memory 列式缓存生效。

```
SQL> SELECT instance_number FROM v$instance;

INSTANCE_NUMBER
---------------
              1

SQL> SELECT /*+ FULL(p) NO_PARALLEL(p) */ COUNT(*) FROM lineorder p;

SQL> @cellstat
NAME                                                              VALUE
----------------------------------------------------------------- ----------
HCC scan cell CUs processed for uncompressed                      1302
HCC scan cell CUs sent uncompressed                               1302
cell IO uncompressed bytes                                        75734226
cell blocks helped by minscn optimization                         166547
cell blocks processed by cache layer                              166547
cell blocks processed by data layer                               622
cell blocks processed by txn layer                                166547
cell blocks returned by data layer                                1924
cell flash cache read hits                                        2619
cell flash cache read hits for smart IO                           2619
cell num smartio automem buffer allocation attempts               1
cell physical IO bytes eligible for predicate offload             1364353024
cell physical IO bytes eligible for smart IOs                     516800512
cell physical IO bytes processed for IM capacity                  85327872
cell physical IO bytes saved by columnar cache                    426377216
cell physical IO interconnect bytes                               3068672
cell physical IO interconnect bytes returned by smart scan        3019520
cell pmem cache read hits                                         6
cell scans                                                        1
cellmemory IM scan CUs processed for capacity                     1302

20 rows selected.
```

关闭实例 1 中的 PDB 以模拟故障,当然也可以关闭整个实例或整个主机。

```
SQL> ALTER PLUGGABLE DATABASE orclpdb1 CLOSE;
Pluggable database altered.

SQL> SHOW PDBS;

    CON_ID CON_NAME                       OPEN MODE  RESTRICTED
---------- ------------------------------ ---------- ----------
         2 PDB$SEED                       READ ONLY  NO
         3 ORCLPDB1                       MOUNTED
```

然后在实例 2 上执行查询,从执行统计信息中可以确认 In-Memory 列式缓存生效,而且绝大多数的统计值与之前实例 1 上的查询结果相同:

```
SQL> SELECT instance_number FROM v$instance;

INSTANCE_NUMBER
---------------
```

```
                     2
SQL> SELECT /*+ FULL(p) NO_PARALLEL(p) */ COUNT(*) FROM lineorder p;

SQL> @cellstat
NAME                                                              VALUE
----                                                              -----
HCC scan cell CUs processed for uncompressed                       1302
HCC scan cell CUs sent uncompressed                                1302
cell IO uncompressed bytes                                     75734226
cell blocks helped by minscn optimization                        166547
cell blocks processed by cache layer                             166547
cell blocks processed by data layer                                 622
cell blocks processed by txn layer                               166547
cell blocks returned by data layer                                 1924
cell flash cache read hits                                         2619
cell flash cache read hits for smart IO                            2619
cell num smartio automem buffer allocation attempts                   1
cell physical IO bytes eligible for predicate offload        1364353024
cell physical IO bytes eligible for smart IOs                 516800512
cell physical IO bytes processed for IM capacity               85327872
cell physical IO bytes saved by columnar cache                426377216
cell physical IO interconnect bytes                             3044096
cell physical IO interconnect bytes returned by smart scan      3019520
cell pmem cache read hits                                             3
cell scans                                                            1
cellmemory IM scan CUs processed for capacity                      1302

20 rows selected.
```

以上实验结果表明，由于存储是共享的，计算层面部件（如实例、节点）的失效只会导致计算能力的下降，但不会影响到对 In-Memory 列式缓存的使用。

10.3.3　In-Memory 列式缓存性能比较

相对于内存而言，In-Memory 列式缓存提供更大的容量，不会受到 RAC 节点故障的影响，不需要重新分布，这也是其相对于内存列式存储的主要优势。但是闪存的性能毕竟不如物理内存，因此在评估使用 In-Memory 列式缓存前，需要首先确认其性能是否能满足需求。

以下通过一个实验来比较 In-Memory 列式缓存和内存列式存储的性能差异。采用的测试语句为 SSB 示例 SQL 1_1，此查询涉及到 LINEORDER 和 DATE_DIM 两张表。

首先建立性能基线，此时内存列式存储和 In-Memory 列式缓存均被禁用。

```
ALTER TABLE lineorder NO CELLMEMORY NO INMEMORY;
ALTER TABLE date_dim NO CELLMEMORY NO INMEMORY;
```

此时的查询时间为 0.3 秒。从执行计划中可以看到使用了布隆过滤器优化，并且启用了智能扫描。在执行统计信息中，可以看到存在大量物理读。

```
SQL> @1_1
Elapsed: 00:00:00.30
```

```
SQL> @../../../showplan

PLAN_TABLE_OUTPUT
----------------------------------------
SQL_ID  62gb0a7n7xvdm, child number 0
----------------------------------------
SELECT /* 1.1 SSB_SAMPLE_SQL */ /*+ MONITOR */   SUM(lo_extendedprice
* lo_discount) AS revenue FROM    lineorder,   date_dim WHERE
 lo_orderdate = d_datekey     AND d_year = 1993    AND lo_discount
BETWEEN 1 AND 3    AND lo_quantity < 25

Plan hash value: 2403472142

-----------------------------------------------------------------------------
| Id  | Operation                   | Name     | Rows  | Bytes | Cost (%CPU)|
-----------------------------------------------------------------------------
|   0 | SELECT STATEMENT            |          |       |       | 45538 (100)|
|   1 |  SORT AGGREGATE             |          |     1 |    28 |            |
|*  2 |   HASH JOIN                 |          |  340K | 9311K | 45538   (1)|
|   3 |    JOIN FILTER CREATE       | :BF0000  |   365 |  3650 |    13   (0)|
|*  4 |     TABLE ACCESS STORAGE FULL| DATE_DIM |   365 |  3650 |    13   (0)|
|   5 |    JOIN FILTER USE          | :BF0000  | 2243K |   38M | 45517   (1)|
|*  6 |     TABLE ACCESS STORAGE FULL| LINEORDER| 2243K |   38M | 45517   (1)|
-----------------------------------------------------------------------------

SQL> @../../../imstats

NAME                                                            VALUE
------------------------------------------------------------ ----------
CPU used by this session                                            57
physical reads                                                  333324
physical reads direct                                           333324
session logical reads                                           333446
session pga memory                                             9919440
table scans(long tables)                                             1

6 rows selected.
```

在存储节点相关的执行统计信息中，没有任何 CELLMEMORY 的信息。说明没有使用 In-Memory 列式缓存。

```
SQL> @cellstat

NAME                                                            VALUE
------------------------------------------------------------ ----------
HCC scan cell CUs processed for uncompressed                    234162
HCC scan cell CUs sent uncompressed                             234096
cell IO uncompressed bytes                                  2731130880
cell blocks helped by minscn optimization                       337086
cell blocks pivoted                                             234162
cell blocks processed by cache layer                            337086
```

```
cell blocks processed by data layer                              333390
cell blocks processed by txn layer                               337086
cell num smartio automem buffer allocation attempts                   1
cell physical IO bytes eligible for predicate offload        2730590208
cell physical IO bytes eligible for smart IOs                2730590208
cell physical IO interconnect bytes                           101755872
cell physical IO interconnect bytes returned by smart scan    101755872
cell scans                                                            1

14 rows selected.
```

接下来测试单独启用 CELLMEMORY 时的性能。设置完属性后，通过全表扫描发布数据：

```
ALTER TABLE lineorder CELLMEMORY NO INMEMORY;
ALTER TABLE date_dim CELLMEMORY NO INMEMORY;
SELECT /*+ FULL(p) NO_PARALLEL(p) */ COUNT(*) FROM lineorder p;
SELECT /*+ FULL(p) NO_PARALLEL(p) */ COUNT(*) FROM date_dim p;
```

确认两张表均没有发布到内存列式存储中：

```
SQL> SELECT COUNT(*) FROM v$im_segments;

  COUNT(*)
----------
         0
```

此时的查询时间为 0.05 秒。执行计划与基线测试类似。从执行统计信息中，可以确认使用了 In-Memory 列式缓存。

```
SQL> @1_1
Elapsed: 00:00:00.05

SQL> @../../../showplan

---------------------------------------------------------------------------------
| Id  | Operation                   | Name     | Rows  | Bytes | Cost (%CPU)|
---------------------------------------------------------------------------------
|   0 | SELECT STATEMENT            |          |       |       | 45538 (100)|
|   1 |  SORT AGGREGATE             |          |     1 |    28 |            |
|*  2 |   HASH JOIN                 |          |  340K| 9311K | 45538   (1)|
|   3 |    JOIN FILTER CREATE       | :BF0000  |   365 |  3650 |    13   (0)|
|*  4 |     TABLE ACCESS STORAGE FULL| DATE_DIM |   365 |  3650 |    13   (0)|
|   5 |    JOIN FILTER USE          | :BF0000  | 2243K|   38M | 45517   (1)|
|*  6 |     TABLE ACCESS STORAGE FULL| LINEORDER| 2243K|   38M | 45517   (1)|
---------------------------------------------------------------------------------

SQL> @cellstat

NAME                                                          VALUE
------------------------------------------------------------ ----------
HCC scan cell CUs columns accessed                             2796
HCC scan cell CUs decompressed                                 2703
HCC scan cell CUs processed for uncompressed                   4353
HCC scan cell CUs row pieces accessed                          2703
```

```
HCC scan cell CUs sent uncompressed                                        3708
HCC scan cell bytes compressed                                         913885081
HCC scan cell bytes decompressed                                       247000499
HCC scan cell columns theoretical max                                      45207
HCC scan cell rows                                                      24917824
cell IO uncompressed bytes                                             255233459
cell blocks helped by minscn optimization                                 345982
cell blocks pivoted                                                         1005
cell blocks processed by cache layer                                      345982
cell blocks processed by data layer                                         1005
cell blocks processed by txn layer                                        345982
cell flash cache read hits                                                 18892
cell flash cache read hits for smart IO                                    18892
cell num smartio automem buffer allocation attempts                            2
cell physical IO bytes eligible for predicate offload                 2834284544
cell physical IO bytes eligible for smart IOs                         1052811264
cell physical IO bytes processed for IM capacity                      2826051584
cell physical IO bytes saved by columnar cache                          12255232
cell physical IO interconnect bytes                                    561387920
cell physical IO interconnect bytes returned by smart scan             561387920
cell scan CUs pcode pred evaled                                               93
cell scans                                                                     2
cell smart IO session cache hits                                               1
cell smart IO session cache lookups                                            1
cell smart IO session cache soft misses                                        1
cellmemory IM scan CUs processed for capacity                               3348
key vector hash cells scanned                                              50345

31 rows selected.
```

最后，看一下单独启用内存列式存储时的性能。为两张表仅设置 INMEMORY 属性，然后发布：

```
ALTER TABLE lineorder NO CELLMEMORY INMEMORY;
ALTER TABLE date_dim NO CELLMEMORY INMEMORY;
EXEC DBMS_INMEMORY.POPULATE('SSB', 'LINEORDER');
EXEC DBMS_INMEMORY.POPULATE('SSB', 'DATE_DIM');
```

此时的查询时间为 0.02 秒。从执行计划中可以确认内存列式存储已使用。

```
SQL> @1_1
Elapsed: 00:00:00.02

SQL> @../../../showplan

---------------------------------------------------------------------------------------
| Id  | Operation                | Name     | TQ    |IN-OUT| PQ Distrib |
---------------------------------------------------------------------------------------
|   0 | SELECT STATEMENT         |          |       |      |            |
|   1 |  SORT AGGREGATE          |          |       |      |            |
|   2 |   PX COORDINATOR         |          |       |      |            |
|   3 |    PX SEND QC (RANDOM)   | :TQ10000 | Q1,00 | P->S | QC (RAND)  |
|   4 |     SORT AGGREGATE       |          | Q1,00 | PCWP |            |
```

```
|*  5 |       HASH JOIN                    |           | Q1,00 | PCWP |          |
|   6 |        JOIN FILTER CREATE          | :BF0000   | Q1,00 | PCWP |          |
|*  7 |         TABLE ACCESS INMEMORY FULL | DATE_DIM  | Q1,00 | PCWP |          |
|   8 |        JOIN FILTER USE             | :BF0000   | Q1,00 | PCWP |          |
|   9 |         PX BLOCK ITERATOR          |           | Q1,00 | PCWC |          |
|* 10 |          TABLE ACCESS INMEMORY FULL| LINEORDER | Q1,00 | PCWP |          |
---------------------------------------------------------------------------------

Predicate Information (identified by operation id):
---------------------------------------------------

   5 - access("LO_ORDERDATE"="D_DATEKEY")
   7 - inmemory("D_YEAR"=1993)
       filter("D_YEAR"=1993)
  10 - inmemory(:Z>=:Z AND :Z<=:Z AND ("LO_DISCOUNT"<=3 AND "LO_QUANTITY"<25 AND
       "LO_DISCOUNT">=1 AND
              SYS_OP_BLOOM_FILTER(:BF0000,"LO_ORDERDATE")))
       filter(("LO_DISCOUNT"<=3 AND "LO_QUANTITY"<25 AND "LO_DISCOUNT">=1 AND
              SYS_OP_BLOOM_FILTER(:BF0000,"LO_ORDERDATE")))

Note
-----
   - dynamic statistics used: dynamic sampling (level=5)
   - automatic DOP: Computed Degree of Parallelism is 12
   - parallel scans affinitized for inmemory
```

在执行统计信息中可以看到大量 IM scan 信息，但没有任何 CELLMEMORY 相关信息。说明内存列式存储已启用，而 CELLMEMORY 未被使用：

```
> @../../../imstats

NAME                                                        VALUE
----------------------------------------------------------  ----------
CPU used by this session                                            15
IM scan CUs columns accessed                                       200
IM scan CUs columns theoretical max                                901
IM scan CUs current                                                 53
IM scan CUs memcompress for query low                               53
IM scan CUs no cleanout                                             53
IM scan CUs pcode pred evaled                                      241
IM scan CUs predicates applied                                     147
IM scan CUs predicates received                                    147
IM scan CUs readlist creation accumulated time                      18
IM scan CUs readlist creation number                                53
IM scan CUs split pieces                                            54
IM scan bytes in-memory                                     1203234226
IM scan bytes uncompressed                                  2287358990
IM scan delta - only base scan                                      53
IM scan rows                                                  24011940
IM scan rows projected                                          479390
IM scan rows valid                                            24011940
IM scan segments disk                                                6
```

| IM scan segments minmax eligible ... | 6 |

总结以上 3 个实验的结果，与基线测试不采用任何 In-Memory 列格式优化相比较，启用 CELLMEMORY 将性能提升了 6 倍，启用 INMEMORY 将性能提升了 15 倍。在测试中，INMEMORY 的性能比 CELLMEMORY 快了 2.5 倍。实际环境中的性能差异可能与此不同，但总的原则是，可以将性能要求最高的数据发布到内存列式存储中，相对较"冷"的数据则发布到位于 Exadata 存储节点上的 In-Memory 列式缓存中。